Introductory Immunobiology

Introductory Immunobiology

Huw Davies
School of Life, Basic Medical & Health Sciences
Division of Life Sciences
King's College
London

CHAPMAN & HALL
London · Weinheim · New York · Tokyo · Melbourne · Madras

Published by Chapman & Hall, 2–6 Boundary Row, London SE1 8HN, UK

Chapman & Hall, 2–6 Boundary Row, London SE1 8HN, UK

Chapman & Hall GmbH, Pappelallee 3, 69469 Weinheim, Germany

Chapman & Hall USA., 115 Fifth Avenue, New York NY 10003, USA

Chapman & Hall Japan, ITP- Japan, Kyowa Building, 3F,
2–2–1 Hirakawa-cho, Chiyoda-ku, Tokyo 102, Japan

Chapman & Hall Australia, 102 Dodds Street, South Melbourne,
Victoria 3205, Australia

Chapman & Hall India, R. Seshadri, 32 Second Main Road, CIT East, Madras
600 035, India

First edition 1997

© 1997 H. Davies

Typeset in 11.5/14 pt Garamond by WestKey Limited, Falmouth, Cornwall

Printed in Italy by Vincenzo Bona, Turin

ISBN 0 412 37240 1

A catalogue record for this book is available from the British Library

Library of Congress Catalog Card Number: 96–84895

Contents

Preface

Why immunobiology?

Immunology is the study of the immune system – the internal defence reactions that protect the body from invading microorganisms and the diseases they cause. Spectacular advances have been made over the last few decades in understanding how the immune system works. There is no doubt that these advances have been made possible by concentrating research on a few species of animals, most notably mouse and man. The main motivation for studying the human system, for example, has been to further the cause of medicine. Indeed, the roots of modern immunology can be traced back to pioneering studies of vaccines against viruses and bacteria. The mouse has become the favoured non-human animal in which to study immunity, both in relation to protection from microorganisms, but also at a more fundamental level. The term 'immunology' has become virtually synonymous with the study of the immune systems of humans and mice. 'Immunobiology' in contrast is a broader field, encompassing the immune systems of all animals. It is the study of the origins and evolution of immune systems in general, and the underlying role that microorganisms play in the process.

The penalty for this focussed effort has been a disproportionately mammalocentric database. A few non-mammalian vertebrates, in particular the chicken, the amphibian *Xenopus* and several species of fishes, are emerging as the model animals of other vertebrate classes. These, and other non-mammalian vertebrates, are now attracting the attentions of steadily increasing numbers of 'comparative immunologists'. By contrast with vertebrates, the study of invertebrate immunity has been a less popular field of research. Yet vertebrates belong to a single phylum and account for less than 1% of the million or so known species of animals. More than three-quarters of known species are arthropods – insects, crustaceans and chelicerates, with insects representing the majority. A picture that emerges is that the solutions to the problem of protecting the body from microorganisms are varied, with vertebrates and invertebrates tending to do things differently.

To reflect the comparative depth of information available, the majority of this book is concerned with the mammalian approach in immunity.

vaccine *n.* a preparation, usually derived from an infectious pathogen, administered to provide protective immunity without causing disease.

microorganism *n.* an organism too small to be seen clearly with the naked eye; often used as a collective term for organisms that colonize host organisms and which may or may not be pathogenic (viruses, bacteria, fungi, protozoa and worms).

Xenopus *n.* a genus of anuran amphibians, including *X. laevis* (the African clawed toad).

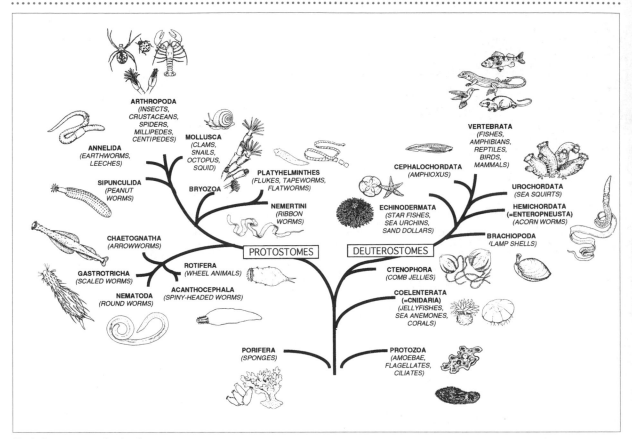

Evolutionary tree of animals. Animals redrawn from Valentine, J.W. (1978) *Scientific American,* **239,** 104–17; phylogenetic relationships interpreted from Figure 1.4 of Nielsen, C. (1995) *Animal Evolution: Interrelationships of the Living Phyla.* Oxford University Press, Oxford UK.

Where possible, however, data from non-mammalian and invertebrate animals have been included for comparison. A tree has been depicted here to help you place different types of animals in relation to each other. Their approaches to immunity are not inferior or any less able to eliminate microorganisms than the more familiar methods of the mammals. Indeed, invertebrate species vastly outnumber mammals and they have successfully exploited virtually every ecosystem. Each solution to the common problem of defence against microorganisms does the job successfully in its own way.

Who is this book for?

This book is an introductory text written mainly for students who are encountering immunology for the first time as part of a broader course of study. This book will also be useful to students already studying immunology to help place their understanding of the human and mouse immune system into a broader perspective.

The objectives of the book have been twofold. Firstly I wanted where possible to place the mammalian immune system into perspective. One way is to build on its underlying biological principles. Often, wherever an aspect of immunobiology overlaps with a more general biological process, this has been drawn out into a 'Box'. Further reading is given in most Boxes to provide a 'non-immunological' view of the subject. The other way to place immunity into perspective is by using the comparative approach. By examining the immune systems of non-mammalian vertebrates it is possible to see a progressive expansion during the course of evolution, by stepwise additions onto a pre-existing system. The 'sophistication' or 'complexity' of the mammalian system does not necessarily equate with 'better', but instead reflects this evolutionary process. The second objective was to keep the book small and readable. Immunology, of human and mouse, has grown into a vast subject that encompasses all sorts of biomolecular disciplines: biochemistry, genetics, cell biology, and molecular biology. It is very easy to lose the overall picture in a sea of detail.

How is this book organized?

Immunology conveniently falls into discrete topics, like 'immunoglobulins', 'the T cell receptor', 'complement', 'inflammation', 'cytokines' and so on. Each area lends itself to a separate chapter. Unfortunately this requires a pre-existing conceptual framework of the system in order to understand how these individual components fit together.

This book has a less fragmentary approach, and starts in Chapter 1 with the actual business of killing microorganisms. Although the solutions are varied, the problem is the same for vertebrates and invertebrates alike. To kill microorganisms without damaging healthy cells in the body, an immune system requires some form of discrimination between self and non-self. This responsibility falls on recognition receptors and molecules, which form the subject of Chapter 2. This chapter includes an overview of the antigen receptors of lymphocytes, and explains how these receptors underlie all the remarkable properties of the vertebrate 'adaptive' immune system.

Chapter 3 examines the cells and tissues of immune systems and the circulatory systems by which they communicate. Particular emphasis is placed on the progression of these tissues during the course of vertebrate evolution. The remaining chapters are largely concerned with the vertebrates. The antigen receptors of mammalian T and B lymphocytes are discussed in greater depth in Chapter 4, and the nature of their ligands in

ligand *n.* the complementary structure bound by a receptor (also called a **counter-receptor**).

Chapter 5. Chapter 6 expands on the ligand of the T lymphocyte antigen receptor and the inheritance of immune responsiveness. Both are the responsibility of the major histocompatibility complex (MHC). In Chapter 7 we step back and examine the other vertebrates, and theories of how the antigen receptors and MHC molecules originated and evolved. Chapter 8 examines the orchestration of cellular activities through signalling cascades and cytokines. We end with a chapter on mechanisms of tolerance, possibly the least understood aspect of vertebrate immunity, but one that plays a major role in discriminating self from non-self.

Acknowledgements

I am very grateful to many colleagues for their help and advice: Harold Baum, Wilson Caparros-Wanderley, Benny Chain, Francis Darwin, Julian Hickling, Colin Howe, Athena Koulourianos, Mohammad Ibrahim, Tim Littlewood, John Murphy, Philip Shepherd, Hans Stauss and Graham Wallace – who gave up their time to read and comment on individual chapters, Paul Hobby who generated the crystal structure pictures, and especially Mike Johnstone, Derek Wakelin and Marcia, who each read the whole book. I am also grateful to Dominic Recaldin for his encouragement, and the production team at Chapman & Hall for making it all possible.

Huw Davies

List of boxes

Protection against pathogens

1

In this chapter

- Parasitism, disease and the need for defence.
- Barriers to infection.
- Inducible defence: blood coagulation, inflammation and effector mechanisms.

Introduction

Virtually every multicellular animal, from the simplest to the most complex, is a potential host for microorganisms. Microorganisms multiply rapidly and any that gain entry to the tissues of a host may be quickly dispersed by its circulatory system. Without means to protect against the entry of microbes, a host would be quickly overwhelmed. In this chapter we will learn about the barriers to infection – genetic, physical and chemical – that prevent microbes gaining entry, and the internal defence mechanisms that have evolved to eliminate microbes that breach the barriers.

> **infection** *n.* invasion of a host by microorganisms and subsequent multiplication.

1.1 The need for defence

Many microorganisms – viruses, bacteria, fungi, protozoa and worms – spend some or all of their life cycle within, or upon, the bodies of other host animals. Many of these merely colonize external epithelial surfaces, or gain entry through orifices to reside on internal surfaces lining the digestive, respiratory or genital tracts. While such a relationship has obvious benefits for the microbe (a stable environment and a secure food supply), their presence usually neither harms nor benefits the host.

However, otherwise harmless microorganisms may become disease-

pathogen *n.* a microorganism that causes disease in its host.

pathogenicity *n.* the relative ability of different species of pathogens to cause disease.

parasite *n.* an organism that lives on or in another (host) organism and which benefits from the relationship at the expense of the host.

causing (or **pathogenic**) should they have the opportunity to enter the tissues or the circulatory system through damaged epithelium. This is because microorganisms can cause cellular damage when they gain entry, often as a result of their metabolic by-products or toxic components in their cell walls (Table 1.1). Disease – a state of illness or reduced competence of the host animal to perform normal activities – has been a major influence in the evolution in hosts of barriers and defence reactions to resist infection.

Other microbes have taken this a step further and have evolved means for actively penetrating a host and colonizing a particular environmental niche, such as inside a particular cell or tissue. The majority of these specialized relationships are **parasitic**, in which the microorganism lives at the expense of the host's welfare. Although these are also a major source of disease, a particularly pathogenic organism is not likely to be successful in the long term if it is dependent on the host for its own survival. Indeed, most successful parasitic microorganisms co-exist with their hosts without causing severe disease, and only do so when the defence reactions of the host are compromised in some way. Not surprisingly, an **endoparasitic*** lifestyle has been accompanied by the evolution by such microbes of a variety of strategies for evading the internal defence reactions of the host. We will encounter examples of a few of these evasion strategies periodically through this book.

1.1.1 Genetic barriers

Resistance to infections and diseases depends on many factors, both inherited and acquired. At the most fundamental level is **constitutional resistance** to disease. Each species of microorganism is adapted to exploit a particular environment offered by the particular species of host in which it lives. Different hosts have different physiologies and different levels of constitutional resistance to disease-causing microbes. For example, mammals such as sheep, mice, guinea pigs, rabbits and monkeys are susceptible to anthrax, a disease caused by a toxin of the bacterium *Bacillus anthracis*, whereas dogs, horses, cats and rats are not (Figure 1.1).

Even within a single species of host, there is variation in the resistance to disease between different individuals. Factors such as age and diet can greatly influence susceptibility to infection by a given microorganism. However, genetic variability between individuals accounts for differing levels of susceptibility and provides opportunities for host populations to adapt to pathogens. Microorganisms have very short generation times and

---

*Another category of parasites, typified by certain arthropods such as fleas, ticks and lice, have become adapted to an external parasitic lifestyle and live on, rather than in, their hosts. These **ectoparasites** do not encounter the internal defence mechanisms of the host and are not discussed any further.

Table 1.1 A few examples of microbes that cause disease in humans

	Brief description	Sources of cell or tissue damage	Examples that cause disease in humans [name of disease(s) in parenthesis]
Viruses	Particles consisting of nucleic acid (DNA or RNA) surrounded by a protein shell. Viruses are absolutely dependent for their existence and replication on penetrating a host cell	• Cause infected cells to fuse	• Human Immunodeficiency Virus infection of helper T lymphocytes (acquired immunodeficiency syndrome – AIDS) • Measles virus infect many cells (measles or rubeola)
		• Cause infected cells to become tumour cells	• Epstein–Barr Virus infect B lymphocytes, nasopharyngeal epithelium (infectious mononucleosis, nasopharyngeal carcinoma, Burkitt's lymphoma) • Human T lymphotrophic leukaemia virus infect T lymphocytes (leukaemia)
		• Burst (lyse) infected cells	• Varicella infection of cutaneous epithelium (chickenpox) • Herpes simplex infection of cutaneous epithelium (cold sores)
Bacteria	True microorganisms capable of independent existance. Bacteria are prokaryotes (unicellular organisms lacking a membrane-bound nucleus or membrane-bound organelles)	• Contain toxic lipopolysaccharides (LPS) in their cell walls (**endotoxins**)	• Many Gram-negative bacteria, e.g. *Salmonella typhimurium* (*Salmonella* gastroenteritis)
		• Liberate soluble toxins (**exotoxins**). **Cytotoxins** (inhibit protein synthesis and cause cell death)	• *Corynebacterium diphtheriae*; diphtheria toxin poisons many cells (diphtheria)
		Neurotoxins (interfere with nerve synapse function)	• *Clostridium botulinum*; botulinum toxin blocks nerve synapse function (botulism) • *Clostridium tetani*; tetanus toxin blocks inhibitory nerve synapses (tetanus)
		Enterotoxins (alter fluid/ion transport across membrane) **Lysins** (disrupt cell membranes causing cell lysis)	• *Vibrio cholera*; cholera toxin causes loss of water and salt through intestinal epithelium (cholera) • *Streptococcus pyogenes* Streptolysin-O lyses erythrocytes and other cells
Fungi	The moulds and yeasts; spore-bearing eukaryotes (organisms whose cells possess membrane-bound organelles).	• Penetrate epithelial tissue with threadlike mycelia	• *Trichophyton* species; different species penetrate skin at different anatomical sites (skin mycoses or ringworm)
Protozoa	Unicellular eukaryotic cells classified according to mode of locomotion (flagellates, cilliates, amoebae and sporozoa)	• Tissue invasion and ulceration	• *Entamoeba histolytica* invades intestinal epithelium (amoebic dysentery) • *Plasmodium* species invade erythrocytes and hepatocytes (malaria)
		• Penetration of cells causing lysis	• *Leishmania* species invade tissue macrophages (leishmaniasis. • *Trypanosoma* species invade muscle and nerve cells (Chaga's disease, sleeping sickness)
Helminth worms: flatworms (flukes and tapeworms) roundworms (nematodes)	Multicellular (metazoan) parasitic worms with complex life cycles often involving more than one host. In humans, worms inhabit sites such as blood vessels, lymphatics, alimentary canal	• Blockage or damage to blood vessels	• *Schistosoma* species (schistosomiasis or bilharzia)
		• Blockage or damage to lymphatics	• *Brugia malayi* (lymphatic filariasis)
		• Obstruction or perforation of intestine	• *Trichuris trichiura* (whipworm infection) • *Taenia saginata* (tapeworm disease, cestodiasis)

Figure 1.1 Schematic representation of constitutional resistance to diseases caused by bacteria in different mammals. (a) Anthrax caused by *Bacillus anthracis*, (b) botulism caused by *Clostridium botulinum*, (c) tetanus caused by *Clostridium tetani*. Reproduced with permission from Rumyantsev, S.N. (1992) *Immunology Today*, **13**(5), 184–7.

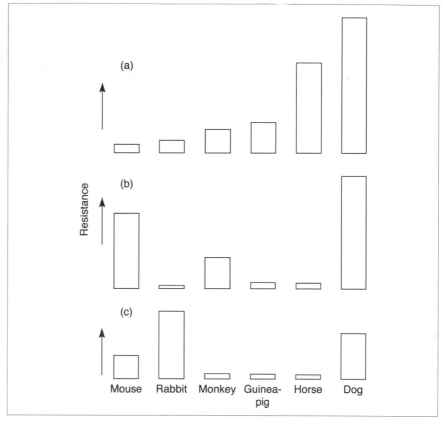

can also adapt rapidly to evolving environments offered by populations of host organisms.

1.1.2 Physical barriers to infection

The epithelial surface surrounding an organism and lining its internal digestive, respiratory and genital surfaces protects the soft tissues. It also performs a variety of specialized physiological functions such as respiration, secretion, and thermal regulation.

Outer epithelial surfaces are also the first line of defence against penetration by microbes. Many are constructed of multiple layers of flattened and closely packed cells (**stratified epithelium**) which provide an impenetrable barrier to most microbes. Different animals augment the efficacy of this barrier by secreting oils or mucus, or hardened cuticle onto their surfaces.

To reduce the possibility of infection, colonization of epithelial surfaces is minimized in several ways. Epithelia are cleared regularly by self-cleansing actions. Constant shedding of the outermost layers of cells (**exofoliation**) of skin and mucosal surfaces removes adherent microorganisms. Similarly, the respiratory tracts of mammals are coated by a blanket of mucus that is constantly swept away by ciliated epithelial cells. Resistance to infection by potential pathogens may also be afforded by the resident population of commensal organisms, with which they must first compete for the available resources.

commensalism *n.* a symbiotic relationship in which one partner (the commensal) benefits without benefit or detriment to the other (the host). See **mutualism** and **parasitism**.

1.1.3 Chemical barriers
Epithelial surfaces are often augmented by chemical barriers. The most ubiquitous is **lysozyme** which is an enzyme that hydrolyses bacterial cell walls and is particularly effective against Gram-positive bacteria. Lysozyme is found in many secretions, including tears and mucus of mammals, the egg whites of birds, and the blood of many invertebrates. This is an ancient form of defence against microorganisms. Another ubiquitous microbicidal agent is **transferrin**. This is an iron-binding glycoprotein that is found abundantly in the mucosal secretions of many vertebrate species. Transferrin exerts its microbicidal effects by sequestering the iron needed by some microorganisms for growth.

1.1.4 Inducible defence is needed if outer barriers are breached
Although specialized endoparasites have evolved ways to actively penetrate epithelial surfaces, other opportunistic microbes may also gain entry through the epithelium of the host when it becomes damaged by injury. For this reason all animals have evolved various forms of active defence reactions that can be mobilized quickly to seal the wound, eliminate any invading microorganisms and repair tissue damage (Figure 1.2). These reactions consist of:

*   **Blood clotting** or **coagulation**.
*   **Inflammation**.
*   **Inducible effector mechanisms** (or **immune responses**).

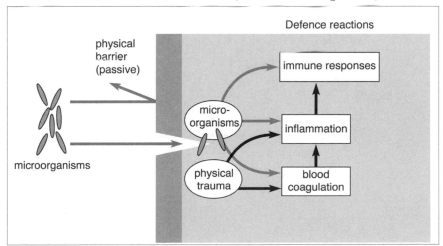

Figure 1.2 The interrelationships between physical barriers and defence reactions.

In general, all three reactions can be elicited by microorganisms, whereas inflammatory reactions and clotting cascades can also be triggered by physical trauma alone. There is considerable overlap between these systems, particularly in the invertebrates.

A common theme we will encounter in inducible defence reactions is the **enzymatic cascade**. Cascades have evolved independently in several different types of defence reactions in vertebrates and invertebrates alike.

A cascade consists of a sequence of enzymatic reactions, each enzyme being activated in sequence by the enzymatic activity of the previous component in the pathway (Figure 1.3). To prevent uncontrolled amplification, an active enzyme is quickly inactivated by other, inhibitory components. Individual components of an enzymatic cascade are usually preformed. This requires that the components are kept apart to prevent the accidental triggering of the cascade. This is achieved in two ways:

- Physical segregation within intracellular vesicles.

- Chemical segregation, by being present in an inactive form. Cleavage exposes the catalytic site of the enzyme, which enables it to activate the next component in the pathway.

Enzymatic cascades have two useful properties in defence reactions. Firstly, the components are preformed (i.e. do not require gene expression) which brings about a rapid response. Secondly, enzymes in a cascade may be able to activate many molecules of substrate before being inactivated. A cascade therefore amplifies the effect of the original stimulus. Thirdly, by-products created when an inactive enzyme is cleaved usually have other

Figure 1.3 Principle of enzymatic cascades in defence reactions. Enzymes of a cascade are segregated, either physically in intracellular vesicles or chemically, by being present in the blood in an inactive form. The surface of a microorganism (the stimulus) causes activation of the first component. One cleavage product possesses enzymatic activity and may activate many molecules of its substrate. The other cleavage product usually has different properties.

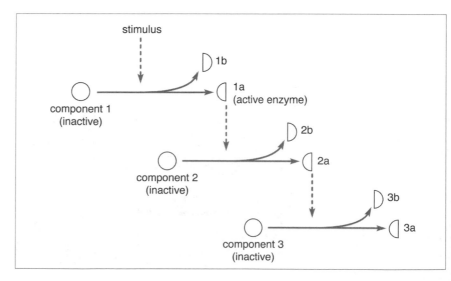

biological properties that influence cell behaviour. In this way a cascade can elicit a variety of responses by different cells.

1.2 Blood coagulation

Blood coagulation, or clotting, is very similar in widely differing animals (Figure 1.4). The response is immediate and is brought about by an enzymatic cascade, resulting in the polymerization of a fibrous, protein-aceous matrix and the entrapment of blood cells.

In all animals, blood coagulation serves two important functions. One is to seal haemorrhages and prevent the escape of blood. Uncontrolled loss of blood can have rapid and severe consequences for transport of oxygen to the tissues. Also, in soft-bodied invertebrates such as annelids, gastro-pods and other invertebrates lacking an exoskeleton, the circulatory fluid also provides hydrostatic pressure required for locomotion. Therefore, any punctures must be sealed rapidly; local muscular contractions immediately serve to staunch the flow of blood while clotting reactions take place. In animals with an inflexible integument, such as the shell of the echinoderm (called the **test**), muscular contractions cannot be used and a rapid clotting reaction is a vital defence in response to injury.

The second function is to assist in the removal of any microorganisms that gain entry through the wound. This is especially important in the invertebrates, in which blood coagulation quickly entraps invading micro-organisms and prevents their dispersal throughout the body. In vertebrates, clotting reactions are also linked to defence against micro-

> **blood coagulation (clotting)** a response to haemorrhaging from damaged blood vessels in which blood is converted to a solid plug usually via an enzymatic cascade.

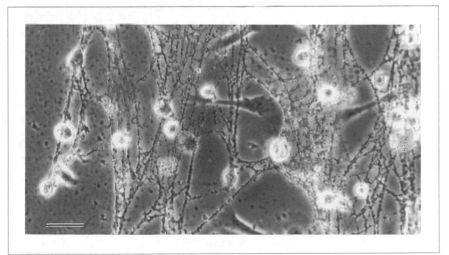

Figure 1.4 Coagulating insect blood. The strands of coagulum connecting the blood cells have been visualized by adding small latex particles which stick to the coagulum. From Davies, D.H. and Vinson, S.B. (1988) *Cell and Tissue Research,* **251**, 467–75.

organisms, although indirectly. Clotting triggers inflammatory reactions, which, as we will see later, play a vital role in the removal of invading microbes.

Among the vertebrates, blood coagulation in humans is very well understood and it is probably quite similar in other vertebrates (Figure 1.5a). The clot is an insoluble protein gel formed by the conversion of soluble fibrinogen in the blood into fibrin. This reaction is catalysed by an enzyme in the blood, thrombin, which is itself activated by cleavage of an inactive precursor, prothrombin. The cleavage of prothrombin occurs on the surface of **platelets**. These are small non-nucleated blood cell fragments that aggregate at punctures in blood vessels. Clots are eventually removed by another enzymatic cascade. In humans, this is performed by the **fibrinolytic system** (Figure 1.5b). The key component is plasminogen which, upon activation by other proteinases, digests fibrin clots into smaller degradation products.

Much of the research on clotting mechanisms in invertebrates has been

platelet *n.* the **thrombocytes** of mammals; non-nucleated cells derived from the fragmentation of megakarocytes that are involved in blood clotting.

Figure 1.5 Enzymatic cascades in humans. (a) *Coagulation cascade.* A pivotal enzyme in the cascade is factor X, because it can be activated in two different pathways. One is via the **intrinsic pathway** which is triggered when blood comes into contact with tissue damaged by trauma or bacterial endotoxins (see Table 1.1), or in contact with glass *in vitro*. In this way, factor XII in the blood binds to the newly exposed negatively charged surfaces and becomes activated. Activated factor XII also cleaves factor XI, which in turn cleaves factor IX. Activated factor IX is then able to form a complex with factor VIII, phospholipid and calcium on the surface of platelets, which together activate factor X. Factor X can also be cleaved by a more rapid **extrinsic pathway,** which augments the intrinsic pathway. Damaged tissues release factor III (tissue factor). In the presence of calcium ions, tissue factor activates factor VII in the blood, and together these are able to cleave factor X on the platelet membrane. (b) *Fibrinolytic cascade.* Clots are solubilized by plasmin, which is activated from an inactive precursor, plasminogen.

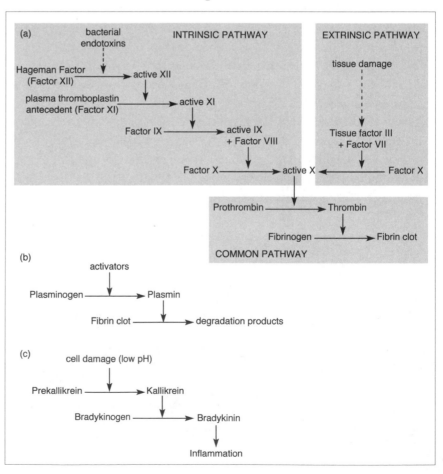

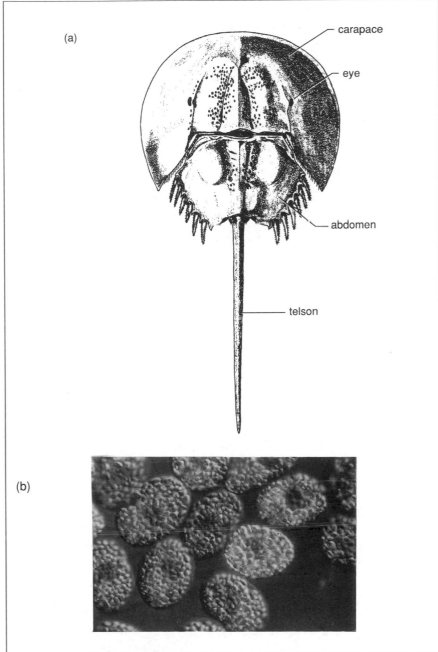

(a)

carapace

eye

abdomen

telson

(b)

Figure 1.5 *continued from opposite page* (c) *Bradykinin* cascade (see Section 1.3). Low pH caused by tissue injury activates an enzyme in the extracellular fluid called **kallikrein.** Activated kallikrein cleaves an inactive precursor of **bradykinin** which, when activated, binds to receptors on mast cells causing degranulation. The bradykinin cascade is also triggered by the intrinsic pathway of the coagulation cascade. Activated factor XII is able to activate prekallikrein into kallikrein, thereby connecting the coagulation response to inflammatory reactions. Redrawn and modified from Klein, J. (1982) *Immunology: The Science of Self-Nonself Discrimination.* J. Wiley, New York.

Figure 1.6 (a) *Limulus polyphemus*, a horseshoe crab. Reproduced from Barnes, R.D. (1974) *Invertebrate Zoology,* 3rd edition, copyright © 1974 by Saunders College Publishing, by permission of the publisher. (b) Blood cells (amoebocytes) of *Limulus.* Courtesy of Dr P.B. Armstrong (Armstrong, P.B. (1977) *Experimental Cell Research,* **107**, 127–38).

carried out with arthropods, in particular *Limulus*, a horseshoe crab (Figure 1.6). *Limulus* possesses only one type of blood cell, whose major function is coagulation of blood (Figure 1.7). Clotting reactions have been studied in other invertebrates. A puncture made in the skin of an insect larva causes labile cells to degranulate at the site of injury and a sticky matrix clots the blood. Blood cells are attracted to the area where they seal the wound, and

Figure 1.7 Clotting cascade of *Limulus*. The amoebocytes are packed with granules which contain the majority of the clotting components, including **coagulogen.** When released in response to stimuli from microorganisms, it is cleaved by activated proteases into **coagulin,** the major component of fibrous clot. Adapted from Figure 3 in Iwanaga, S., *et al.* (1994) *Annals of the New York Academy of Sciences,* **712**, 102–16.

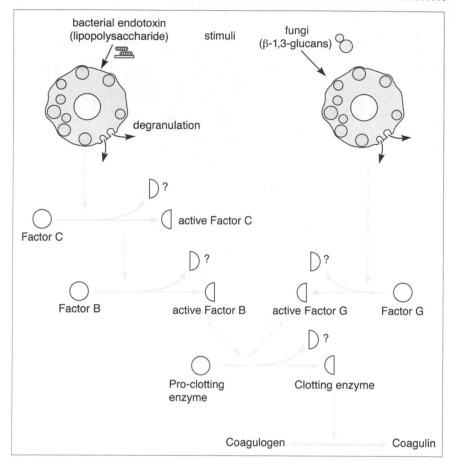

haemocyte, amoebocyte *n.* a blood cell of insects and other invertebrates.

a brown pigment, **melanin,** is deposited. It is likely that tissue exposed by the injury also triggers the clotting system, although the details of the insect cascade and how it is triggered are not clear.

1.3 Inflammation

inflammation *n.* localized response to tissue injury and invading microorganisms, characterized by infiltration by phagocytes, removal of foreign cells and debris, and tissue repair.

Inflammation is a complex physiological response that occurs in the epithelium. In mammals, inflammation is triggered in response to physical trauma (such as cuts, bruises or burns), microorganisms and a variety of other chemical irritants. Inflammation serves two functions: firstly to quickly infiltrate the local area with cells to repair any connective tissue damage; secondly, to allow the influx of cells and soluble components of the immune system required to eliminate any microorganisms that have gained entry through the damaged tissue. While this process has been studied most closely in mammals, the clotting reactions of invertebrates also fulfil these two functions and should be regarded as a kind of inflammatory reaction.

1.3.1 Inflammation in mammals

Within a few minutes of physical trauma to the skin, called the acute phase, three physiological changes occur (Figure 1.8).

- Local capillaries increase in diameter (**vasodilation**), causing the tissue to become reddened and warm.

- The stretching of local blood vessels causes them to become more permeable to fluid and protein, which leak into the tissue.

- Blood cells called **neutrophils** infiltrate the tissue. The function of these cells is to ingest any invading microorganisms by phagocytosis (see Section 1.4.1).

Inflammation is triggered by the release of preformed components from the granules of **mast cells** in the damaged skin. Mast cells are fragile and degranulation can be triggered directly by the trauma. Also, degranulation is triggered by the debris produced by damaged cells which lowers the pH of the extracellular fluid and activates the bradykinin cascade (Figure 1.5c). The end-product of this enzymatic cascade, bradykinin, binds to receptors on mast cells causing degranulation.

Mast cells release several mediators that trigger the acute phase inflammatory reaction. **Histamine** activates vascular endothelial cells, causing vasodilation and increased vascular permeability (something that does this is **vasoactive**). Bradykinin itself is also vasoactive. Histamine also induces the expression of **adhesion molecules** by vascular endothelial cells. This increases the adhesiveness of the inside, or luminal, surface of blood vessels for passing neutrophils. They then migrate out of the blood supply and into the tissue by pushing between the endothelial cells. This process is called **extravasation** and we will be examining the adhesion molecules that mediate the process later. Several chemical attractants, or **chemotactic factors**, are also released by mast cells, which attract infiltrating neutrophils to the site of injury.

Phagocytic cells in the skin called **tissue macrophages** also become engaged in the clearance of microorganisms. These cells also release digestive enzymes to assist in the breakdown of the material. The accumulation of fluid and cells in inflamed tissue causes the familiar swelling. During this phagocytic activity, macrophages become stimulated and release other important inflammatory components of their own (Table 1.2). These cause many local and whole-body (systemic) effects associated with the acute phase of inflammation.

vasodilation enlargement of blood vessels.

neutrophil n. a class of phagocytic blood cell in vertebrates.

phagocytosis n. the uptake of large particles (microorganisms, other cells, etc.) into a cell.

phagocyte n. a cell capable of **phagocytosis**.

extravasation n. the passage of leucocytes through a blood vessel wall into surrounding tissues by migration between the vascular endothelial cells (also known as **diapedesis** or **transendothelial migration**).

chemotaxis n. movement of motile cells to or from the source of a chemical stimulus.

Figure 1.8 Overview of the acute phase of inflammation in mammals. Trauma triggers mast cells in the tissue to release preformed vasoactive agents from granules. These cause (1) vasodilation, (2) increased vascular permeability and (3) increased adhesiveness of vascular endothelium for phagocytes.

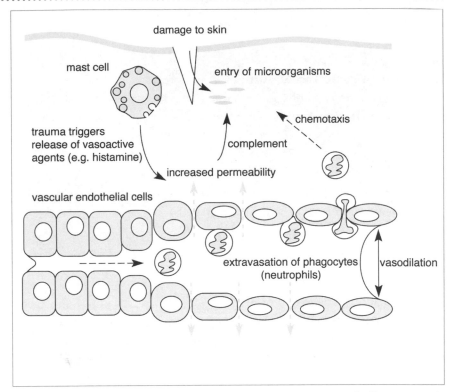

Locally, these effects include the increased expression of adhesion molecules on vascular endothelial cells, and increased vascular permeability. Systemic effects include elevation of the body temperature (**fever**). This is caused by hormones released by the hypothalamus in the brain. This has beneficial effects in fighting infections as it suppresses the growth of many microorganisms adapted to growing optimally at normal body temperatures. Fever is a common response to infection throughout the vertebrates. Endotherms (mammals and birds) rely on physiological responses to raise their body temperatures, whereas ectotherms (fish, amphibians and reptiles)

Table 1.2   Inflammatory mediators produced by activated macrophages

	tumour necrosis factor-α (TNF-α)	interleukin-1 (IL-1)	interleukin-6 (IL-6)
Local effects			
increase permeability of vascular endothelium	+	+	+
increase the adhesiveness of vascular endothelium for neutrophils	+	+	−
Systemic effects			
induce fever	+	+	+
induce liver hepatocytes to produce acute phase proteins	+	+	+

which are dependent on the external environment for their heat content, raise their body temperatures by behavioural means. Liver hepatocytes also respond by releasing **acute phase proteins** into the blood. These include C-reactive protein, serum amyloid A and mannose-binding protein, which all dramatically increase in concentration in the blood during the acute phase. Several of these proteins bind to polysaccharides on the cell walls of microbes and promote their removal by phagocytes bearing appropriate receptors.

The duration of the acute phase may last from a few hours to several days depending on the severity of the injury or the kinds of microbes that are present. Some bacteria, for example, are coated in a glycocalyx that protects them from phagocytic cells. These are particularly difficult to remove and the inflammatory response enters a longer term, or chronic phase reaction. During the chronic phase there is a gradual shift in the type of cells infiltrating from the blood, from neutrophils to **monocytes** and **T lymphocytes**. When circulating monocytes enter the inflamed tissues they mature into more tissue macrophages. These assist in the clearance of microorganisms and cellular debris. Because of the large numbers of phagocytic cells now involved, healthy bystander cells are often damaged by the amount of extracellular digestive enzymes. The dead cells and cellular debris accumulate with fluid to form **pus**.

monocyte *n.* a leucocyte in the blood that is the precursor of tissue macrophages.

T lymphocytes, in contrast, are non-phagocytic cells, although they are involved in effector functions in various other ways. We will encounter these cells a great deal in future chapters. T lymphocytes release inflammatory mediators that are the hallmarks of a chronic phase inflammatory reaction. These include **interferon-γ (IFN-γ)** and **tumour necrosis factor-α (TNF-α)** which both stimulate macrophages and neutrophils to become more potent at ingesting and destroying microorganisms.

In order to limit tissue damage and to allow cells to repair damage, other diffusible substances calm the inflammatory reaction. Among these is **transforming growth factor-β (TGF-β)** which is secreted by many cell types including epithelial cells and fibroblasts. A main effect is the inhibition of lymphocyte activity. It is also important in wound healing as it stimulates fibroblasts to grow and produce extracellular matrix. Over a period of several days or weeks the tissue is repaired by fibroblasts and other epithelial cells, and by extracellular matrix deposited by fibroblasts. This may result in scarring. Any particularly resistant microorganisms that remain at this stage are not eliminated but instead become sequestered inside a barrier of phagocytes and lymphocytes called **granulomatous tissue.**

1.3.2 Inflammation in invertebrates

Invertebrate inflammatory processes have been studied mostly in the arthropods. These appear to have a less complex system mediated by fewer components. However, these animals lack the vasculature that is present in vertebrates and have no need for vasoactive substances. Nevertheless, in common with the vertebrates, they depend on phagocytes to kill microorganisms and prevent the infection spreading around the body, and several invertebrates also have opsonic and chemotactic inflammatory mediators (see below).

Homologues of the mammalian macrophage-derived inflammatory mediators (Table 1.2) have been found in the blood of several invertebrates, including the annelids, echinoderms and urochordates. Their main role is to stimulate haemocyte proliferation and enhance phagocytic activity. These invertebrate molecules act on both mammalian and invertebrate cells, suggesting that receptors on mammalian cells can bind the invertebrate molecules. This indicates that the mammalian and invertebrate inflammatory mediators may be related and perhaps descended from common ancestral molecules in existence before the divergence of invertebrates and vertebrates.

Summary

- Microorganisms that live in or on hosts may cause disease if they are allowed to enter the hosts' tissues. These microorganisms are called pathogens. These have been a major influence leading to the evolution of inducible defence reactions.

- External physical and chemical barriers provide the first line of defence against entry of microorganisms. These may be breached if the barriers are damaged.

- Damage to the epithelial barriers initiates clotting reactions to seal haemorrhages, inflammatory reactions to expedite repair of damaged tissue, and to trigger immune effector mechanisms if microorganisms have gained entry.

- Inflammation in mammals involves mast cell degranulation and the release of vasoactive substances. These cause vasodilation, which increases blood flow and vascular permeability, and adhesiveness of vascular endothelial cells for phagocytic blood cells.

1.4 Effector mechanisms of immune systems

A third group of defence reactions are also triggered by microorganisms that breach the outer physical and chemical barriers and gain entry to the body. These reactions kill or neutralize the microbes, and are largely the responsibility of the **immune system**. As might be predicted, there are numerous examples of endoparasites that have evolved counter-strategies to escape elimination by the immune system. In general, these strategies are used by highly specialized microorganisms that have adapted to a particular environment inside the body of the host, rather than by the opportunists that enter through damaged skin.

In the mammals it has been convention to divide responses by the immune system into **cell-mediated** and **humoral** immunity, according to whether immunity to microbes could be transferred from one mouse to another in either the cellular or fluid component of the donor's blood, respectively. In reality, all responses require both cellular and humoral components, and the terms are rather misleading. Also, there is considerable debate over whether such terms apply to invertebrates, or indeed whether there is an invertebrate 'immune system' at all. For the purposes of this book, any reactions that kill or neutralize microbes will be considered part of an immune system. Some of the more commonly used models for the study of invertebrate immunity are listed in Table 1.3.

The problem of killing or neutralizing microorganisms has been solved in a variety of different ways in different animals. Some are more complex than others but they are all adequate to protect the particular animals within which they operate. Vertebrates and invertebrates have, on the whole, evolved different ways of doing things. It is important to remember that the divergence of the protostomes (which is the lineage containing the major invertebrate groups of arthropods, annelids and molluscs) and the deuterostomes (echinoderms, hemichordates, protochordates and chordates) occurred over 500 million years ago, and these two lineages have been evolving separately ever since. Only a few molecules seem to be involved in the immune systems of both lineages, and may therefore predate the divergence. These are considered more ancient and primitive features of host defence. Most features are peculiar to one or other lineage and so probably arose independently since the divergence.

Irrespective of the animal we examine, responses by the immune system have two key events. The first is the recognition of the microorganisms, followed by the second step – triggering of **effector mechanisms** – which effects the killing or neutralization of the invading microbes. Although this

immune system *n.* the cells and tissues that collectively recognize and eliminate invading microorganisms, parasites, tumour cells, etc., from the bodies of vertebrates and invertebrates.

cell-mediated immunity *n.* originally used to describe immunity that could be adoptively transferred to irradiated recipient mice by cells and not antibodies; now used to describe responses by effector T lymphocytes (i.e. cell-mediated cytotoxicity or delayed-type hypersensitivity), and sometimes includes NK cells and macrophages.

humoral immunity *n.* originally used to describe immunity that could be adoptively transferred to irradiated recipient mice by antibodies rather than by cells; antibody-mediated immunity.

Table 1.3

Arthropoda	Annelida	Mollusca	Echinodermata	Urochordata
Limulus polyphemus, Tachypleus tridentatus (horseshoe crabs)	*Lumbricus terrestris, Eisenia foetida* (earthworms)	*Mytilus edulis* (mussel)	*Strongylocentrotus purpuratus* (sea urchin)	*Botryllus, Clavelina picta, Ciona intestinalis* (tunicates)
Pacifastacus leniusculus Astacus astacus (crayfish)		*Crassostrea* spp. (oysters)		
Carcinus maenas (crab)		*Helix* spp., *Planorbarius corneus , Biomphalaria glabrata, Lymnea stagnalis* (snails)		
Locusta migratoria (locust)				
Hyalophora cecropia, Bombyx mori (silkmoths)				
Manduca sexta (tobacco hornworm) *Heliothis* spp. (noctuid moths)				
Drosophila melanogaster (fruitfly)				
Periplaneta americana, Blatta orientalis, Blaberus spp. (cockroaches)				
Sarcophaga (fleshfly)				

endocytosis *n.* the process by which eukaryotic cells ingest extracellular components by the invagination and pinching off of the plasma membrane to form an endocytic vesicle (**endosome**); can be divided according to the size of the endosome into **phagocytosis** and **pinocytosis**.

has been achieved in many different ways (summarized in Figure 1.9), all the mechanisms are based on two basic strategies.

- Disruption of the cell wall or membrane of the microorganism. This may occur either within intracellular compartments of phagocytes, or extracellularly by the action of soluble membrane destabilizers. Most effector mechanisms operate in this way.

- Enclosure of the microorganisms in a multicellular sheath of cells to sequester them from the rest of the body. This is a more common effector mechanism in the invertebrates.

1.4.1 Phagocytosis

Phagocytosis is a type of active uptake by cells (endocytosis) by which cells ingest particulate material such as cellular debris or bacteria. This is the most primordial and important of all cell-mediated effector mechanisms. Phagocytosis probably has its evolutionary roots in the feeding mechanisms of free-living single-celled organisms, such as that seen in

	phagocytosis	opsonins	encapsulation/ nodulation	agglutination	membrane disruption	
					antimicrobials	assembly of pores in membranes
simple invertebrates						
sponges	amoebocytes	?	no	no	?	?
coelenterates	amoebocytes	?	amoebocytes	?	?	?
protostomes						
molluscs	haemocytes	lectins	haemocytes	lectins	lysins, molluscan antibiotics	?
annelids	coelomocytes	lectins	coelomocytes	lectins	lysozyme lysins	?
crustaceans	haemocytes	lectins, components of PO pathway, limulin	haemocytes	components of PO pathway	components of PO pathway	?
insects	haemocytes	lectins, components of PO pathway	haemocytes	components of PO pathway	lysozyme, cecropins, attacins, defensins, components of PO pathway	?
deuterostomes						
echinoderms	amoebocytes	lectins, complement-like components	coelomocytes	mucoid substances	?	?
tunicates	leucocytes	lectins	leucocytes	lectins	?	?
jawless fish	leucocytes (macrophages granulocytes)	complement components, C-reactive protein	no	lectins	defensins	?
jawed vertebrates	leucocytes (macrophages granulocytes)	complement components, C-reactive protein antibodies	no	immune complexes	defensins lysozyme	perforins, complement membrane attack complex

Figure 1.9 A summary of known effector mechanisms found in different animals. PO, = prophenoloxidase.

Figure 1.10 Phagocytosis as a feeding mechanism. Scanning electron micrograph of the protozoan amoeba, *Acanthamoeba*, feeding on *Pseudomonas fluorescens*, a common soil bacterium. This amoeba has trapped numerous bacteria over its surface which, when the amoeba moves off, will be swept posteriorly into a cluster and ingested. Micrograph courtesy of T.M. Preston, University College, London [Preston, T.M. and King, C.A. (1984) *Journal of General Microbiology*, **130**, 1449–58].

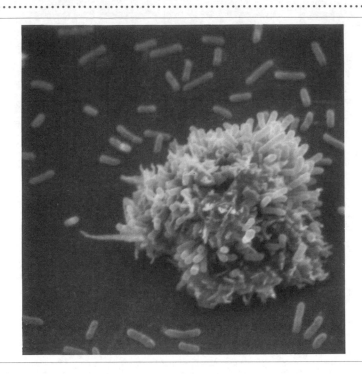

modern amoebae (Figure 1.10). In simpler multicellular invertebrates, like sponges and coelenterates, phagocytosis is retained as the major means for capturing food. Sponges are colonies of essentially autonomous cells with specialist functions (Figure 1.11). Within the cells of the sponge are wandering phagocytic cells called **amoebocytes** which ingest larger particulate material for food reserves. Even in the simplest true multicellular animals (the metazoans), phagocytosis continues to perform a feeding function. For example, in turbellarians (flatworms), a simple blind sac serves as a digestive cavity. The wall of the cavity is a single layer of phagocytic and excretory cells. Fragments of food broken down by the excreted enzymes are ingested by the phagocytes where further digestion occurs.

In more complex metazoans we see a reduced involvement of phagocytes in digestive processes. This corresponds with the possession of a true digestive tract, consisting of cells specialized for nutritive absorption, and enclosing a lumen where enzymatic degradation of food occurs. In essence, the digestive enzymes previously contained within digestive vacuoles inside the phagocyte are now contained within the lumen of the gut. Such an arrangement is more efficient and allows the gut to sustain a larger organism.

The hostile environment created by having digestive enzymes within the gut lumen has necessitated the use of protective mucus, which limits the effectiveness of phagocytosis. Free-living phagocytes are retained in these

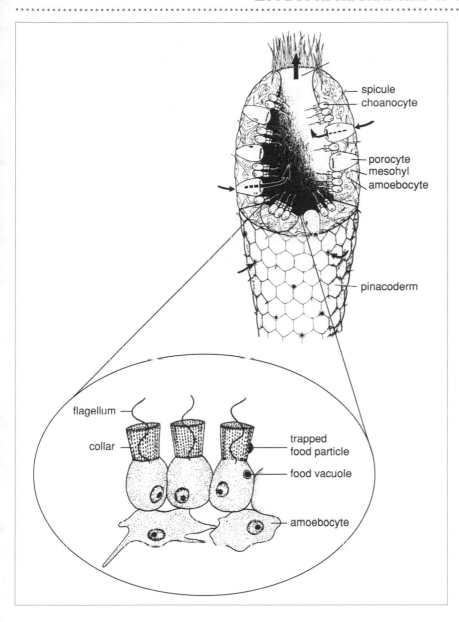

spicule
choanocyte

porocyte
mesohyl
amoebocyte

pinacoderm

flagellum

collar

trapped
food particle

food vacuole

amoebocyte

Figure 1.11 Organization of an ascanoid sponge. The colony is organized into a system of chambers, through which water is circulated by flagellated cells called **choanocytes** lining the chambers. The circulating water supplies oxygen and particulate nutrients to the colony, and removes waste material. Arrows indicate the direction of water flow. The gelatinous protein matrix contains skeletal spicules and wandering phagocytic cells called ameobocytes. Particulate material is filtered by choanocytes that are then ingested by the amoebocytes. Reproduced from Barnes, R.D. (1974) *Invertebrate Zoology,* 3rd edition, copyright © 1974 by Saunders College Publishing, by permission of the publisher.

animals, although their function is in defence against microorganisms.

In all cases the mechanics of the process are similar (Figure 1.12). Phagocytosis is triggered by the attachment of the particle to be ingested to the plasma membrane of the phagocytic cell. This is followed by an active process in which the phagocyte plasma membrane increases the area of contact with the particle. Finally, when the particle is entirely enclosed, the invaginated plasma membrane pinches off to release the particle-containing vacuole into the cytoplasm of the cell. This category of endocytic compartment, or **endosome**, is called a **phagosome**.

endosome (endocytic vesicle) *n.* an intracellular vesicle produced by invagination of the plasma membrane (see **endocytosis**).

Figure 1.12 The sequence of events in phagocytosis.

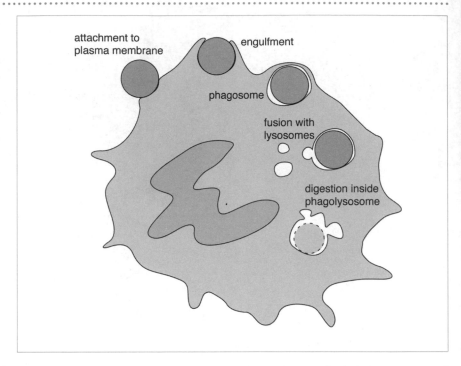

lysosome *n.* a membrane-bound organelle of eukaryotic cells that contains digestive enzymes at acidic pH; the endpoint of the endocytic pathway in which extracellular material is ingested and broken down.

lytic *a.* a toxin, virus or other agent that causes cell lysis (haemolytic, cytolytic).

Vesicles (**lysosomes**) containing toxic compounds, lytic enzymes and proteinases fuse with the phagosome to yield a **phagolysosome**, and the contents are killed and/or broken down. The components in the lysosomes of mammalian macrophages and neutrophils are listed in Table 1.4. These include reactive oxygen intermediates, toxic nitrogen intermediates, defensins (see below) and a cocktail of proteinases and other enzymes that

Table 1.4 Toxic components of mammalian macrophages and neutrophils that mediate killing and digestion of ingested microorganisms

Microbicidal agents	reactive oxygen intermediates:
	superoxide radical (O_2^-)
	hydrogen peroxide (H_2O_2)
	singlet oxygen ($^1O_2$)
	hydroxyl radical (OH^-)
	toxic nitrogen intermediates:
	nitric oxide (NO)
	nitrite (NO_2^-)
	nitrate (NO_3^-)
	defensins (Section 1.4.3)
Enzymes	lysozyme
	phospholipase A
	RNase
	DNase
	proteinases

help break down the ingested material. The reactive oxygen intermediates are produced by a process called the **respiratory burst,** and are very toxic to microorganisms. There is evidence that phagocytic blood cells of crustaceans and tunicates can also produce a respiratory burst.

The efficiency of phagocytosis is enhanced in several ways. One is if attachment is mediated by receptors on the membrane of the phagocyte that bind to complementary structures on the surface of the particle (Figure 1.13). In larger metazoan animals, phagocytes are almost entirely dependent on receptor-mediated uptake, although in their absence low-level uptake may continue to occur through mechanisms based on physical properties such as electrostatic charge. Receptor-mediated uptake is usually achieved by the deposition of a substance called an **opsonin** onto the particle to be ingested. Most opsonins we know of have been found in mammalian blood. Several are generated by the complement cascade in response to microorganisms (see below). Antibodies, which we discuss in future chapters, are also opsonins. Mammalian phagocytes have receptors for these opsonins which enhance their phagocytic activities.

A ubiquitous opsonin is C-reactive protein. As we saw earlier, this appears in the serum of vertebrates in the acute phase of inflammatory reactions, particularly in response to components in the cell walls of bacteria. C-reactive protein has been found in most vertebrates, including

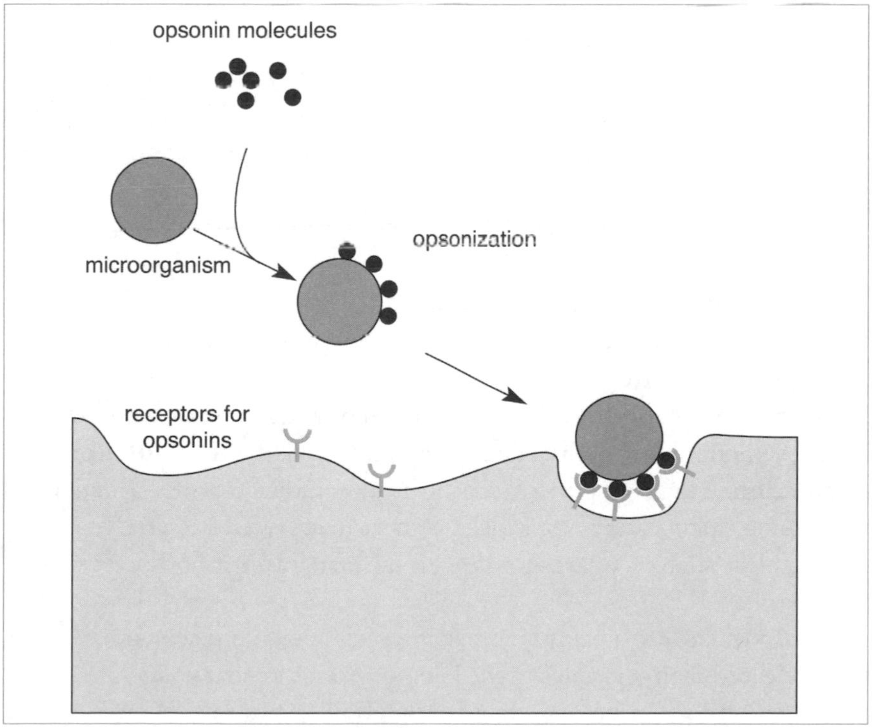

tunicates *n. plu.* sub-phylum of chordates (urochordates) consisting of sea squirts.

opsonin *n.* a substance that attaches to microorganisms and promotes their phagocytosis by cells bearing receptors for the opsonin.

opsonization *n.* the process in which a microorganism or antigen is coated with opsonins prior to ingestion by phagocytic cells.

opsonin molecules

microorganism

opsonization

receptors for opsonins

Figure 1.13 The enhancement of phagocytosis by opsonization.

mammals, amphibians and fishes. C-reactive protein has several useful properties. In all species it binds to polysaccharides in the cell walls of many microorganisms and acts as an opsonin. In mammals, C-reactive protein also triggers the alternative complement pathway which results in the deposition of a complement opsonin, C3b (see below). C-reactive protein has also been found in the horseshoe crab, *Limulus* (termed **limulin**), where it also appears to opsonize microorganisms, suggesting this represents an ancient form of defence.

The efficiency of phagocytosis is also increased if phagocyte movement is not random, and directed instead toward the microorganisms. As we saw exemplified by the activities of mammalian neutrophils in inflamed tissue, this is achieved by the release of locally acting chemotactic factors. Directed movement toward the source of a chemotactic factor, called **chemotaxis**, is achieved by the migration of the phagocyte in the direction of increasing concentration of the chemotaxin. Once again, several chemotactic factors are generated by the activated complement cascade of mammals (see below).

The mammalian tissue macrophage is a typical phagocytic cell. Normally in a state of quiescence, macrophages can be rapidly activated by various stimuli to become more mobile and with more potent microbicidal properties. These stimuli act through receptors on the surface of the cell, and include receptors for opsonins, chemotactic factors, the C-termini of antibody molecules (see Chapter 2), some extracellular matrix proteins, and various diffusible signalling agents. A particularly potent stimulator of mammalian macrophages is IFN-γ which, as we learned earlier, is secreted by some T lymphocytes. Activation enhances the phagocytic and microbicidal activities of macrophages, and stimulates them to release their own factors that have bactericidal or inflammatory properties.

1.4.2 Encapsulation and nodulation

In simpler invertebrates, cell-mediated defence against invading microorganisms is mounted entirely by phagocytosis. In more complex invertebrates, notably the arthropods, annelids and molluscs, phagocytosis is supplemented by multicellular responses called **encapsulation** and **nodulation**. These responses can be readily induced when a paticulate suspension, such as carbon particles or heat-inactivated bacteria, or a large foreign body such as a parasitoid egg, is injected into the body cavity of an insect larva.

Small particles are quickly clumped in a sticky coagulum, which is thought to be released from granules in the haemocytes. A similar coagulum is also deposited on the surface of larger foreign bodies within a few minutes of

parasitoid *n.*
Hymenopteran wasps that are partly free-living and partly parasitic. Adults lay their eggs on or in the larvae of other (host) insects which are consumed by the parasitoid larvae.

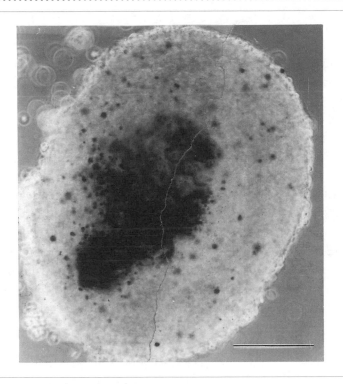

Figure 1.14 Nodulation in insects. A nodule formed 3 days after the experimental injection of carbon particles into larvae of the moth *Heliothis*. The particles have become enclosed in a multicellular capsule of blood cells. Bar = 0.2 mm. Reproduced from Davies, D.H. and Vinson, S.B. (1988) *Cell and Tissue Research,* **251**, 467–75.

injection. Over the course of the next few hours, the foreign material is completely enclosed by numerous haemocytes which, over the following few days, gradually consolidate to produce a cohesive capsule (Figure 1.14). In general, the response to numerous small particles is called **nodulation** and the response to the larger foreign bodies that cannot be ingested by single phagocytes is called **encapsulation**. Nodulation occurs particularly in response to large numbers of particles that cannot be eliminated by phagocytosis alone. These reactions serve primarily as a barrier to sequester the invading microorganisms or parasites, which may be subsequently killed by toxic products released by the blood cells or by simple starvation or asphyxiation.

encapsulation *n.* a defence reaction of many invertebrates in which macroscopic foreign bodies gaining entry to the coelom are enclosed and sequestered in a multicellular sheath of haemocytes.

1.4.3 Microbicidal agents

Many different kinds of antibacterial compounds increase in concentration in the blood of invertebrates and vertebrates after infection. Many of these are low molecular polypeptides, smaller than 100 amino acids long, and which operate by damaging the membranes of microorganisms. These include **cecropins**, **attacins** and **defensins**.

In invertebrates, inducible antibacterial peptides form a major line of defence against infection. Cecropins, named after the silk moth *Hyalophora cecropia* in the larvae of which they were first discovered, have been found in many species of insects. These compounds have antibacterial activity against a broad spectrum of both Gram-positive and Gram-negative bacteria and are thought to form channels in bacterial membranes. Another family of insect proteins are the **attacins**, which are active mainly against Gram-negative bacteria. Cecropins and attacins operate synergistically and together play a more significant role in controlling infections than lysozyme.

Defensins are also low molecular weight antimicrobial polypeptides that are active against bacteria, fungi and viruses. Defensins are found within granules of phagocytic blood cells, including macrophages and neutrophils of mammals. These defensins are active in endosomes wherein they operate to disrupt the membranes of ingested microbes (Table 1.4). Defensins are also found in the blood and tissues of humans, normally at low concentrations but which rise in response to heavy infections with microorganisms. Defensins are also found in intestinal cells of mice and humans, called Paneth cells, which secrete the defensins onto the epithelial surface of the intestine.

Defensin-like peptides have also been discovered in the blood of several species of insects, and are similar in structure to the membrane-disruptive venom of scorpions. Unlike mammalian defensins, insect defensins are strictly bactericidal and have no activity for fungi or viruses. The structure of insect defensins also differs markedly from that of mammalian defensins and it seems unlikely that these two families evolved from a common ancestor. It is more likely that mammals and arthropods acquired these compounds independently.

Interferons are diffusible factors released by cells infected with viruses. They have been found in most vertebrates, including fish, although they seem absent from invertebrates. These confer a state of enhanced viral resistance to uninfected cells nearby (see Chapter 8).

1.4.4 The phenoloxidase cascade of invertebrates

The phenoloxidase cascade is another invertebrate effector mechanism, studied mainly in arthropods, echinoderms and gastropods, and triggered by components in the cell walls of microorganisms. A variety of products with microbicidal, chemotactic and opsonic activities are generated. The key component, phenoloxidase, oxidizes phenol derivatives to quinones, which are converted into melanin by a series of non-enzymatic reactions. Melanin is widely used as a pigment. It is also able to bind to the amino groups on proteins and is used by arthropods to toughen their outer cuticles. It is now appreciated that melanin, quinones and other phenolic components have powerful microbicidal properties.

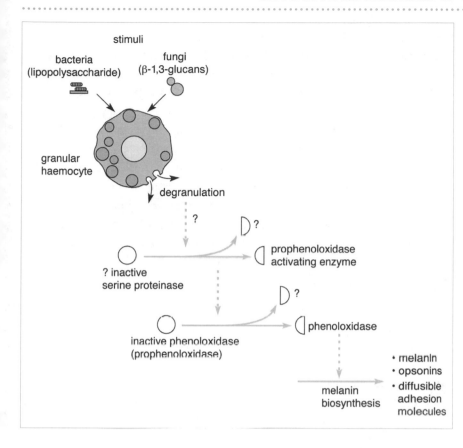

Figure 1.15 The prophenoloxidase pathway of the crustacean *Pacifastacus* (crayfish). Phenoloxidase is an enzyme that oxidizes phenol derivatives to quinones which are used in melanin synthesis. Phenoloxidase is stored as an inactive precursor, prophenoloxidase, in secretory granules of granular blood cells. It is activated by cleavage by prophenoloxidase-activating enzyme, a proteinase that is also sequestered in granules. These are released into the blood when degranulation of the granular cells is triggered by components in the cell walls of microorganisms. Several other components have been identified. These include other enzymes, β-1,3-glucan-binding proteins, and proteins with opsonic and chemotactic properties. Adapted from the work of Söderhäll, K., *et al.* (1994) *Annals of the New York Academy of Sciences,* **712**, 156–61.

In the crayfish, phenoloxidase is located in the plasma as an inactive prophenoloxidase form (Figure 1.15). It is cleaved by the activation of proteinases, which are themselves activated by lipopolysaccharide of bacteria and certain polysaccharides (mannans and glucans) in the cell walls of yeast and fungi. In other crustaceans, such as crabs, activators of the phenoloxidase cascade also cause the production of factors with chemotactic, opsonic and encapsulation-promoting properties for the blood cells.

In insects, phenoloxidase is located within granules of haemocytes rather than in the blood. These are stimulated to degranulate by components in the cell walls of microorganisms, which become coated and entrapped in a coagulum generated by the cascade. Other haemocytes are recruited into the area by chemotactic factors generated, which then proceed to encapsulate the agglutinated mass. Melanin is deposited in the core of nodules and capsules, which contributes to the killing of the microorganisms at the centre. In smaller insects such as diptera, nodules and capsules are made entirely from deposited melanin.

plasma *n.* the liquid part of body fluids (blood, lymph, milk).

lipopolysaccharide (LPS) *n.* one of the components of Gram-negative bacteria.

1.4.5 The complement cascade of vertebrates

The **complement cascade** is found only in vertebrates. It is analogous to the phenoloxidase cascade of invertebrates in many respects: it is triggered by components in the cell walls of microbes, and its main function is to generate opsonins and chemotactic factors. However, these two cascades show no homology and it is likely that they evolved independently.

Complement is an enzymatic cascade of approximately 30 cell-surface and serum proteins found in vertebrates. The complement system of mammals, particularly of humans, is the best understood (Box 1.1). Its most important function is the generation of opsonins and chemotactic factors that help in the removal of invading microbes by phagocytosis. An additional, but less important, product that can be produced by the cascade is the assembly of a **membrane attack complex.** This forms pores in the membrane of susceptible microorganisms which may disrupt the membrane, causing cell death.

BOX 1.1 The mammalian complement cascade

Complement is an enzymatic cascade of approximately 30 cell-surface and serum proteins. The nature of antibodies and the thioester bond are discussed in the next chapter. You may wish to read this Box after you have read Chapter 2.

The alternative pathway (Figure 1a, left)

Central to this pathway is an abundant serum protein called **C3.** When the cascade is activated, enzymes called **C3 convertases** are generated that cleave C3 into two fragments, C3a and C3b. C3 has an internal **thioester bond** which becomes exposed on the larger C3b fragment when C3 is cleaved. The thioester bond of C3b, which is very unstable when it becomes exposed, will readily form covalent bonds indiscriminately with any amino or hydroxyl group of a nearby amino acid (this is explained in more detail in the next chapter). If this binding occurs on the surface of a cell, the alternative pathway can be triggered, whereas if such a substrate is unavailable the thioester bond reacts harmlessly with water (hydrolysis).

Once C3b is bound to a cell surface, a component called **factor B** associates to

Figure 1 Overview of the mammalian complement cascade. The shaded bars represent membrane, to indicate which components are membrane bound. Enzymatic cleavages are depicted by the dashed arrows. (a) The alternative and classical pathways. Modified with permission from Kinoshita, T. (1991) *Immunology Today,* **12**(9), 291–5. (b) The lytic pathway.

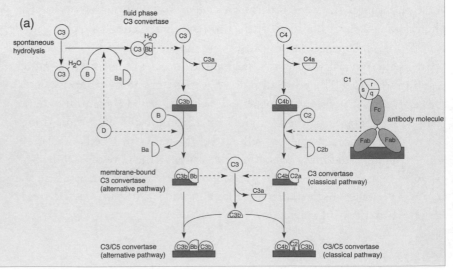

produce C3bB. This is then cleaved by another serum component, **factor D,** to yield a membrane-bound **C3 convertase**. A single molecule of C3 convertase can proceed to cleave several hundred C3 molecules, which causes a positive feedback loop and an explosive amplification of the original signal. Because reactive C3b molecules are unstable in water they cannot diffuse far before being rapidly hydrolysed. For this reason, C3b molecules that bind to the membrane tend to cluster around the membrane-bound C3 convertase.

The alternative pathway is constantly being triggered because the thioester bond inside serum C3 undergoes spontaneous hydrolysis. This produces C3-H$_2$O which has C3b-like properties. Like C3b, C3-H$_2$O can react with factors B and D to produce a soluble, or fluid-phase, C3 convertase. Alternatively, the thioester bond of C3 binds with a hydroxyl group (-OH) on cell membranes instead of with water, which also reacts with factors B and D to produce a membrane-bound C3 convertase (not shown). By either route, additional C3 can then be cleaved into C3a and C3b.

A regulatory system, consisting of inactivators in the serum and on cell surfaces, blocks the amplification loop and prevents healthy cells in the body triggering the cascade. Two membrane receptors, complement receptor 1 (**CR1**) and membrane cofactor protein (**MCP**), or a serum component, **factor H**, can each bind C3b, while a serum proteinase, **factor I**, degrades it into C3bi and soluble C3f. Factor I continues to degrade C3bi into smaller fragments, C3c and C3dg. Moreover, the C3 convertase (C3bBb) is rapidly degraded unless it is stabilized by **factor P**. This is accelerated by decay-accelerating factor (**DAF**), a membrane-attached molecule that blocks C3b from associating with factor B and so decreases the net amount of C3b deposited. Together, these factors prevent the positive feedback loop from occurring and maintain the spontaneous cleavage of C3 to a constant but low, 'tickover' level. No such inhibitory mechanisms are found in the cell walls of most microorganisms such as bacteria. Their surfaces allow the amplification loop to occur and the rapid deposition of C3b.

C3a and C3b mediate several important reactions (Table 1.5 in the text). C3b is also involved in advancing the cascade, by combining with C3 convertase (C3bBb) to produce a **C5 convertase**. This leads into the lytic phase of the cascade, culminating with the assembly of the transmembrane membrane attack complex from components C5 to C9 (see after the classical pathway).

The classical pathway (Figure 1a, right)

This pathway is triggered by the surfaces of microorganisms indirectly, via the binding of antibodies to foreign surfaces. We will examine the molecular details of antibodies in a later chapter, but in short, antibody molecules are produced only in vertebrates and act as the recognition molecules for triggering certain effector mechanisms. Antibodies

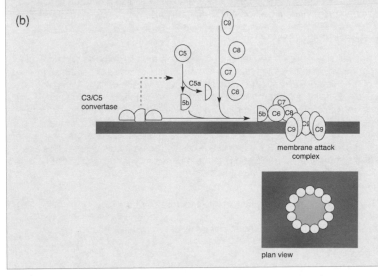

BOX 1.1 *continued*

mark foreign surfaces for activation by the classical pathway, but can also opsonize microorganisms and enhance uptake by phagocytes bearing receptors for the Fc domains of antibodies. Antibodies also agglutinate microorganisms into clumps called immune complexes which enhances their entrapment and clearance by phagocytes.

The first component in the cascade is a complex of serum complement proteins called C1, which binds to the Fc region of antibodies in immune complexes. The molecular interaction between C1 and the Fc region will be discussed in more detail when we examine the structure of antibodies (Chapter 4). In so doing, C1 acquires enzymatic activity. The next component in the sequence of events is C4. This is cleaved by activated C1 into two smaller products, C4a and C4b. These two fragments are homologous to the C3a and C3b cleavage products of C3 in the alternative pathway (indeed the genes of C3 and C4 seem to have arisen by the duplication of a common ancestral gene). Like C3a, the smaller C4a fragment initiates inflammatory reactions because it is both a chemotactic factor for phagocytes and an anaphylatoxin (see p.29) that causes mast cells to degranulate.

Like C3b, the larger C4b fragment has an exposed and highly labile thioester group able to form covalent bonds with any nearby surface and further the cascade. Several C4 molecules can be cleaved by a single activated C1, which may deposit many C4b molecules on the foreign surface around the C1 complex. This serves to amplify the original signal. Also like C3b, C4b is opsonic and enhances phagocytosis of the microorganism by phagocytes bearing the CR1 complement receptor.

There are also similarities in the inhibitors of these components. C4b is rapidly degraded into by the serum enzyme, factor I, unless the next component in the cascade, C2, is recruited. If C4b is deposited on a cell surface it engages C2 and becomes protected from factor I. Recall that factor I also cleaves C3b into C3d. Factor I requires cofactors, either factor H or the CR1 complement receptor, before it will cleave C4b. DAF, located on the surface of endothelial cells, prevents C4b from associating with C2. Decay-accelerating factor also prevents C3b associating with factor B in the alternative pathway. Together, these inhibitory factors protect cells in the body from damage caused by activation of the complement cascade.

If C4bC2 complexes form in close proximity to an activated C1 complex, the C2 is cleaved by C1 into two fragments. Cleaved C2 is an active proteinase, with the active site residing in the larger C2b fragment. The smaller C2a fragment remains non-covalently associated to C2b. This activated C4bC2 complex is the C3 convertase (or more precisely, the C3/C5 convertase) of the classical pathway. This splits serum C3 into C3a and C3b. As we learned earlier, C3b becomes the focus for the assembly of more C3/C5 convertase, triggering an uncontrolled amplification through the alternative pathway. The C4bC2 complex has a tendency to dissociate spontaneously, which helps limit any damage to self cells. The dissociation is also promoted by DAF and an additional factor in the blood called C4-binding protein.

The lytic pathway (Figure 1b)

This comprises the downstream sequence of events common to both the classical and alternative pathways. In this pathway the C5–C9 membrane complex assembles in the membrane of the microorganism. C5 is cleaved by the active site of C2 within the C4bC2 complex. This is the last enzymatic reaction of the cascade, with the subsequent steps occurring as a chain reaction. C5a, another anaphylatoxin, diffuses away. C5b remains bound to the C4bC2 complex and then binds C6, which in turn binds C7, which in turn binds C8. This causes damage to the membrane. Several copies of the final component, C9, enlarge the hole by forming a hollow tube. This kills the microorganism, either by destabilizing the membrane or by inducing osmotic lysis.

Further reading

Kinoshita, T. (1991) Biology of complement: the overture. *Immunology Today,* **12,** 291–5.
Frank, M.M. and Freis, L.F. (1991) The role of complement in inflammation and phagocytosis. *Immunology Today,* **12,** 333–6.

Table 1.5 Biological properties of complement cascade cleavage products

Product	Property	Notes
C3a	anaphylatoxic chemotactic	causes mast cells and basophils to release vasoactive agents such as histamine attracts phagocytes
C3b	opsonic	bound by CR1 (CD35[a]) on phagocytes CR1 also involved with regulatory factor I in degrading C3b
	antigen trapping	CR1 receptor on follicular dendritic cells in lymphoid follicles (see Chapter 3)
C3bi	opsonic	bound by CR3 (CD11b/CD18) and CR4 (CD11c/CD18) on phagocytes and natural killer cells
C4a	anaphylatoxic	causes mast cells and basophils to release vasoactive agents such as histamine
C4b	opsonic	Bound by CR1 (CD35) on phagocytes CR1 also involved with regulatory factor I in degrading C4b
C5a	anaphylatoxic haemostatic chemotactic others	causes mast cells and basophils to release vasoactive agents such as histamine causes platelet aggregation attracts phagocytes increases expression of complement receptors (CR3 and CR4) on neutrophils

[a] The cluster of differentiation (CD) system of leucocyte markers is covered in detail in Appendix 1. CR, complement receptor.

The details of the mammalian complement cascade are considerably more complex than the phenoloxidase cascade. The cascade is triggered by foreign surfaces in two ways, leading to two different but homologous pathways called the **alternative pathway** and the **classical pathway**. Each pathway converges on a pivotal enzyme at the centre of both, called **C3 convertase**. In the process, a variety of components are generated (Table 1.5). C3b is an opsonin, which is recognized by receptors on the surface of neutrophils, and macrophages. C3a diffuses into the surrounding tissues and causes mast cells to degranulate and release histamine (i.e. it is an **anaphylatoxin**). As we saw earlier, histamine is vasoactive and has several effects on local blood vasculature. C3a is also a chemotactic factor, which together with chemotactic factors released by mast cells, guides phagocytes to the site of infection. Thereafter, both pathways can use a common sequence of components that culminate in the assembly of the membrane attack complex. This terminal sequence of events which can be reached by either stimulus is called the **lytic pathway**.

1.4.6 Cell-mediated cytotoxicity

Cell-mediated cytotoxicity is performed by cytotoxic or 'killer' cells that directly lyse other cells in the body that have become infected by intracellular parasites, or cells that have become tumour cells. The process requires direct physical contact between the cytotoxic cell and the target cell. The molecular details of the killing mechanism are best understood in mammals.

Several types of blood cells are equipped with this effector function. The best studied are **cytotoxic T lymphocytes** and **natural killer** (or **NK**) **cells**. The method used by these two types of effector cells to recognize an

cytotoxic *n.* an action that destroys cells.

cytotoxic T lymphocyte (CTL) *n.* a subset of T lymphocytes that kill target cells, usually tumour cells or cells infected with viruses.

natural killer (NK) cell *n.* a type of lymphocyte that lacks antigen receptors and displays cell-mediated cytotoxicity (used to be called null cells).

Figure 1.16 Sequence of events in the lysis of a transformed target cell by a mammalian cytotoxic effector cell (NK cell or cytotoxic T lymphocyte).

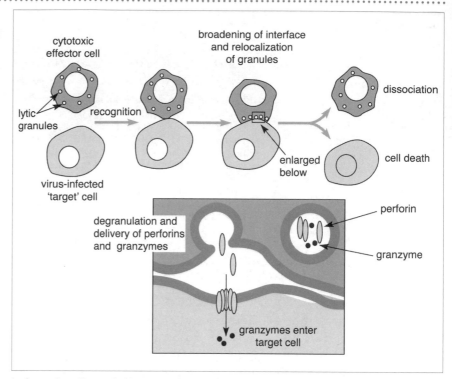

infected cell is different, although the effector mechanism is similar. Cytotoxic cells kill their target cells by first making contact (Figure 1.16) whereupon granules within the cytotoxic cell move to the region of contact and release their contents onto the target cell surface. The cells then dissociate. After dissociation the cytotoxic cell can recycle to engage several other target cells.

The granules contain several active components, including **perforins** and **granzymes.** Perforins are subunits that assemble in the target cell plasma membrane as 16 nm pores. Perforin is similar to the C9 component of the membrane attack complex of complement and the two may have evolved from a common ancestral gene. Granzymes are a family of at least eight proteinases and esterases, which enter the cytoplasm of the target cell through the perforin pores. This triggers a cascade of events leading to 'suicide' of the target cell (see Box 9.2).

perforin *n.* a protein released by **cytotoxic T lymphocytes** and NK cells that polymerizes with other perforin molecules to produce pores in the membranes of target cells.

Summary

• If microorganisms gain entry at sites of injury, inducible effector mechanisms are elicited. These often involve the perturbation of the membranes of living cells, either within phagocytes, or by soluble components that operate to disrupt membranes extracellularly.

- Phagocytosis is the most primordial of cellular effector mechanisms. Wandering phagocytic cells are found in invertebrates and vertebrates alike. The efficiency of phagocytosis is enhanced when (1) opsonins are deposited on the microbial surface and (2) chemotactic factors guide the migration of phagocytes to the site of infection.

- A variety of inducible microbicidal peptides have been found in invertebrates and vertebrates. These act to destabilize microbial cell walls.

- Several inducible effector mechanisms are enzymatic cascades. These allow (1) a rapid mobilization of the effector mechanism, (2) amplification of the original signal and (3) the simultaneous production of other components (such as inflammatory mediators, opsonins and chemotactic factors).

- In many invertebrates, blood clotting and melanin biosynthesis are enzyme cascades that generate opsonins and chemotactic factors. Invading microorganisms are first agglutinated with a sticky coagulum. These are then encapsulated in a sheath of haemocytes to form nodules and capsules.

- Vertebrates uniquely have complement cascades. Opsonins, chemotactic factors and inflammatory agents are by-products of the complement cascade. The end-product is the assembly of pores (the membrane attack complex) in the membrane that causes lysis of microorganisms.

- Vertebrates also have cytotoxic cells (NK cells and cytotoxic T lymphocytes) which lyse tumour cells and cells infected with viruses. This is achieved by the release of perforins that assemble in the target cell membrane to form pores. Perforin and C9 of the complement membrane attack complex are structurally similar.

Further reading

Asman, R.B. and Mullbacher, A. (1984) Infectious disease, fever and the immune response. *Immunology Today*, **5**(9), 268–71.

Baumann, H. and Gauldie, J. (1994) The acute phase response. *Immunology Today*, **15**(2), 74–80.

Ganz, T. and Lehrer, R.I. (1994) Defensins. *Current Opinion in Immunology*, **6**, 584–9.

Hoffmann, J.A. and Hetru, C. (1992) Insect defensins: inducible antibacterial peptides. *Immunology Today,* **13**(10), 411–15.

Liu, C.-C. *et al.* (1995) Perforin: structure and function. *Immunology Today,* **16**(4), 194–201.

Podack, E.R. and Kupfer, A. (1991) T cell effector functions: mechanisms for delivery of cytotoxicity and help. *Annual Review of Cell Biology,* **7**, 479–504.

Smyth, M.J. and Trapani, J.A. (1995) Granzymes: exogenous proteinases that induce target cell apoptosis. *Immunology Today,* **16**(4), 202–6.

Strand, M.R. and Pech, L.L. (1995) Immunological basis for compatibility in host parasitoid relationships. *Annual Review of Entomology,* **40**, 31–56.

Stee, D.M. and Whitehead, A.S. (1994) The major acute phase reactants: C-reactive protein, serum amyloid P component and serum amyloid A protein. *Immunology Today* **15**(2), 81–8.

Review questions

Fill in the blanks

A relationship in which a microorganism lives in or on a host organism at the expense of the host is a _____[1] relationship. Many such microorganisms can cause a reduction in the host's capacity to perform normal activities, called _____[2]. Microorganisms that do this are called _____[3]. The first line of defence in blocking invasion comprises epithelial layers, which are augmented by _____[4] barriers and microbicidal agents, such as _____[5] which digests the cell walls of Gram-positive bacteria. Microorganisms can gain entry when the epithelial surface is damaged. Blood coagulation, triggered by physical trauma to the skin, serves to prevent _____[6] and limits the dispersal of invading microorganisms. Inflammation is also triggered by physical _____[7] and by microorganisms that may gain entry. In mammals, inflammation is triggered by degranulation of _____[8] cells in response to tissue _____[9]. This causes the release of _____[10] substances such as histamine that cause increased _____ _____[11]. This allows _____[12] of the tissues with cells and fluid. The first cells to arrive are _____[13] which are guided to the site of infection by _____[14] factors. These and the tissue _____[15] are phagocytic cells that remove microorganisms and cellular debris by _____[16]. The latter cells release inflammatory mediators, _____[17], _____[18] and _____[19], which together cause local and systemic effects. Phagocytosis is an _____[20] mechanism of the immune system. Another is the _____[21] cascade, which generates a variety of opsonins and _____[22] factors. A third is cell-mediated _____[23], as performed by _____ _____[24] cells and _____[25] T lymphocytes. Both these cells use _____[26] and _____[27] to kill target cells. _____[28] is related to the C9 component of complement.

Short answer questions

1. Select from the following a term that most closely matches the descriptions listed below relating to different defence reactions: the respiratory burst, histamine, C-reactive protein, serum amyloid A, defensins, phagocytosis, opsonization, interferon-γ, nodulation, encapsulation, lysozyme, chemotaxis, phagolysosome, interleukin-1.

    (a) A ubiquitous enzyme that digests peptidoglycan in the cell walls of bacteria.

    (b) A multicellular effector mechanism mounted by insects against metazoan endoparasites.

    (c) A feeding mechanism of amoebae and the means by which neutrophils defend the body against bacteria.

    (d) An inflammatory mediator produced by activated macrophages.

    (e) A mammalian acute phase protein that opsonizes microorganisms.

    (f) An intracellular compartment of macrophages in which ingested material is broken down.

    (g) The process by which toxic reactive oxygen intermediates are produced in macrophages and neutrophils.

    (h) The process by which motile cells migrate towards the source of a chemical substance.

    (i) A potent enhancer of macrophage activity that is released by T lymphocytes.

    (j) A vasoactive substance released by mast cells.

2. What are the main inflammatory mediators released by mammalian macrophages, and what are their main local and systemic effects?

3. Which of the following defence reactions are found in vertebrates, invertebrates or both?

    (a) Phagocytosis.

    (b) Nodulation.

    (c) Lysis by natural killer cells.

(d)   Complement.

(e)   Encapsulation.

(f)   The phenoloxidase cascade.

(g)   Blood coagulation.

(h)   Attacins.

(i)   Interferons.

4.   Read Box 1.1 and then say if the following statements are true or false. If you think the answer is false, give your reasons.

(a)   C4a and C4b are produced by the cleavage of C4.

(b)   C4a is opsonic.

(c)   C4b is opsonic.

(d)   C4 is part of the classical pathway.

(e)   The alternative pathway starts with the spontaneous cleavage of serum C3.

(f)   The most important function of complement is the generation of the membrane attack complex.

(g)   Decay-accelerating factor is a membrane-attached molecule that blocks the formation of C3 convertase.

5.   Which kind of effector mechanism(s) in mammals would be appropriate for the following types of microorganisms?

(a)   A virus in the intracellular stage of its life cycle.

(b)   A bacterium that has gained entry to the tissues.

(c)   A protozoan parasite living in hepatocytes in the liver.

Recognition of pathogens

2

In this chapter

- Innate recognition.
- Introduction to the structure and genetics of antigen receptors.
- The consequences of rearrangement of antigen receptor genes.

Introduction

All immune responses have two key components. As we saw in the previous chapter, one of these components is the effector mechanism that destroys the microorganism. Because effector mechanisms are potentially harmful to the healthy cells of the host, this activity must be confined to invading microorganisms. The other feature of an inducible defence reaction, therefore, is a capacity to recognize the microorganism and distinguish it from normal cells in the body.

Recognition is performed by a variety of molecules and receptors that operate freely in solution or at the surface of cells. In this chapter we will learn that essentially two, fundamentally different strategies have evolved for recognizing and discriminating between self and non-self. These are by innate recognition receptors and antigen receptors. We will examine innate recognition first.

> **antigen receptor**
> cell-surface T lymphocyte receptor ($\alpha\beta$ or $\gamma\delta$) or immunoglobulin on B lymphocytes (IgM, IgD, IgG, IgE or IgA).

2.1 Innate recognition

Innate recognition refers to all forms of self/non-self discrimination, other than that mediated by antigen receptors. Many innate receptors have evolved to specifically bind to components in the cell walls of microorganisms. The same components are found in different species of micro-organisms, thereby giving innate receptors a broad spectrum of activity. Innate receptors are abundant in all animals and the effector mechanisms they trigger are quickly mobilized. Broadly speaking, two different means of innate recognition have evolved.

2.1.1 Animal lectins

lectin *n.* proteins found in plants and animals that bind to sugar residues and can agglutinate cells; several animal lectins perform innate recognition of microbes.

First, are the **lectins** [Latin *legere*, to select]. These are a diverse family of carbohydrate-binding proteins that are ubiquitous throughout the plant and animal kingdoms. Lectins were first discovered through their ability to clump together cells, such as mammalian erythrocytes or bacteria (a process called agglutination). This occurs because carbohydrates are present on the surfaces of eukaryotic and prokaryotic cells.

In eukaryotic cells, newly synthesized proteins destined for the plasma membrane, or for release into the extracellular fluid, are usually first glycosylated by the addition of carbohydrate, as chains of saccharides (Figure 2.1). Different polysaccharides confer different surface topographies to proteins, and a diverse array of protein structures can be produced able to interact specifically with other macromolecules. Polysaccharides are also found in the cell walls of prokaryotic microorganisms. Some are unique to microbes, such as lipopolysaccharide of bacteria. Normally, these allow microbes to adhere to host tissues, although their presence also provides the basis for innate recognition by soluble animal lectins (see below).

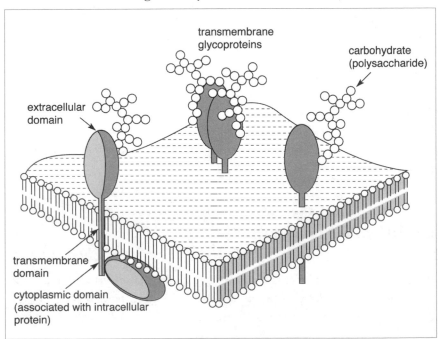

Figure 2.1 Glycosylated cell-membrane proteins.

In plants, the real functions of lectins are not known, while lectins in animals are known to participate in a wide range of cellular activities. Many animal lectins are integral membrane proteins that are involved in providing the specificity of cell-to-cell interactions in developmental processes such as the fertilization of eggs, and cellular reorganization during embryogenesis and metamorphosis.

Despite a common carbohydrate-binding property of animal lectins, there is no general pattern to their structure. Sufficient information is available for vertebrate lectins to classify them into three distinct groups – the **C-type lectins**, the **S-type lectins** and the **pentraxins** (Table 2.1).

In vertebrates, several of these are responsible for triggering defence reactions in response to microbes. The collectins, a subclass of the C-type lectins, bind to a range of specific carbohydrates on microorganisms commonly found in blood and lung fluid. These are opsonins that promote the removal of microbes by phagocytes bearing collectin receptors. They can also trigger activation of the complement cascade via the classical pathway. Pentraxins such as C-reactive protein and mannose-binding protein are acute phase proteins that also opsonize microorganisms, and trigger the complement cascade via the alternative pathway.

Acute phase proteins:
Section 1.3 Inflammation

Soluble lectins have also been detected in the blood of a large number of different invertebrates, suggesting this is a particularly ancient form of defence. A few representatives from different invertebrate phyla have been purified and characterized. Of those, some can be assigned to the classification based on vertebrate lectins. C-type lectins are found in arthropods, tunicates and echinoderms, whereas S-type lectins have been found in nematode worms, sponges and tunicates. **Limulin** (from *Limulus*) and lectins from some tunicates are similar to the vertebrate acute phase proteins. Others, such as **echinonectin** (from echinoderms), do not fit with any of the groupings. Moreover, a much larger number of invertebrate lectins that await characterization have yet to be assigned.

A key question is whether these invertebrate lectins are also involved in defence reactions. There is ample evidence that they agglutinate erythrocytes and microorganisms such as bacteria *in vitro*. Agglutinating activity has been found in the blood of nematodes, arthropods, annelids, molluscs, tunicates and echinoderms. If agglutination of invading microorganisms by lectins also occurs *in vivo*, this would prevent their dispersal throughout the body and allow them to be cleared by phagocytes more efficiently. Lectins from some molluscs such as the oyster and the mussel have also been shown to have opsonic activities *in vitro*, and they increase in concentration in the blood in response to injections of foreign particles. It is likely that, in these animals at least, the lectins are involved in host defence.

in vitro experiments conducted with living cells outside the organism.

in vivo within the living organism.

Table 2.1 Categories of animal lectins

Group	Subgroup	Examples	Characteristics
C-lectins	II	Liver asialoglycoprotein receptor Macrophage galactose receptor Lymphocyte IgE-receptor (CD23) Lectins on NK cells	• Membrane-bound lectins found in many vertebrates. Involved in endocytosis
	III (Collectins)	Mannose binding protein Lung surfactant proteins Bovine conglutinin	• Soluble lectins found in mammals and birds Involved in innate defence
	IV (Selectins)	L-selectin P-selectin E-selectin	• Membrane-bound adhesion molecules Involved in leucocyte traffic
S-lectins	Prototypic (Galaptins)		• Found in many animals; sponges, various fish, amphibians (*Xenopus*), birds and mammals. Function unknown. • 14k Da homodimers with single CRD*
	Chimaeric	Elastin/laminin-binding protein	• Found in mammals only. Function unknown • 29–35 kDa monomers
	Tandem repeat		• Found in nematode worm (*C. elegans*) and rat intestine. Function unknown. • 32 kDa lectin with tandem repeat of CRD*
Pentraxins		C-reactive protein Serum amyloid protein	• Inflammatory acute phase proteins found in mammals. • C-reactive protein also in *Xenopus*, *Limulus*

* CRD = carbohydrate recognition domain
Insufficient data is available for most invertebrate lectins to assign to this classification

The receptors that trigger other invertebrate effector mechanisms (encapsulation, nodulation, blood clotting and the phenoloxidase cascade) are not well defined, although lectins are potential candidates.

Cell-surface lectins may also be important for non-self recognition by vertebrate NK cells. These cells are a group of large granular white blood cells whose function is to kill tumour cells or virus-infected cells in the body. They have been found in mammals, birds and a few other vertebrates, although it is unclear whether invertebrates have an equivalent cell type. NK cells are derived from the same stem cells as lymphocytes during the production of blood cells, although they lack antigen receptors and recognize their target cells by innate means.

NK cells have a typical broad spectrum of activity. A variety of different tumours, and cells infected with a range of different viruses, can be killed by NK cells. The 'NK receptor' that mediates this innate recognition has been rather elusive to define, and the identity of its ligand on target cells is also unclear although there is mounting evidence that the recognition is mediated in part by C-type lectins on the NK cell.

NK cells also readily kill microorganisms that have been tagged by antibodies (Figure 2.2). As we will learn later, the main function of antibody molecules is to attach to microorganisms and trigger immune effector mechanisms. The C-terminus of antibody molecules is called the **Fc domain**. NK cells bear **Fc receptors** that enable these cells to attach to and lyse antibody-coated cells. This type of effector mechanism is called **antibody-dependent cell-mediated cytotoxicity** (or **ADCC**).

2.1.2 The thioester bond

The other main form of innate recognition is based on the thioester bond. There are several proteins in human serum that belong to a family of related proteins that possess an internal thioester bond, called the α_2-**macroglobulin family.** These include C3 and C4 of the complement cascade, and α_2-macroglobulin.

As we learned in the previous chapter, the C3 and C4 components of complement can bind to the surface of microbes forming the focus for the deposition of opsonins, and the generation of chemotactic factors and inflammatory mediators. Binding can occur because a highly reactive carbonyl group is produced when the thioester bond is exposed by cleavage (Figure 2.3). This hydrolyses harmlessly in water, but it will also readily form bonds with amino or hydroxyl groups on nearby amino acids. In itself, this has no discriminatory ability, and bonds form as readily to self and non-self proteins alike. The presence of inactivating proteins on self cells and in the serum (factor I, decay accelerating factor) ensures that the complement

stem cell *n.* a cell with the capacity for constant self-renewal and to undergo differentiation.

Fc receptor (FcR) *n.* cell-surface receptor that binds to the Fc region of immunoglobulin.

antibody-dependent cell-mediated cytotoxicity (ADCC) a form of killing of foreign cells coated by bound antibody and mediated by cytotoxic cells bearing Fc receptors (neutrophils, NK cells, macrophages).

thioester *n.* (or thiol ester) an ester with sulphur instead of oxygen.

Figure 2.2 Schematic representation of two routes for cytotoxicity by natural killer (NK) cells.

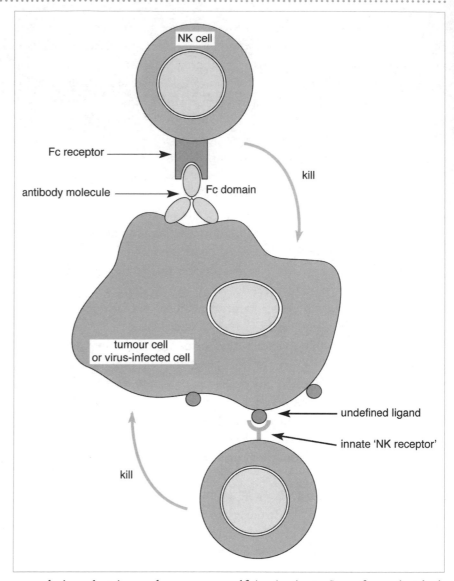

cascade is only triggered upon non-self (or 'activator') surfaces that lack these inhibitors. Discrimination between self and non-self in the complement system is, therefore, based on the presence or absence of inhibitory molecules.

Another molecule in mammals that has a thioester bond is the α_2-macroglobulin molecule itself. This is a proteinase inhibitor that captures and sequesters a broad spectrum of different proteinases, and is thought to assist in the removal of the enzymes from the blood. The blood of *Limulus*, the horseshoe crab, also contains α_2-macroglobulin, which like its mammalian counterpart, has an internal thioester bond and inactivates proteinases. It also appears to play a role in defence in *Limulus*. Although invertebrates

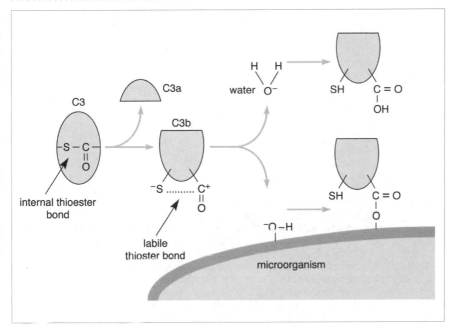

Figure 2.3 The binding of C3b to the surface of a microorganism. C3 produces C3a and C3b fragments by spontaneous cleavage. A labile thioester bond is exposed, which may either form a covalent bond with -OH (or -NH) groups on nearby proteins, or react harmlessly with water. Adapted from Sim, R.B., *et al.* (1981) *Biochemical Journal*, **193**, 115–27.

lack the complement system of mammals, α_2-macroglobulin in this animal does appear to have a direct cytolytic effect against foreign cells *in vitro* and may function in an analogous way to the C3 component of complement in the recognition of non-self surfaces.

The cell walls of bacteria are complex structures that provide mechanical protection and allow the cell to attach to surfaces in their environment. They also provide protection for bacteria that live inside host organisms that are subjected to the effector mechanisms mounted by the host's immune system.

All bacteria have a layer of **peptidoglycan** (also known as **murein**) that surrounds their plasma membranes. Peptidoglycan is a gel-like polymer of the sugar derivatives, *N*-acetylmuramic acid (NAM) and *N*-acetylglucosamine (NAG), plus several different

BOX 2.1 Bacterial cell walls protect against phagocytes and complement

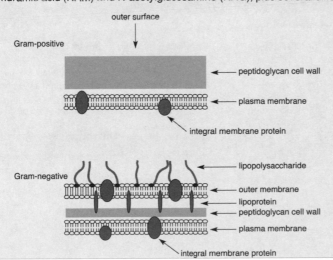

Figure 1.

amino acids. Long polysaccharide chains are formed of alternating NAM and NAG subunits, which are cross-linked by short, peptide side chains (see Figure 1). Peptidoglycan is vulnerable to digestion by lysozyme which cleaves the bond connecting NAM and NAG.

The cell walls of Gram-positive bacteria are composed mainly of a thick layer of peptidoglycan. This provides good protection against the formation of the C5–C9 membrane attack complex of complement, although these organisms are relatively susceptible to digestion by lysozyme. In contrast, Gram-negative bacteria have a much thinner peptidoglycan coat which is surrounded by an additional lipid bilayer that is anchored to the peptidoglycan by lipoproteins. Embedded in the outer lipid membrane are chains of **lipopolysaccharide** (LPS). These are molecules constructed of lipid and carbohydrate. The composition of the carbohydrate varies considerably between different species of bacteria. LPS both helps stabilize the outer membrane and, as we saw in the first chapter, is an endotoxin and accounts for some of the symptoms to hosts that arise in hosts with Gram-negative infections.

Many Gram-positive and Gram-negative bacteria also have a thick layer of polysaccharides overlying the cell wall called a **capsule.** A capsule is a well-organized layer that is not easily washed off. The main pathogens that cause pneumonia and meningitis in humans, including *Streptococcus pneumoniae*, *Klebsiella pneumoniae*, *Haemophilus influenzae* and *Neisseria meningitidis*, all have polysaccharide capsules. Studies of *Streptococcus pneumoniae* in mice show that the capsule provides protection against phagocytosis. Normally mice are killed by infection, but when the capsule is removed experimentally the bacteria lose their virulence. Capsules also provide good protection against the membrane attack complex and some are also poor activators of the alternative complement pathway because they lack surface structures that will stabilize C3a. *Streptococcus* has a capsule of hyaluronic acid, which is a normal component of mammalian connective tissue. This bacterium may avoid recognition by mimicry of the host tissues.

Some bacteria, instead of resisting phagocytosis, have instead exploited the phagocyte as a place to live. Among these are the human pathogens that cause Legionnaire's disease (*Legionella pneumophila*), tuberculosis (*Mycobacterium tuberculosis*), typhoid fever (*Salmonella typhi*) and listeriosis (*Listeria monocytogenes*). When these organisms breach the physical and chemical barriers of the host they are ingested by phagocytic macrophages into phagosomes. As we saw in the first chapter, this is a hostile, acidic environment containing degradative enzymes, defensins and toxic oxygen and nitrogen intermediates. Various strategies have evolved for surviving in this. *Legionella*, *Mycobacterium* and *Salmonella* alter the nature of the membrane of the phagosome, making it more resistant to fusion with lysosomes. *Mycobacterium* also is able to raise the pH of the phagosome. *Listeria* instead erupts from phagosome by using a lipase and enters the comparative safety of the cytosol of the cell.

Further reading

Finlay, B.B. and Falkow, S. (1989) Common themes in microbial pathogenicity. *Microbiological Review*, **53**(2), 210–30.

Falkow, S. (1991) Bacterial entry into eukaryotic cells. *Cell,* **65**, 1099–102.

Andrews, N.W. and Webster, P. (1991) Phagolysosomal escape by intracellular pathogens. *Parasitology Today,* **7**(12), 335–40.

Moulder, J.W. (1985) Comparative biology of intracellular parasitism. *Microbiological Review,* **9**(3), 298–337.

Cooper, N.R. (1991) Complement evasion strategies of microorganisms. *Immunology Today,* **12**(9), 327–31.

Summary

- Innate recognition receptors discriminate self from non-self in different ways. Many soluble animal lectins have evolved to bind to structural features shared by many microorganisms that are absent from self, such as the lipopolysaccharide of bacteria.

- NK cells can kill target cells by virtue of their own ability for innate recognition, thought to be mediated in part by membrane-bound lectins. NK cells are also able to kill cells tagged by antibodies (antibody-dependent cell-mediated cytotoxicity).

- The thioester bond has also been adopted as a means of triggering effector mechanisms, although discrimination is achieved by having inhibitors of the effector mechanism on self-surfaces.

2.2 The adaptive immune system

In the previous section we saw two basic mechanisms by which innate discrimination of self and non-self is achieved. A feature of innate recognition is the ability to bind to the components common to different microorganisms. This gives each innate receptor a broad spectrum of target pathogens. A measure of the importance of innate-type recognition systems is ubiquity – innate recognition is found in all animals.

However, vertebrates are unique among animals as they have evolved an additional way of recognizing microbes, which is the basis of what is commonly called the **adaptive immune system**. The adaptive system uses many of the same effector mechanisms as the innate system. The real distinction between the two lies in how non-self is first recognized.

The following is an overview of the adaptive system. It is intended to provide a foundation for more detailed discussion in future chapters. At this stage we are concerned with the roots that underlie all that is special about the vertebrate adaptive system. Two main hallmarks distinguish it from innate kinds of immunity.

- Firstly, recognition is performed by a receptor that exists in billions of different forms in an individual. This enormous diversity endows an animal with the ability to recognize any microorganism. Even newly evolved microorganisms that have not been encountered before, and which have had no opportunity to influence the evolution of the host, can be recognized.

- Secondly, the adaptive system retains a **memory** of each particular microorganism to which it has been exposed. Memory allows the adaptive system to eliminate the same microorganism more effectively upon subsequent exposure. This phenomenon only applies to microbes encountered before and it is therefore specific for the original microorganism. The innate system, in contrast, responds to a pathogen with the same reaction time irrespective of the number of previous exposures.

The receptor that is responsible for these remarkable properties is the **antigen receptor**. These are found only on lymphocytes. Although the function of antigen receptors is to recognize microorganisms and trigger effector mechanisms, it is conventional to term the ligand to which antigen receptors bind to simply as **antigen**, regardless of its origin. Many different substances can be an antigen, including protein, polysaccharide and nucleic acid.

There are two general categories of lymphocytes, according to the type of antigen receptor they possess – **T lymphocytes** and **B lymphocytes**. The antigen receptor of B lymphocytes is also called **immunoglobulin**. Both types of antigen receptor operate at the lymphocyte surface. In many respects they are like receptors for other extracellular molecules, such as hormones or adhesion molecules. The interaction between receptor and ligand is specific, arising from a precise fitting together of complementary surfaces. Receptors are linked, via intracellular communication pathways, to the genes that control behaviour. Receptors thus enable cells to perceive and respond in predefined ways to stimuli from the outside world. In the case of the antigen receptor, the binding of antigen triggers lymphocytes to become active in an immune response, a complex process usually referred to as **lymphocyte activation.**

As we will see, the antigen receptors of T and B lymphocytes are genetically and structurally related. However, there is an important distinguishing feature. After activation by antigen, B lymphocytes start to manufacture large quantities of immunoglobulin molecules which are then released into the blood. These soluble forms of immunoglobulin are commonly called **antibodies**.

2.2.1 Antigen receptors show enormous diversity

The antigen receptors of T and B lymphocytes differ from other kinds of receptors because the nature of their ligands is unpredictable. How do antigen receptors recognize a foreign antigen that enters the body *before* its

memory *n.* in immunological parlance, the enhanced immune reactivity mounted against an antigen encountered previously (see **secondary response**).

antigen *n.* any substance that is bound specifically by immunoglobulin or T lymphocyte (antigen) receptor.

immunoglobulin (Ig) *n.* antigen receptor of B lymphocytes; forms the antigen-binding component of the **B lymphocyte receptor complex** and **antibody** when released as soluble molecules.

Receptors and signalling: Chapter 8

antibody *n.* soluble form of immunoglobulin; membrane-bound immunoglobulin is the antigen-binding component of the B lymphocyte receptor complex.

conformation is encountered? For many years it was thought that antibody molecules folded around the antigen, which acted as a template. In this way the antibody was thought to acquire an imprint of the antigen. This was called the **instructive theory**. However, the conformation of an antibody is fixed and the particular antigen it binds to is an intrinsic property. For example, if antibodies that bind to a particular antigen are reversibly unfolded (denatured) and then allowed to refold into their native shape, they regain the same antigen specificity as before.

An alternative idea, embodied by the **clonal selection hypothesis**, proposes that antigen receptors pre-exist in many different forms, each one carried by a different lymphocyte. Only those lymphocytes carrying the receptor that is complementary for a particular antigen are stimulated to respond. It is now accepted that this hypothesis is correct. If mammalian lymphocytes are separated into individual cells and each is allowed to proliferate *in vitro*, each population (or **clone**) is found to have a different antigen receptor. It has been estimated that the number of different antigen receptors in a typical mammal is around 10^9, thereby endowing it with the capacity to recognize an equally vast array of complementary antigens. (Actually, as we will see later, antigen receptors recognize only a part of the antigen, called an **antigenic determinant** or **epitope**.) It is this array of different receptors, called the **repertoire**, that prepares the system to recognize any antigen that may be encountered during the lifetime of the animal. Before we examine how this is achieved, and its consequences, we will briefly examine the structure of these receptors.

2.2.2 The structure of antigen receptors

Antigen receptors are polypeptide chains organized into discrete, globular **domains** (Figure 2.4). At the lymphocyte cell surface, antigen receptors are not single chains but exist as heterodimers of chains (i.e. two dissimilar chains). That of the T lymphocyte is constructed from two chains, α and β, or γ and δ, held together by a disulphide bond. Each chain is encoded at a different locus in the genome. Immunoglobulin is also constructed of two chains – heavy (or H) and light (or L). In mammals there is one H chain locus, and two light chain loci that encode different L chains, called κ or λ. At the cell surface, the heterodimers of one H and one κ or λ chain are linked into a pairs to produce a four chain structure. Each type of receptor is anchored into the plasma membrane of the lymphocyte by transmembrane regions in the receptor polypeptides.

T and B lymphocyte antigen receptors are homologous, that is, they have a structural similarity that originates from common ancestry. The basic structural unit is called the **immunoglobulin fold**. Each domain of an

clonal selection the proliferation in response to stimulation by antigen, of lymphocytes with the corresponding antigen specificity.

clone *n.* a population of cells derived from a single cells; *v.* to produce a set of identical cells or DNA molecules from a single cell or copy.

epitope *n.* the part of an antigen recognized by an antigen receptor; sometimes also called an **antigenic determinant**.

repertoire *n.* the combined assortment of antigen receptors present in the pool of mature T and/or B lymphocytes.

domain *n.* a structurally defined section of a protein, often compact and globular.

dimer *n.* a protein made of two polypeptide chains (or subunits).

Figure 2.4 Schematic representation of the T and B lymphocyte antigen receptors.
 Both receptors are heterodimers (non-identical polypeptide chains). T lymphocyte receptors are constructed from αβ or γδ chains, those on B lymphocytes (immunoglobulin) from heavy (H) and light (L) chains. Each chain is encoded at a different genetic locus. Two different loci encode immunoglobulin L chains giving rise to chains called κ or λ. The immunoglobulin is a pair of identical heterodimers. Interchain disulphide bonds represented by black bars; variable domains are shaded darker blue (see text).

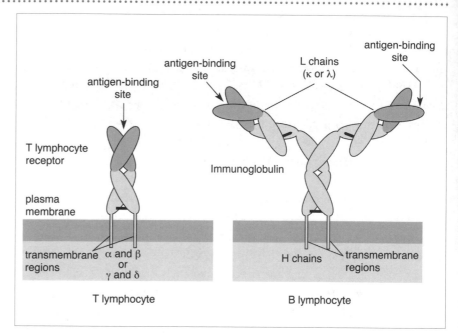

heterodimer *n.* a protein consisting of two dissimilar polypeptide chains.

disulphide bond S-S bond formed between two cysteines within a polypeptide or between two polypeptides, important in maintaining three-dimensional structure.

immunoglobulin fold *n.* a globular **domain** structure of approximately 110 amino acids in length folded into two β-pleated sheets of three or four β-strands each held together by an intradomain disulphide bond.

antigen receptor contains a single immunoglobulin fold. It consists of a sandwich of two anti-parallel β-pleated sheets, formed by the looping back and forth of the polypeptide chain, held together by a disulphide bond inside the domain (Figure 2.5).

Virtually all of the differences between the antigen receptors of different T or B lymphocyte receptors are localized in the domains furthest from the membrane. These are called the **V** (or **variable**) domains, and they form the antigen-binding part of the molecule. The remaining domains are called **C** (or **constant**) domains. This concentration of variability in the V domains is illustrated by the variability plot in Figure 2.5. This plot was generated by comparing the amino acid sequences of 15 immunoglobulin light chains from different clones of B lymphocytes. Peaks in the plot identify positions in the primary amino acid sequence that may be occupied by different amino acids. The higher the peak, the more variability is found at that position among the different clones. The amino acids in between the peaks are more conserved.

2.2.3 Diversity is generated by gene rearrangement

The origin of antigen receptor diversity, and the feature that sets antigen receptors apart from all other receptors, lies in their genes. The amount of genetic information that would be required if each receptor in the mammalian repertoire was encoded by a separate gene would be several orders of magnitude larger than the total number of genes actually present in the genome.

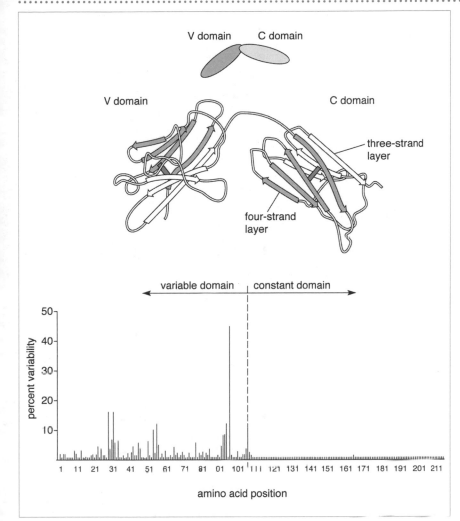

variable domain | constant domain

amino acid position

Figure 2.5 Structure of an immunoglobulin L chain. The light chain consists of two domains: a variable (V) domain and a constant (C) domain. At the top is the schematic of the light chain used in Figure 2.4. In the centre is a ribbon diagram of the light chain. Each domain contains an immunoglobulin fold structure, consisting of a sandwich of two sheets of three or four β-strands each, held by an intradomain disulphide bond (dark blue). At the bottom is a variability plot of the two domains generated from the amino acid sequences of 15 different immunoglobulin light chains from different clones of B lymphocytes. The percent variability at each position is calculated from the formula (number of different amino acids at the position/frequency of most common amino acid at that position) x 100. Notice that the variability between the clones is localized in the variable domain. Ribbon diagram reprinted with permission from Edmundson, A.B. *et al.* (1975) *Biochemistry,* **14** (18), 3953–61, copyright 1975 American Chemical Society; data for variability plot from Klein, J. (1990) *Immunology.* Blackwell Scientific Publications, Oxford.

The explanation came when the first antigen receptor genes were isolated from clones of B lymphocytes. It was noticed that their organization was different than in other cells in the body. This reorganization is achieved by a type of DNA **recombination**, often referred to as antigen receptor gene **rearrangement**. Figure 2.6 shows a schematic representation of how the process of rearrangement of antigen receptor genes can be observed. DNA in the non-rearranged configuration is obtained from germline cells (strictly speaking, from the fertilized egg, although in practice, a non-lymphocyte cell from the same individual can be used). DNA is also obtained from a clone of lymphocytes. The DNA from both sources is then cleaved into fragments with a restriction endonuclease, and the fragments probed for the receptor gene. In the germline configuration, the probe hybridizes with two separate fragments whereas in mature lymphocytes it recognizes only one. This indicates that the functional antigen

recombination *n.* any exchange or integration of one DNA molecule into another.

rearrangement *n.* a type of DNA **recombination** seen in antigen receptor genes.

germline *n.* gametes (sperm and ova) or the cells that give rise to them.

Figure 2.6 Somatic rearrangement of antigen receptor genes. Antigen receptors are encoded by different gene segments that are separated from each other in the germline DNA. When the lymphocytes mature, the gene segments are moved ('rearranged') into close proximity, as depicted in (a). This can be demonstrated by digesting germline and lymphocyte DNA with a restriction endonuclease (b). The gene segments that encode the receptor are on the same DNA fragment after rearrangement whereas they are in different fragments when in the germline configuration. In practice this is shown by resolving the fragments by electrophoresis and probing the fragments with radiolabelled mRNA expressed by the receptor gene (c).

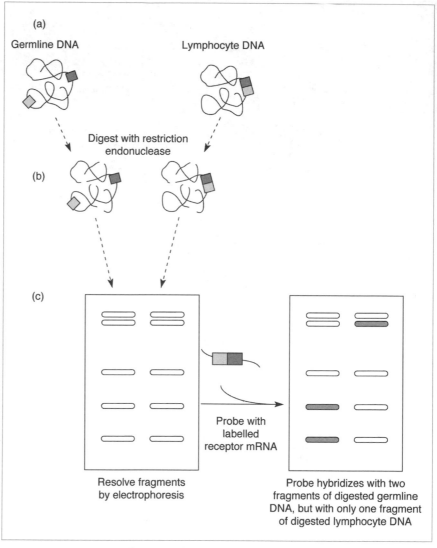

(a)

Germline DNA   Lymphocyte DNA

Digest with restriction endonuclease

(b)

(c)

Probe with labelled receptor mRNA

Resolve fragments by electrophoresis

Probe hybridizes with two fragments of digested germline DNA, but with only one fragment of digested lymphocyte DNA

somatic cell *n.* any cell in the body other than a germline cell.

receptor gene is actually created from separate pieces, or **gene segments,** which are located in different parts of the genome before rearrangement occurs. It is now known that the gene segments in the germline configuration do not possess all the necessary information to be expressed into proteins by themselves, but only do so when joined together to become a single, rearranged gene.

Unlike most forms of molecular heterogeneity, which is inherited in the germline, antigen receptor diversity is generated after fertilization and is therefore described as **somatic** (non-germline) **diversity**. Although every other nucleated cell in the body has antigen receptor genes, they remain in a non-rearranged, or germline, configuration. Non-rearranged antigen receptor genes are inactive and unable to be transcribed and translated into protein. Only in lymphocytes does the rearrangement of antigen receptor genes occur.

Rearrangement of antigen receptor gene segments is depicted in very general terms in Figure 2.7. In the germline configuration, the gene segments are organized into discrete clusters along the chromosome. One cluster contains **variable (V) gene segments** and the other contains **joining (J) gene segments.** Some antigen receptor genes also have a third cluster of segments called **diversity (D) gene segments** (not shown in the figure). The segments within each cluster are genetically related, but each encodes a slightly different sequence of amino acids. In this configuration the gene is inactive and not translated into the receptor proteins.

In the mature lymphocyte, the gene that encodes this domain is constructed from a single V gene segment and a single J gene segment, each

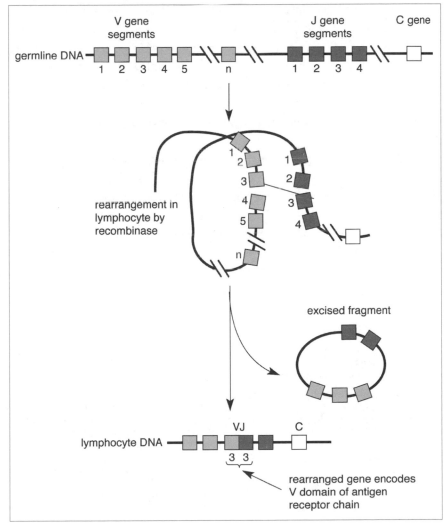

Figure 2.7 Schematic view of rearrangement of antigen receptor V and J gene segments. An antigen receptor locus contains many non-functional variable (V) gene segments and joining (J) gene segments, clustered into families. The segments within a family have different, but highly homologous, DNA sequences. In lymphocytes only, a functional gene is produced by the physical joining of one gene segment from each family and the removal of intervening DNA. Some receptor chains involve a third gene segment chosen from a family of diversity (D) gene segments (see Chapter 4 for more details). In this way, the location of the chosen segments changes within the chromosome from the original, germline configuration to a new, rearranged configuration in the mature lymphocyte.

selected from the many available segments. As shown in Figure 2.7, an enzyme called **recombinase** removes the intervening DNA between the segments and they become ligated into a single functional gene. As a general rule, one V domain in the receptor heterodimer is constructed from VJ gene segments, the other from VDJ segments.

Precisely which gene segment from each cluster is used for the final rearranged receptor gene is an essentially random process. As a consequence, a different combination of rearranged V(D)J genes are generated in each lymphocyte. Further diversity arises during the rearrangement process because of the imprecise joining of the gene segments. This adds or removes genetic information at the junction and amplifies the diversity by several orders of magnitude. For example, even if the same pair of V and J segments were juxtaposed in two separate rearrangement events, the protein produced by these two rearranged VJ genes would be slightly different at the region encoded by the junctions of the gene segments. As a result of these mechanisms, the size of the repertoire is vast. Mammals typically have the capacity to generate more different antigen receptors than the number of lymphocytes they possess. In effect, each receptor is unique.

DNA recombination of one sort or another occurs in all living organisms (see Box 2.2). Most are accidental events and have no known function in the life of the organism. Several cause gene damage, whereas others create useful mutations that are propagated into future generations by natural selection. Occasionally, DNA recombination has been harnessed and programmed into normal developmental pathways. Notable are some protozoan parasites that use this as a means of changing their surface structures as a defence against the immune system.

However, the process is at its most sophisticated and intricate in the vertebrate lymphocyte. We will examine the mechanics of the process in detail in Chapter 4. Here, meanwhile, we will examine in outline the *consequences* of antigen receptor gene rearrangement, for it underlies all the remarkable hallmarks of the adaptive immune response.

mutation *n.* a change in the sequence of DNA caused by errors in DNA replication or chemicals, radiation, etc.

Antigenic variation: Box 4.3

BOX 2.2 Recombination and gene conversion	Genes move (or transpose) from one location to another within the genome of prokaryotic and eukaryotic cells by one of two general mechanisms.

- **Recombination.** This is a 'cut and paste' process, in which genetic information is exchanged between homologous DNA molecules during nuclear division (Figure 1).
- **General recombination** covers any exchange or insertion of one DNA molecule into another to produce recombinant DNA. This occurs more readily between two similar (or homologous) DNA sequences than between two heterologous (or dissimilar) sequences.
- **Replicative transposition.** Here the gene is copied into another DNA molecule, either directly or via a mRNA intermediate, which then recombines elsewhere in the genome. Unlike the former case, the replicative process duplicates or copies the original gene to the new site and the original remains *in situ*.

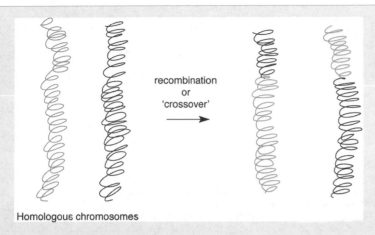

Figure 1 Recombination between two homologous chromosones.

Homologous chromosomes

General recombinations are particularly important in immunobiology for several reasons. Firstly, antigen receptors are generated by a form of generalized recombination of antigen receptor gene segments called **rearrangement** (Chapter 4). Secondly, inserting DNA at homologous sites (homologous recombination) has been exploited in the laboratory as a means for interrupting specific genes in the genomes of experimental animals. This technology has had a major impact on our understanding of the development of lymphocyte repertoires (Chapter 9). Thirdly, recombinations that occur between homologous chromosomes during meiosis, called **crossovers**, have several important consequences.

- Crossovers generate new combinations of genes in the major histocompatibility complex, or MHC, of vertebrates (Chapter 6).
- Unequal crossovers are a major source of gene duplication and divergence leading to the evolution of gene superfamilies.
- Crossovers can be used as an experimental tool for mapping the relative positions of genes linked together on the same chromosome and have been particularly important for mapping the genes in the MHC (Chapter 6).
- Crossovers are an opportunity for a replicative transposition of DNA called **gene conversion**. This is a major source of new alleles in the MHC (Chapter 7). A similar process occurring during mitosis is the means by which the lymphocytes of birds and some mammals generate antigen receptor diversity (Chapter 4) and how some pathogens change their surface antigens to evade host immunity (see below).

A model for recombination between two homologous double-stranded DNA molecules is

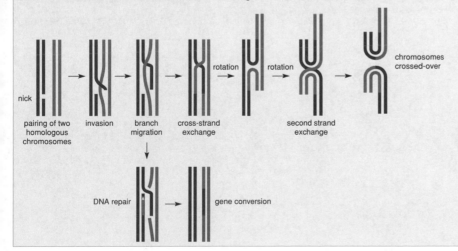

Figure 2 Schematic view of crossover and gene conversion at meiosis.

BOX 2.2 *continued*

shown in Figure 2. The event is initiated by a nick in one strand releasing a free end that can 'invade' the homologous partner and hybridize with the complementary strand. This is followed by a rapid migration of the paired region for some distance from the original point of crossover. A double-stranded region is formed between the homologous, but not identical, DNA strands called a heteroduplex. This swapping over may create a new DNA sequence, and is a major source of genetic variability in organisms. A reciprocal strand exchange can lead to an intermediate stage in crossing-over called a cross-strand exchange. To complete the crossover, the remaining two strands must be broken and the ends exchanged. This is achieved by a series of rotations of the cross-strand structure.

Exchanges between homologous DNA at the same locus on each chromosome leads to an equal or reciprocal exchange of DNA, whereas crossovers between homologous sequences (or repeated sequences) at different loci lead to non-reciprocal exchange (Figure 3). Non-reciprocal exchanges are a major route to the duplication of genes and the evolution of families of related genes like the immunoglobulin superfamily. This is discussed further in Chapter 7.

An important form of replicative transposition is called **gene conversion**. Classically, gene conversion was described as an apparent 'conversion' of a gene during meiosis from one allele to the other (see Figure 2). This is caused by normal DNA repair mechanisms that operate near to the sites of crossovers. One mechanism is shown in the figure. The gap created by branch elongation is filled in using the unpaired strand as the template, while the unpaired strand on the homologous chromosome is degraded. Alternatively, one strand of the heteroduplex may be replaced by DNA repair enzymes using the other strand as a template. Either way, one segment of DNA is copied over another thereby giving the impression of a gene 'converting'. The size of DNA that can be converted varies enormously, from a few base pairs to whole genes.

More recently, the term 'gene conversion' has come to embrace any replicative transposition that causes sections of genetic information to be copied into new locations, during meiosis or mitosis. In particular, this has been found to occur in the immunoglobulin genes of birds, rabbits and sheep, during B lymphocyte proliferation. Gene conversion is also the means by which the protozoan parasite *Trypanosoma brucei* constantly changes the **variable surface glycoproteins** it carries on its surface in order to evade recognition by the human immune system. Each variant of the surface glycoprotein is encoded in part by a different, but closely related, copy of the glycoprotein gene. These copies are clustered upstream of a single actively transcribed locus. A major mechanism for antigen switching is the replacement of the copy at the transcribed locus with a different copy by gene conversion during cell division.

Further reading

Thompson, C.B. (1992) Creation of immunoglobulin diversity by intrachromosomal gene conversion. *Trends in Genetics,* **8**(12), 416–22.

Van Der Ploeg, L.H.T., Gottesdiener, K. and Lee, M.G.-S. (1992) Antigenic variation in African trypanosomes. *Trends in Genetics*, **8**(12), 452–7.

Lieber, M.R. (1991) Site-specific recombination in the immune system. *FASEB Journal,* **5**, 2934–44.

Figure 3 Equal (a) and unequal (b) crossovers by homologous recombination.

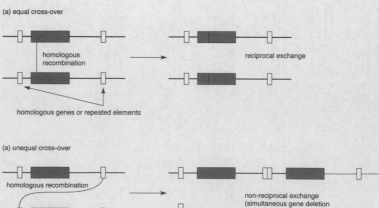

2.2.4 Antigen specificity

The first major consequence of antigen receptor gene rearrangement is that each clone has an antigen receptor with a uniquely shaped antigen-binding site. This dictates which complementary surface it will bind. Like other kinds of receptors, each antigen receptor has an exquisite sensitivity to the conformation of its complementary antigen. It has a much lower affinity for other, even similarly shaped, antigens. This means that each clone of lymphocytes has its own **antigen specificity**.

> **antigen specificity** the limited recognition of a particular epitope, or group of structurally related epitopes, by a lymphocyte clone.

2.2.5 Clonal selection

A second major consequence of antigen receptor gene rearrangement is the clonal expression of each different receptor. As a result, the number of lymphocyte clones able to recognize a particular antigen is only a fraction of the total repertoire. This in turn leads to a cascade of other consequences.

Clonal expansion is necessary for a response

Because of the rarity of responders that recognize a particular antigen in the repertoire, lymphocyte clones must proliferate and develop into fully functional cells before an effector mechanism can be mounted. This process is entirely antigen driven (Figure 2.8). Only those few lymphocyte clones bearing the appropriate receptor are activated by antigen. The remainder of the repertoire remains at rest; indeed it is thought that the majority of mature lymphocytes are never stimulated during the lifetime of an animal. This antigen-dependent activation of appropriate responder clones is called **clonal selection**. Notice how this stands in contrast to non-clonal innate effector cells, in which there is no need for clonal expansion.

The primary response is slow

As a direct result of the low responder frequency, the first few days of an immune response to antigen involve a time-consuming expansion of the relevant lymphocyte clones. During this time there is a window where pathogenic organisms are held at bay only by the mechanisms of innate immunity. Indeed, many of the symptoms of infection by a pathogen (such as fever and inflammation) are the outwardly visible responses of the innate immune system.

Lymphocyte encounter with antigen is non-random

There are also physiological consequences of a low lymphocyte responder frequency. It is likely that if left to random circulation, the encounter between an antigen and the responder lymphocytes would be very inefficient. All vertebrates have therefore evolved tissues where antigens that gain

Figure 2.8 Schematic view of clonal selection. The lymphocyte clone(s) bearing receptors complementary to a specific antigen becomes activated. The clone proliferates and daughter cells undergo differentiation into effector and memory phenotypes. The other lymphocytes in the repertoire remain inactivated by the antigen.

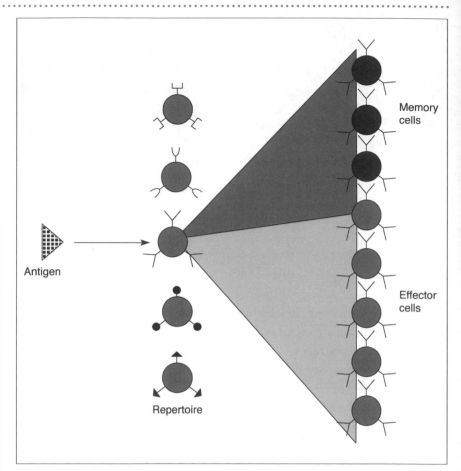

Memory cells

Antigen

Effector cells

Repertoire

lymphoid tissue *n.* tissues in which immature lymphocytes develop (**primary lymphoid tissue**) or in which mature lymphocytes undergo antigen-dependent activation (**secondary lymphoid tissue**).

secondary lymphoid tissue *n.* tissue in which lymphocytes encounter antigen and undergo antigen-dependent activation; **lymph node**, **spleen**, **Peyer's patch**, **tonsil**.

Secondary lymphoid tissues: Section 3.6

costimulation a second signal (or 'signal 2') in addition to that received through the antigen receptor (or 'signal') required for full activation of lymphocytes; lymphocytes may be paralysed by antigen in the absence of costimulation (**anergy**).

access to the body are deposited and through which lymphocytes constantly recirculate. These tissues, termed **secondary lymphoid tissue**, increase the likelihood of the encounter between antigen and antigen receptor occurring. They also provide the stringent cellular microenvironment required for the activation of lymphocytes and maturation into effector cells. As a general rule, the complexity of these tissues in vertebrates increases with phylogenetic progression. Thus fish have the simplest lymphoid tissues and mammals have the most complex.

Primary responses have high stringency

Lymphocytes that have not yet encountered their specific antigen are **naïve** lymphocytes. Activation of a naïve lymphocyte is a two signal process, requiring one signal delivered through the antigen receptor by the binding of antigen, and a second, simultaneous **costimulatory** signal delivered through a different receptor. This costimulatory signal is usually only available in the cellular microenvironment found within the secondary lymphoid tissue.

After receipt of both signals a naïve lymphocyte then undergoes a programmed series of events that include clonal expansion (proliferation) and maturation into effector and memory cells.

Secondary responses have low stringency

Another hallmark of the adaptive immune response is memory. While the primary response is slow, the response to the same antigen upon a subsequent exposure is much more rapid. A memory response can usually eliminate invading microorganisms before any disease symptoms develop. This effect is caused by **memory cells** that are produced during the proliferation and expansion of lymphocyte clones during the primary exposure to the antigen (Figure 2.8).

Memory lymphocytes are long-lived cells that remain in the body long after the antigen has been eliminated. Upon subsequent exposure to the antigen they respond more rapidly than before. This is in part because the previous expansion of the clone means there is a higher responder frequency, but also because memory lymphocytes can be reactivated more readily than naïve lymphocytes.

Tolerance to self: the basis of self/non-self discrimination

As we learned above, the production of antigen receptors is an essentially random genetic process. As such the antigen receptor has no inherent capacity to distinguish between self and non-self, and many are produced that recognize self-antigens. In order to discriminate, the adaptive system first produces a full repertoire of antigen receptors, and then subsequently removes the self-reactive lymphocytes from the repertoire, either by causing them to die, or by paralysing them, or actively suppressing their activities. As a result, the majority of lymphocytes are purged from the system. This is described as maintaining a state of **tolerance** to self-antigens.

2.3 Evolutionary aspects

Innate recognition receptors are found in both protostomes and deuterostomes (see Figure in the preface), suggesting that innate recognition must have been in existence before the divergence of these two lineages. Some of these molecules have been well conserved since the divergence. C-reactive protein and other lectins, thioester bond-containing molecules such as α_2-macroglobulin, and proteins containing an immunoglobulin fold, are found in both in both lineages.

secondary response a rapid humoral or cell-mediated response mediated by memory B or T lymphocytes, that is stimulated upon the second exposure to a particular antigen (also known as **memory** or **anamnestic** responses).

memory cells *n.* differentiated T and B lymphocytes that remain after antigen-dependent clonal selection and which maintain immunological **memory**.

Memory cells: Section 8.1.6

tolerance *n.* a state of antigen-induced non-responsiveness of lymphocytes achieved by **clonal deletion**, suppression or **anergy**.

Tolerance: Chapter 9

The immunoglobulin superfamily: Section 3.8.1

In contrast, clonally distributed antigen receptors encoded in rearranging genes are found only in vertebrates. Indeed, antigen receptors and lymphocytes have been found in representatives of all jawed vertebrate classes. As yet, there is no evidence for antigen receptors or the recombinase enzymes in any invertebrate. Thus, innate recognition appears to have existed before the 'invention' of antigen receptors by vertebrates, which have been incorporated onto a pre-existing innate system. There are other lines of evidence for this.

2.3.1 Innate and adaptive recognition systems use the same effector mechanisms

Several of the effector mechanisms of the innate system are also used by the adaptive system. For example, NK cells and cytotoxic T lymphocytes are effector cells that both kill virus-infected cells or tumour cells. As we learned above, the NK cell uses innate recognition to identify target cells. The response they mount is rapid but this is unchanged upon subsequent exposure to the same virus or tumour cell. Cytotoxic T lymphocytes, in contrast, use antigen receptors to recognize virus-derived or tumour cell-specific antigens on the surface of the target cell. These cells exhibit immunological memory and respond more rapidly to viral infections and tumours encountered before. Despite these different modes of recognition both cells share common membrane glycoproteins required for signalling across the plasma membrane, and they both use perforins to kill target cells. NK cells may represent a more primitive and ancestral form of cell-mediated cytotoxicity, from which the cytotoxic T lymphocyte lineage later evolved in the vertebrates.

Similarly, the complement cascade of vertebrates can be triggered by both innate receptors and antigen receptors. The components of the alternative and classical pathways are homologous and evolutionarily related. It is thought that the original role of complement in primitive vertebrates was the generation of opsonins and inflammatory mediators, and the antibody-dependent classical pathway and the lytic pathway evolved later. This is because antibodies and the classical pathway are absent from more primitive, jawless vertebrates (lampreys and hagfish) and some cartilaginous fish, suggesting they arose later in the vertebrate lineage. In contrast, C3, which is the first component in the innate pathway, has been found in representatives of all major vertebrate groups, including the hagfish and lampreys.

Examples such as these suggest the adaptive system has been built upon the pre-existing system, and the two now operate alongside each other.

2.3.2 Vertebrates are dependent on adaptive immunity for survival

It is doubtful whether the vertebrates as we know them today could have evolved without antigen receptors. The enhanced level of protection that antigen receptors have provided has contributed to the evolution of modern vertebrates into larger and more complex organisms than invertebrates, able to have longer life spans and fewer progeny. It is thought that these body plans provide a richer variety of potential environments for colonization by microorganisms. These body plans may never have evolved if protected by innate mechanisms alone. Indeed, although innate immunity is retained in modern vertebrates, these animals are now entirely dependent on their adaptive systems to protect them from pathogens. For example, inherited mutations in recombinase genes that prevent the rearrangement of antigen receptor genes sometimes arise in mammals. Such animals have no lymphocytes and are very susceptible to life-threatening opportunistic infections and cancers.

Summary

- In addition to innate recognition receptors, vertebrates have evolved antigen receptors. These are found only on T and B lymphocytes. Antigen receptors belong to the immunoglobulin superfamily of molecules which was already in existence at the divergence of invertebrates and vertebrates.

- Vertebrates have evolved a way of rearranging the germline configuration of the antigen receptor genes to create an enormous diversity of receptors (the repertoire) from a relatively small amount of genetic information.

- Antigen receptors are clonally distributed, with each clone of lymphocytes having a unique receptor with a particular antigen specificity. This also means the number of lymphocyte clones that can respond to a given antigen is low (Table 2.2).

- A primary response to antigen necessitates expansion of the responder clones into effector cells and memory cells. During this lag phase, innate mechanisms are the main source of protection. Effector cells die after the antigen is cleared from the body whereas memory cells remain.

- A secondary exposure to the same antigen re-activates the memory

Table 2.2   Differences between recognition by innate and antigen receptors

	Innate receptors	Antigen receptors
Distribution	non-clonal	clonal
Responder frequency	high	low
Recognition	broad spectrum	specific
Primary response	rapid	slow
Secondary response	rapid	rapid
Memory	no	yes

lymphocytes, which results in a more rapid response than the first exposure.

• In contrast, innate receptors are non-clonally distributed. The higher responder frequency ensures a rapid response as clonal proliferation is unnecessary, although memory cells do not seem to be produced. Invertebrates rely entirely on innate kinds of recognition for defence.

The future

• Despite some evidence that lectins and a few other soluble components are involved in effector mechanisms in invertebrates, how innate non-self recognition is achieved in invertebrates is largely unknown.

Further reading

Holmskov, U., Malhotra, R., Sim, R.B. and Jensenius, J.C. (1994) Collectins: collagenous C-type lectins of the innate immune defence system. *Immunol Today,* **15**(2), 67–74.

Drickamer, K. and Taylor, M.E. (1993) Biology of animal lectins. *Annual Review of Cell Biology,* **9**, 237–64.

Plasterk, R.H.A. (1992) Genetic switches: mechanism and function. *Trends in Genetics,* **8**(12), 403–6.

Review questions

Fill in the blanks

Discrimination in the _____ [1] immune system is performed by antigen receptors on _____ [2]. Antigen receptors are generated by an essentially random process of _____ _____ [3]. This endows each lymphocyte with a unique antigen-_____ [4]. The combined effect of a large population of lymphocytes is a _____ [5] capable of anticipation of any antigen. The numbers of lymphocyte clones able to recognize a particular antigen are _____ [6], which necessitates clonal _____ [7] before a response can be mounted. This means the adaptive system is _____ [8] to respond to a primary encounter with antigen. A residual population of _____ [9] lymphocytes persist after the response subsides which can respond more quickly upon subsequent exposure. The randomness of receptor generation produces _____ [10] receptors that must be purged from the system.

Short answer questions

1. What is a lectin? Give examples of soluble animal lectins involved in innate recognition in mammals.
2. Describe the thioester bond-containing proteins involved in recognition in immune systems.
3. Name two ways NK cells can recognize target cells.
4. Provide a succinct definition of clonal selection.
5. What is the immunoglobulin superfamily?
6. What is tolerance and why is it necessary?
7. What is the difference between DNA recombination, general recombination and rearrangement?

3   Cells and tissues

In this chapter

- Blood cells and haemopoietic tissues in different vertebrates.
- The function and evolution of the lymphatic system and lymphoid tissues.
- Lymphocyte traffic and adhesion molecules.

Introduction

In multicellular animals less than a few millimetres in thickness, diffusion alone is adequate for the transport of oxygen and nutrients through tissues. Organisms any larger than this depend on fluid that is circulated around the body by muscular contraction. It is within these **circulatory systems** that cells of the immune system move and maintain immune surveillance.

During the course of evolution, the circulatory fluid of vertebrates and some invertebrate animals has become confined to vessels. The majority of invertebrates have open circulatory systems in which fluid is pumped around by one or more heart-like muscular blood vessels. Elsewhere in the body the fluid is in direct contact with the tissues. In only a few invertebrates such as annelid worms is the blood enclosed in a rudimentary vascular system. All vertebrates possess a closed system of vessels containing blood that is distributed around the body by a heart (**cardiovascular system**). Open blood circulation is uncommon in vertebrates but is still in evidence, for example, in the peripheral circulation of hagfish. Some fishes and all amphibians, reptiles, birds and mammals also have an additional circulatory system called the **lymphatic system** which drains fluid from the tissues and returns it to the blood.

Vessels allow precise control over where circulatory fluid is supplied, although vasculature poses unique problems for cells of the immune system that need to be aware of antigens in the tissues and to gain access to them. Circulatory systems have changed considerably during vertebrate evolution, particularly after the emergence of aquatic vertebrates onto land. These changes have had significant consequences for the organization of cells and tissues of the immune systems of different vertebrates, particularly between fishes and terrestrial vertebrates.

In this chapter we will examine first the blood cells that mediate defence reactions and then focus on the physiology of the vertebrate system. Different vertebrates will be compared to see how these tissues have evolved. We will also examine the movement of cells of the immune system between the circulatory systems and the tissues, and the cell-surface adhesion molecules that allow this to occur.

3.1 Invertebrate blood cells

Most of what we understand of invertebrate defence reactions has come from the study of coelomate invertebrates – animals with a fluid-filled cavity between the gut and the outer body wall called the **coelom**. A wide variety of different species have been studied (see Table 1.3 in Chapter 1), but most of what we know comes from molluscs, annelids, and arthropods – particularly insects and crustacea. In these animals, defence against invading microorganisms is mediated by cells contained within the coelom called **coelomocytes** (particularly in annelids), **amoebocytes** or **haemocytes** (in molluscs and arthropods).

The number of blood cell types varies among different species. The horseshoe crab *Limulus*, for example, has only one blood cell type (see Figure 1.6 in Chapter 1), which may reflect a primitive status of this animal. These cells are packed with granules which contain the components required for blood coagulation. Coagulation plays a dominant role in the defence against microbes in *Limulus*. In addition to arresting bleeding from wounds, it serves to prevent bacteria from dispersing throughout the body.

The majority of other invertebrate species have multiple types of blood cells, each with specialized roles in defence reactions. Classification of the different types is based on morphology and function, although because of the diversity of invertebrates, it has been notoriously difficult to standardize classification schemes across different phyla. Insect larvae have been particularly well studied. *Drosophila*, for example, has only two haemocyte types – **plasmatocytes** that are phagocytic and **crystal cells** that are involved in melanization. In contrast, the moth *Heliothis* has at least five morphologically distinct haemocyte types (Figure 3.1). The most abundant are plasmatocytes and **granular cells**. In this species, granular

morphology *n.* the physical shape or form of an organism; also applied to cells.

granular cells *n.* a class of invertebrate (esp. insect) blood cells with distinctive cytoplasmic granules.

Figure 3.1 The haemocytes of the moth *Heliothis* shown attached to a glass microscope slide. Bar = 20 μm. (1) Prohaemocyte. (2) Plasmatocyte, shown spread over the glass surface *in vitro*. In the haemocoel these cells are normally round or spindle-shaped. Plasmatocytes are the capsule-forming cells *in vivo*. (3 and 4) Granular cells, shown freshly removed from the haemocoel and after being allowed to spread for 30 min, respectively. Notice the fine processes in the unspread cell. These cells are phagocytic and release coagulum. (5) Spherule cell. (6) Oenocytoid. From Davies, D.H., *et al.* (1987) *Journal of Insect Physiology,* **33**, 143–53.

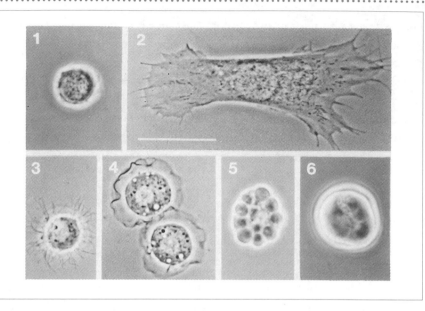

cells are phagocytic and also responsible for producing the coagulum for entrapping particulate material. When examined *in vitro* these cells have a hairy appearance which may enhance their ability to entrap particulate material in the blood. The plasmatocytes have neither of these properties in this species. Instead, the main role of these cells is the formation of the multicellular sheath in nodulation and encapsulation reactions. *In vitro* these cells also spread out over a microscope slide (panel 2 in Figure 3.1). **Oenocytoids** are fragile cells that appear to burst in response to trauma and trigger the melanization of the blood. The functions of the other cell types, the **spherule cells** and **prohaemocytes**, remain enigmatic. Other invertebrates have one or more cells that look like these, or at least have similar properties.

BOX 3.1 Synopsis of vertebrate evolution

Fish are the oldest group of vertebrates, with fossils dating from the Ordovician period. The first to appear were armoured, jawless (agnathous) animals, collectively called ostracoderms (bony-skinned). Their modern relatives are the hagfishes and lampreys, which are considered still to be among the most primitive of all vertebrates. These modern Agnatha are non-armoured, eel-like fish without paired fins. Both also have a round mouth; hence they have collectively been termed cyclostomes. Hagfish are predatory and rasp the flesh from dead and moribund fish. Lampreys have an unusual life cycle featuring a filter-feeding larval stage called an ammocete, which metamorphoses into an adult. Adult lampreys are usually ectoparasitic and suck blood of other fish. Despite common anatomical features these two modern agnathans have a long history of separate evolution. Recent studies suggest that the lampreys are more closely related to the jawed-mouth (or gnathostome) vertebrates than either are to hagfish.

The first jawed fish diverged into three main lineages: the cartilaginous fish or Chondrichthyes (the sharks and rays) and two lines of bony fish or Osteichthyes. Of these two lines of bony fish the Actinopterygii are represented by most modern fish, of which the teleosts form the majority. The other, the Sarcopterygii, have only a few modern representatives, such as the coelacanth and lungfish (Dipnoi), although these are the groups from which the tetrapods (amphibians, reptiles, birds and mammals) evolved. Today fish are the largest group of vertebrates with some 20 000 species. Interest in fish immunobiology is stimulated in part by their ancestral nature and also by economic interest in fish aquaculture.

The first tetrapods, which evolved from Osteichthyes, were amphibious. These emerged onto land about 350 million years ago. Today, amphibians are represented by three orders, the limbless Apoda, the tailed Urodeles (salamanders and newts) and the tailless Anura (frogs and toads). These animals retain a semi-aquatic existence and return to water to breed and lay their eggs. The transition from aquatic to terrestrial existence has been accompanied by profound changes in physiology.

Reptiles are the closest living relatives of the ancestors of birds and mammals. Three main lineages of reptiles have emerged. The Synapsida, although now extinct, gave rise to the mammals. Diapsida, consisting of archosaurs (crocodiles, and the now extinct dinosaurs, and the lineage from which the birds are believed to have evolved), and lepidosaurs (snakes and lizards). In reality birds are more akin to modern reptiles than mammals, with the Crocodilia being the closest living relative of birds. Thirdly are the Anapsids, represented mainly by modern Chelonia (tortoises and turtles). Despite their pivotal position in the ancestry of vertebrates, very little is known about reptilian immune systems. Reptiles, and the birds and mammals that have evolved independently from them, are amniotes. Their embryos are surrounded by an amniotic membrane which encloses amniotic fluid. This has enabled amniotes to develop entirely independently of an aquatic environment.

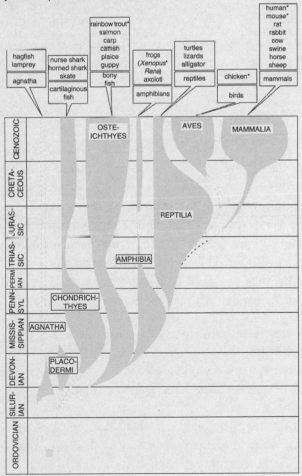

Figure 1 Modified version of Romer's (1966) phylogeny of the vertebrates. The width of the branches indicates the relative abundance of species at certain times. Above are some of the more commonly used animals of each class of vertebrate for immunological study. Particularly well studied model animals are marked (*). Mammals, particularly mouse and human, have received most attention. Commercially important species of fish, bird and mammals are also comparatively well studied. Redrawn and modified from Romer, A.S. (1966) *Vertebrate Paleontology*. The University of Chicago Press, Chicago and London.

BOX 3.1 *continued*

Like fish and amphibians, modern reptiles are all ectotherms and are dependent on the external environment for their heat content. In contrast, mammals and adult birds are endothermic, and derive all or most of their heat content from endogenous metabolic heat rather than from the external environment. Heat generated endogenously allows the body temperature to be regulated more accurately and a stable body temperature can be maintained. This has meant that mammals and birds can better exploit colder environments, higher altitudes or become nocturnal, than ectotherms. All contemporaneous reptiles are ectotherms so it is thought that endothermy evolved independently in birds and mammals after their divergence from their respective reptilian ancestors.

The mammals are descendants of the now extinct synapsid reptiles. Synapsids diverged from other reptile stocks over 300 million years ago. Three main lineages of mammals have emerged: monotremes, marsupials and eutherian mammals. Monotremes, represented today only by the platypus and echidnas, have retained the reptilian characteristic of laying eggs rather than giving birth to live young (called **viviparity**). Marsupials (wallabies, opossums, bandicoots), after a very brief gestation, incubate their young externally, usually in a pouch. Eutherian mammals form by far the largest group of modern mammals and include ruminants, swine, equines, carnivores, rodents, rabbits, primates and bats. Unlike the marsupials these retain their developing foetus in utero for a characteristically long time. By definition mammals provide their young with milk.

Further reading

Forey, P. and Janvirer, P. (1993) Agnathans and the origin of jawed vertebrates. *Nature,* **361**, 129–34.

Kemp, T.S. (1992) *Mammal-like Reptiles and the Origins of Mammals.* Academic Press, London.

3.2 Vertebrate blood cells

Vertebrates, which are all members of a single phylum, also have a heterogeneous mixture of blood cells (Figure 3.2). However, because their properties in immune defence reactions are, on the whole, very different from those of invertebrates, it is almost impossible to draw functional parallels between vertebrate and invertebrate cells with any certainty. Three major types are recognized in vertebrates: the red blood cells or **erythrocytes**, the **thrombocytes** (also called **platelets** in mammals) and a heterogeneous mixture of white blood cells or **leucocytes**.

- Erythrocytes are respiratory cells that contain haemoglobin. They function exclusively in the cardiovascular circulation and carry oxygen from the respiratory surfaces (gills or lungs) to the tissues where it is exchanged for carbon dioxide. Vertebrate erythrocytes differ widely in appearance according to species. In most species, erythrocytes are nucleated although mammals are unusual among the vertebrates as the nucleus is extruded from immature red cells (erythroblasts) during the final stages of their maturation in the bone marrow.

thrombocyte *n.* the non-mammalian platelet, involved in blood clotting.

leucocyte, leukocyte *n.* white blood cell; in mammals comprise **granulocytes and mononuclear cells.**

bone marrow connective tissue inside hollow bones; often contains haemopoietic tissue.

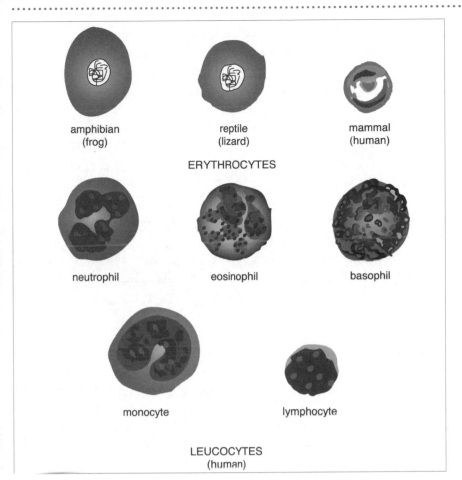

Figure 3.2 Vertebrate blood cells. Redrawn from Fawcett, D.W. (1986) *Bloom and Fawcett: A Textbook of Histology*. W.B. Saunders, Philadelphia, PA.

- Thrombocytes are cells that seal punctures in the vasculature and activate the blood-clotting cascades during haemorrhages. Mammalian platelets are small non-nucleated cells that are derived by the fragmentation of large precursor cells called megakaryocytes.

- Leucocytes are the cells of the immune system. These are a heterogeneous collection of blood cells, and each type is specialized to perform different functions in adaptive and innate responses. The numbers of leucocytes in the body is very plastic – in conditions of infection by microorganisms the number of leucocytes in the blood increases.

Unlike those of erythrocytes and platelets, leucocyte functions take place in the tissues rather than the circulatory systems. Leucocytes that are in the circulation at any one time are in transit from one tissue site to another. In order to enter the tissues leucocytes are able to attach to the endothelial cells inside the blood vessels and squeeze between them out of the blood. This process is called **extravasation**. As we will see, this is a non-random process that directs the entry of different leucocytes into different tissues. Shown in Figure 3.3 is a leucocyte migrating through a blood vessel, in this case a lymphocyte crossing at specialized vascular endothelial cells within lymphoid tissues (see later).

Leucocytes are broadly classified into **granulocytes** and **mononuclear cells**, according to their appearance. Mammalian leucocytes are also distinguished according to the expression of particular cell-surface molecules that are identified using specific antibodies. Over 100 of these **phenotypic markers** (called CD molecules) have been identified on leucocytes of mouse and human. In comparison, relatively few antibodies are available that can be used for distinguishing phenotypes of non-mammalian leucocytes and fewer still for invertebrates.

The descriptions that follow apply to human leucocytes, although the morphology and functions of leucocytes are well conserved in other vertebrates.

3.2.1 Granulocytes

As the name suggests, these cells have prominent granules in their cytoplasm, which are actually vesicles containing a cocktail of different enzymes. Granulocytes also typically have a lobate nucleus, which was erroneously interpreted in early histological studies as the presence of multiple nuclei. Granulocytes are subclassified into **neutrophils**, **basophils** and **eosinophils** according to the histochemical properties of their granules.

In humans, neutrophils are the commonest leucocyte in the blood (Figure 3.4). The main function of these cells is to migrate to sites of inflammation and ingest invading microorganisms by phagocytosis. Neutrophils are able to recognize and ingest pathogens using innate recognition mechanisms independently of antigen receptors, although these cells bear Fc receptors and can ingest microbes opsonized by antibodies.

Basophils are the least numerous leucocyte type in blood. Their granules discharge locally acting inflammatory mediators such as histamine and it is thought that basophils promote acute inflammation.

The function of eosinophils is less clear, but they are phagocytic and may play a role in immunity to parasites and worms.

3.2.2 Mononuclear cells

The other major morphologically distinct blood cells are mononuclear cells. These were so called because their nuclei are round and do not have the multinucleate appearance of granular cells. Mononuclear cells are subclassified into **monocytes** and **lymphocytes**.

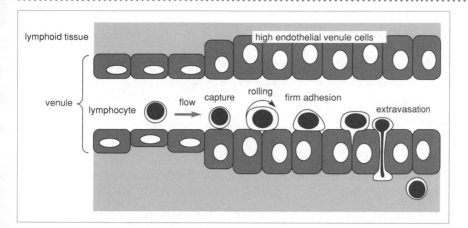

Figure 3.3 A lymphocyte leaving the blood at high endothelial venules. Adapted from Girard, J.-P. and Springer, T.A. (1995) *Immunology Today*, **16**(9), 449–57.

Monocytes

Monocytes in the blood are precursors of connective tissue phagocytes called **macrophages**. Monocytes enter the tissues by migrating through the walls of capillaries wherein they undergo maturation into a macrophage. Macrophages are amoeboid cells that move through tissues ingesting cellular debris. They are attracted to sites of inflammation by chemotaxis where they participate in clearing microorganisms. The type of macrophage a monocyte becomes is determined by the anatomical site of the tissue it infiltrates, and phagocytic macrophages found at different sites are given different names (Table 3.1).

Most of the work performed on mammalian macrophages has been *in vitro* with peritoneal exudate macrophages. These normally migrate into the peritoneal cavity in response to invading microorganisms. This can be induced experimentally in mice by injecting irritants that induce inflammation, such as starch grains or paraffin oil, and then rinsing out the cells with saline a few days later.

> **macrophage** *n.* a large phagocytic mononuclear leucocyte found in the tissues of vertebrates.

Lymphocytes

Lymphocytes are at the heart of the adaptive immune system of vertebrates. These cells can be classified into functional and phenotypic subpopulations called B lymphocytes and T lymphocytes. As we learned in Chapter 2, lymphocytes have the unique ability to recognize non-self components

Table 3.1 Human macrophages

Name	Anatomical site
tissue macrophages	skin
Kuppfer cells	liver
microglial cells	central nervous system
osteoclasts	bone

plasma cell *n.* a fully differentiated B lymphocyte engaging in antibody synthesis.

helper T lymphocyte *n.* a subset of T lymphocytes that elaborate cytokines and provide costimulatory signals for B lymphocytes; helper T lymphocytes are CD4$^+$ and response to antigenic peptides in the context of class II MHC molecules on antigen-presenting cells.

cytotoxic T lymphocyte (CTL) *n.* a subset of T lymphocytes that kill target cells, usually tumour cells or cells infected with viruses; classical CTL are CD8$^+$ and kill target cells bearing antigenic peptides in the context of class I MHC molecules; a minority of CTL are CD4$^+$ and restricted by class II MHC molecules.

Figure 3.4 Proportions of human leucocytes in the blood.

using clonally distributed antigen receptors. If this occurs, a lymphocyte is stimulated to proliferate and mature into a cell with specific effector or regulatory properties. Unlike other leucocytes, B and T lymphocytes have the unique ability to leave the blood vasculature within secondary lymphoid tissues, such as the spleen and lymph nodes, where they encounter deposited antigens (see below).

In their pre-activated, naïve state there is no morphological difference between the two categories of lymphocyte (Figure 3.5). Each has a nucleus surrounded by a thin rim of overlying cytoplasm. Upon activation by antigen, B lymphocytes transform into antibody-producing factories called **plasma cells**. A plasma cell has a very distinctive swollen cytoplasm packed with rough endoplasmic reticulum. This is because these cells are heavily engaged in protein synthesis and secretion.

The functions of T lymphocytes are more diverse, although most can be subclassified into either **helper T lymphocytes** or **cytotoxic T lymphocytes**. Activated cytotoxic T lymphocytes are able to recognize and kill other cells in the body that have become transformed by virus infection or by becoming tumour cells. Cytotoxic T lymphocytes have distinctive lytic granules that are released onto the surface of the transformed cell. Cytotoxic T lymphocytes also express a cell-surface protein, called **CD8**, that is absent from helper T lymphocytes. The function of the CD8 molecule is to assist the antigen receptor when the T lymphocyte is activated by antigen, and it is often referred to as a **coreceptor**. In contrast, helper T lymphocytes have a different coreceptor called **CD4**. Expression of CD4 and CD8 is mutually exclusive on mature T lymphocytes. These cells have *either* the CD4$^+$/CD8$^-$ *or* the CD4$^-$/CD8$^+$ phenotype.

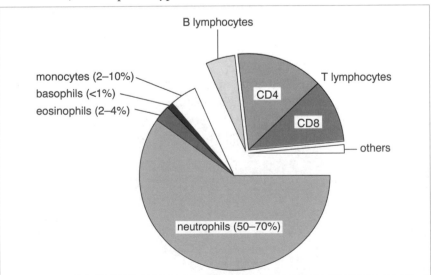

Activated $CD4^+$ (helper) T lymphocytes perform a variety of effector functions. Many of these functions are mediated by soluble signalling molecules called **cytokines** that cause other leucocytes to become effector cells. In particular, helper cells provide signals required by other lymphocytes – some helper cells assist activated B lymphocytes to differentiate into plasma cells and memory cells, whereas other helper cells assist activated cytotoxic T lymphocytes to become fully active effector cells. Other cytokines enhance the phagocytic behaviour of macrophages. We will learn more about T lymphocyte 'help' and the role of cytokines in Chapter 8. $CD4^+$ T lymphocytes are also the major lymphocyte type involved in **delayed-type hypersensitivity** (**DTH**) responses, which are antigen-specific responses in the skin.

An additional population of lymphocytes are $\gamma\delta$ **T lymphocytes**. As we will learn in the next chapter, $CD4^+$ and $CD8^+$ T lymphocytes have an antigen receptor constructed from two polypeptide chains, called α and β. The $\gamma\delta$ T lymphocytes have instead an antigen receptor constructed of γ and δ chains. The precise function of $\gamma\delta$ cells remains enigmatic, although they probably represent a line of defence against microorganisms that gain entry to the gut or lungs. In most mammals studied, the normal anatomical site for $\gamma\delta$ cells is the epithelial tissues of the intestine, the skin and in the lung, where they are particularly abundant. These cells are usually outnumbered by $\alpha\beta$ cells, although in ruminants, $\gamma\delta$ cells are more numerous – possibly because infection across mucous membranes of the stomach poses the greatest infection threat to these animals. Mouse $\gamma\delta$ T lymphocytes differ from the more common $\alpha\beta$ cells in two respects. Firstly, they do not usually bear the coreceptor molecules CD4 or CD8. Second, they appear to be able to recognize free antigen, rather like antibodies do, rather than on the surface of an antigen-presenting cell (see Chapter 5).

Dendritic cells

Dendritic cells (Figure 3.6) are an additional category of cells that may be related to the monocyte/macrophage system. These cells recirculate between the peripheral tissues and the lymphoid tissues via the circulatory systems, and play a vital role in collecting antigen and presenting it to T lymphocytes in the lymphoid tissues. Dendritic cells have the unique capacity for activating naïve T lymphocytes as they present antigen, and are regarded as the sentinels at the interface between the antigenic universe and the adaptive immune system. During recirculation, dendritic cells undergo changes in activity or maturation and these are given different names according to their anatomical location (Table 3.2).

cytokine *n.* a short-range extracellular signalling molecule that influences the growth, differentiation or behaviour of cells.

delayed-type hypersensitivity (DTH) a reaction by memory $CD4^+$ T lymphocytes in response to re-stimulation by antigen in the cutaneous epithelium; manifested as an inflammatory reaction that peaks 2–3 days after application of the antigen to the skin.

dendritic cell *n.* antigen-presenting cells of haemopoietic origin with potent capacity to activate naïve T lymphocytes; these cells are circulatory and are found in the blood, skin (**Langerhans cell**), lymph (**veiled cell**) and secondary lymphoid tissues (**interdigitating dendritic cell**).

Table 3.2 Dendritic cells

Name	Anatomical site
Langerhans' cells	skin
Veiled cells	lymph
Interdigitating dendritic cells	lymphoid tissues (lymph nodes)
Dendritic cells	blood

haemopoiesis *n.* the formation of blood and red and white blood cells from stem cells.

haemopoietic stem cell cell in haemopoietic tissue with capacity for self-renewal and to differentiate into any of the mature blood cells.

3.3 Haemopoiesis

Most mature blood cells have a relatively short life span and are constantly renewed. In vertebrates, blood cells arise from a single immature cell type called a **haemopoietic stem cell** [Greek *haima* = blood + *poietkos* = producing].

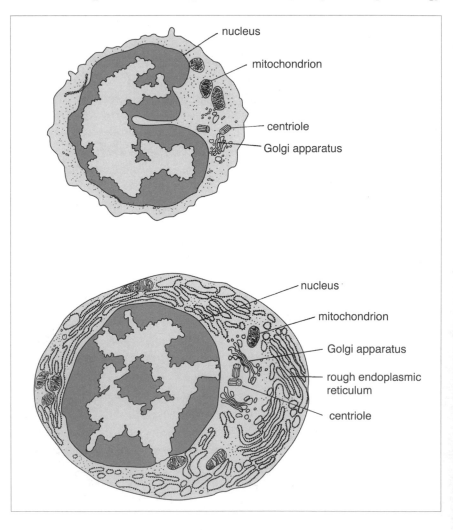

Figure 3.5 Morphology of B lymphocytes before and after activation by antigen. (a) Unactivated lymphocyte as seen by transmission electron microscopy. The cell is characterized by a dense nucleus and a relatively small volume of cytoplasm. Unactivated T and B lymphocytes are indistinguishable by morphology alone. (b) A plasma cell. B lymphocytes that become activated by antigen differentiate into antibody-secreting plasma cells. Notice that the cytoplasmic volume has increased and is rich in rough endoplasmic reticulum. Redrawn from Burkitt, H.G., Young, B. and Heath, J.W. (1993) *Wheater's Functional Histology*. Churchill Livingstone, Edinburgh.

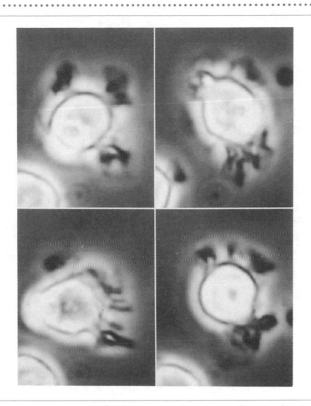

Figure 3.6 Dendritic cell from human blood, shown moving on a glass surface. Courtesy of Dr M. Binks, University College London.

Figure 3.7 shows the lineages that descend from the haemopoietic stem cell of a typical mammal.

By definition a stem cell has the potential to develop into any one of several different kinds of more mature cell types, whilst also being able to produce more stem cells. At the root is the stem cell with the greatest potential, called a multipotential or **pluripotent** stem cell. This can give rise to a number of different 'committed' stem cells able to evolve into one or more specific cell types, in addition to renewing itself. The subsequent, unidirectional maturation from a committed stem cell into a mature cell is called **differentiation**, and is accompanied by characteristic changes in cellular morphology, function and the expression of specific genes. As a committed stem cell differentiates, the potential number of different mature cell types it can generate steadily decreases, so that a multipotential stem cell gives rise to oligopotential progenitors, which eventually differentiate into monopotential progenitors. Which particular differentiation pathway a cell selects is determined by extracellular signals the cell receives.

The process of blood cell formation has been studied most thoroughly in mammals. The mammalian pluripotent and committed haemopoietic stem cells are called **colony forming units**. This is because haemopoietic

pluripotent stem cell a stem cell able to give rise to the greatest number of different 'committed' **stem cells**.

differentiation *n.* development of cells with specialized functions from unspecialized precursor or stem cells.

colony forming unit types of haemopoietic stem cells committed to differentiation along different lineages, so called for their ability to populate haemopoietic tissues, esp. the spleen, to form colonies.

stem cells are able to migrate into haemopoietic tissues and multiply into colonies. For example, when bone marrow-derived stem cells of a mouse are injected into an irradiated recipient mouse (treatment that kills proliferating cells, including the recipient's own haemopoietic cells) the donor's bone marrow cells repopulate the recipient's spleen. These cells begin to proliferate and differentiate, forming colonies. Within each colony are found erythrocytes, granulocytes and megakaryocytes. The bone marrow-derived progenitor of these cells is therefore called a colony forming unit-spleen (or **CFU-S**).

The CFU-S gives rise to three kinds of committed stem cells (Figure 3.7). These are the progenitors of erythrocytes (**CFU-E**), the progenitors of megakaryocytes (**CFU-M**) and the progenitors of granulocytes/monocytes (**CFU-GM**). The signals that determine how a CFU-S becomes committed to each of these three pathways are not well understood. Subsequent differentiation of CFU-E into erythrocytes is stimulated by erythropoietin, a hormone that is produced in response to decreases in the amount of oxygen in the environment. The differentiation of CFU-M is stimulated by a hormone called thrombopoietin, whereas CFU-GM is stimulated by a cytokine called granulocyte/macrophage colony stimulating factor (**GM-CSF**). The CFU-GM then differentiate into monopotential progenitors of granulocytes, monocytes and eosinophils with stimulation from **G-CSF**, **M-CSF** and **E-CSF**, respectively.

The pluripotent CFU-S is not capable of producing lymphocytes. These are generated from a different pluripotent stem cell, CFU-L. These emerge from haemopoietic tissues and colonize **primary lymphoid tissues** in order to undergo further differentiation into mature T or B lymphocytes (see below). Both CFU-S and CFU-L arise from a single type of pluripotent stem cell, called CFU-M,L, from which two lineages descend: the **lymphoid** lineage (lymphocytes) and the **myeloid** lineage (the other blood cell types).

3.3.1 Haemopoietic tissue

In vertebrates, the first haemopoietic stem cells appear in the mesoderm of the embryonic yolk sac. These self-renew and differentiate into blood cells within specialized tissues called **haemopoietic tissues**. In mammals, this role is performed by the embryonic liver, then the embryonic spleen, and ultimately the spleen and bone marrow in the newborn animal. In some mammals, such as humans, the haemopoietic function of the spleen subsides and is lost by the time of birth.

The anatomical location and phylogenetic relationships of haemopoietic and lymphoid tissues of different vertebrates are shown in Figures 3.8 and 3.9. The majority of vertebrates produce blood cells in the spleen.

colony stimulating factors a group of growth factors that stimulate the differentiation of different colony forming units.

primary lymphoid tissue *n.* anatomical site where antigen receptor gene rearrangement occurs (see also **thymus, bone marrow** and **bursa of Fabricius**).

lymphoid *a.* tissues or cells pertaining to lymphocytes.

myeloid pertaining to the bone marrow; tissue in which haemopoiesis occurs in vertebrates; cells of haemopoietic origin other than lymphoid cells.

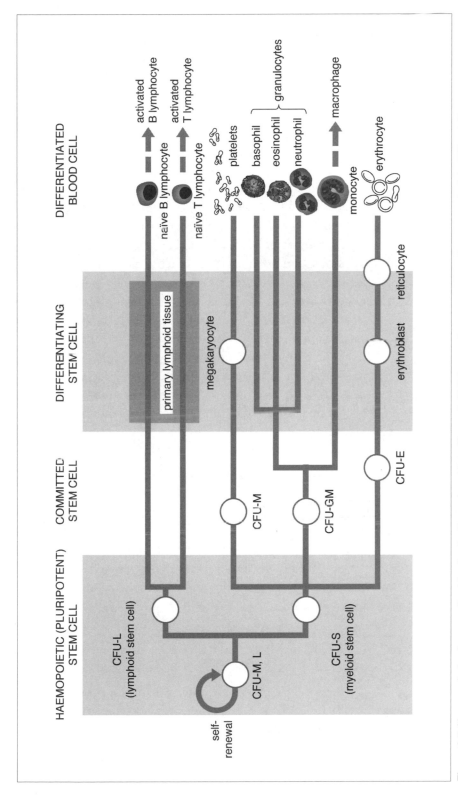

Figure 3.7 Haemopoiesis of mammalian blood cells. Reproduced from Turner (1994) *Immunology – a comparative approach*, John Wiley & Sons.

Agnatha *n.* class of primitive vertebrates lacking a jaw; represented today by hagfish and lampreys.

typhlosole *n.* an invagination of the stomach of larval lampreys that contains haemopoietic and lymphoid cells, together called a protospleen.

Exceptions include those animals that lack a true spleen (the Agnatha). However, larval lampreys have a primitive 'protospleen' of haemopoietic and lymphoid cells located within an invagination of the stomach called the typhlosole. Haemopoietic tissues are also found in the pronephros (head kidney). The haemopoietic function of the protospleen and pronephros regresses after metamorphosis and in adult lampreys spleen-like tissue is instead found associated with the intestine. In the hagfish, which is considered more primitive than the lamprey, haemopoietic tissue is less well defined and scattered in islands in the gut wall and in the pronephros. Other exceptions include some mammals, such as humans, in which the spleen is haemopoietic only during embryonic development. Even in humans the spleen may resume its haemopoietic function in response to certain diseases.

In most amphibians, and in reptiles and mammals, haemopoietic tissue is also housed in bone marrow. Hollow bones and bone marrow evolved when vertebrates first emerged onto land and changed from an aquatic to a terrestrial way of life. Hollow bones are presumably an adaptation to locomotion on land, as they are likely to better resist compression stresses at constant bone mass. Consequently, jawed fishes and urodele amphibians lack the same organized bone marrow of terrestrial vertebrates, and instead have haemopoietic tissue in the kidney and spleen. Birds are also hollow-boned although, presumably through adaptation to flight, the space is occupied by ramifications of the air sacs and haemopoiesis occurs instead in the spleen.

Summary

- There are three classes of vertebrate blood cells: erythrocytes that carry haemoglobin, thrombocytes which are involved in blood coagulation, and leucocytes that are involved in the innate and adaptive immune systems.

- Vertebrate blood cells are produced from haemopoietic stem cells, a self-renewing population that also undergoes differentiation into different committed stem cells in haemopoietic tissue.

- Vertebrate haemopoietic tissues change as the embryo grows, starting with the yolk sac, followed by liver and spleen, and ultimately in terrestrial animals, in the spleen and bone marrow.

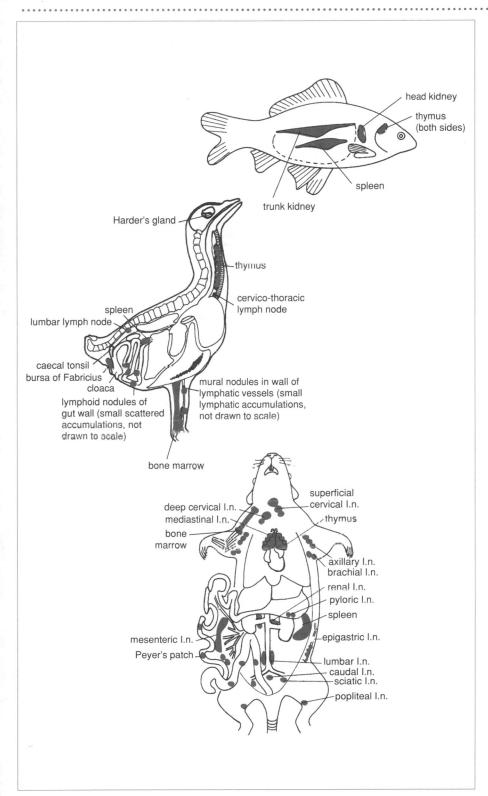

Figure 3.8 The main haemopoietic and lymphoid tissues in carp, chicken and rat. Carp reproduced from Turner, R.J. (ed.) (1994) *Immunology. A Comparative Approach.* J. Wiley & Sons, London; chicken and rat redrawn and modified from Manning, M.J. and Turner, R.J. *Comparative Immunobiology.* Blackie & Son, London.

	agnatha	cartilaginous fish	bony fish	amphibians	reptiles	birds	mammals
Secondary lymphoid tissues — multifollicular (with germinal centres)	no	no	no	no	no	lymph nodes (some species)	lymph nodes, Peyer's patches, appendix, tonsils, spleen
— solitary follicles	no	no	no	no	gut, jugular bodies, lungs	gut lungs	gut lungs
— diffuse	gut gills	gut gills	gut gills	gut gills/lungs	gut lungs	gut lungs	gut lungs
— haemopoietic	kidney (typhlosole of larval lamprey)	spleen kidney	spleen	spleen	spleen	spleen	spleen
Primary lymphoid tissues — B lymphocytes	?	spleen	spleen	spleen (bone marrow)	spleen bone marrow	bursa	(spleen) bone marrow
— T lymphocytes	?	thymus	thymus	thymus	thymus	thymus	thymus
Haemopoietic tissues	kidney, gut (typhlosole of larval lamprey)	spleen kidney	spleen kidney	spleen (bone marrow)	spleen bone marrow	spleen bone marrow	(spleen) bone marrow

Figure 3.9 Phylogenetic relationships of lymphoid tissues. The complexity of lymphoid tissues increases from fish to mammals. In agnathan vertebrates, lymphoid tissue is diffuse and mostly admixed with haemopoietic cells. Diffuse lymphoid tissues are also associated with the typhlosole, along the gut and the gills of larval lampreys, the latter possibly operating as a rudimentary thymus. Adult hagfish are devoid of aggregations of lymphocytes altogether. A dichotomy of lymphoid cells into T and B lymphocytes has not been demonstrated in Agnatha. In contrast, all jawed (gnathostome) vertebrates possess true primary and secondary lymphoid tissue. The thymus is conserved throughout the gnathostome vertebrates as the primary lymphoid tissue for T lymphocytes. Haemopoietic tissue often serves as the equivalent for B lymphocytes. In land-living vertebrates, this is mainly the bone marrow, although in fish and some amphibians the spleen performs this function. The role of bone marrow as a major site of haemopoiesis and B lymphocyte maturation emerges for the first time in the Anura (frogs and toads). Birds have evolved a unique organ dedicated for B lymphocyte maturation called the bursa of Fabricius. Antigens and immune complexes are trapped in secondary lymphoid tissue. Blood-borne antigens are removed by secondary lymphoid tissues admixed with haemopoietic tissues. In fish these are located in the spleen, head kidney and trunk kidney. Diffuse lymphoid tissues are also associated with the digestive tract and gills although fish do not have solitary lymphoid follicles or multifollicular structures. The lymphoid tissues of the simplest amphibians (Apoda and Urodela) are much like those of fish, with the major lymphoid tissues being the thymus and the spleen. Solitary lymphoid follicles are absent from fish and urodele amphibians (newts and salamanders), but appear for the first time in anurans (frogs and toads) and are present in all other tetrapods. Multifollicular lymphoid tissues emerge in marsupial and eutherian mammals. Figure from *Functional Anatomy of the Vertebrates: an evolutionary perspective*, by Warren F. Walker, Jr., copyright © 1987 by Saunders College Publishing, reproduced by permission of the publisher.

3.4 The lymphatic system

As blood is pumped around the vertebrate body, fluid seeps through the thin walls of capillaries and into the interstitial spaces between the cells of tissues. To drain away this fluid and to return it to the blood, an additional circulatory system of **lymphatic vessels** has evolved.

There is considerable debate whether fishes have a true lymphatic system. Because the body of a fish is supported by water there is little gravitational force to be overcome by the heart and blood can be pumped at low pressure. Consequently, relatively little fluid escapes from the blood and tissue drainage is performed mainly by the veins of the cardiovascular system. In general, terrestrial vertebrates have higher pressure cardiovascular systems necessary to compensate for the gravitational forces imposed by living out of water. It is in these animals that a true lymphatic system is seen. Such systems are thought to have arisen first in amphibians (Figure 3.10) and the lymphatic systems of different tetrapods are generally similar.

The lymphatic vessels originate in the tissues as thin, blindly ending ducts. These have very thin walls through which interstitial fluid moves by diffusion. Several blind vessels converge onto a single wider vessel, which joins up with other vessels. Like streams emptying into a river, lymphatic vessels become progressively larger as they eventually converge into one large duct. The drained fluid, or **lymph**, is then discharged into the venous blood, usually near the heart where the pressure is lowest.

Lymph is pushed through the lymphatic system by the contractions of surrounding body musculature. In the majority of vertebrate classes, this flow is pushed in one direction by regions of contractile lymphatic vessels called **lymph hearts** (Figure 3.10), although mammals do not have lymph hearts. Mammals instead maintain a unidirectional flow of lymph by using regularly spaced one-way valves. As we will see, mammals have exploited the lymphatic circulation as a means for recirculating lymphocytes, dendritic

lymph *n.* the clear fluid that collects between tissue cells and which is drained away by a network of vessels called the **lymphatic system**.

tetrapods *n. plu.* vertebrates with four limbs (includes birds).

lymph hearts *n.* small contractile areas in the lymphatic systems of non-mammalian vertebrates that push lymph in one direction.

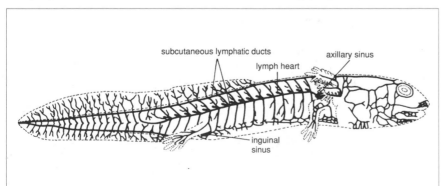

subcutaneous lymphatic ducts   axillary sinus
lymph heart
inguinal sinus

Figure 3.10 The lymphatic system of a urodele amphibian (salamander). Reproduced from Walker, W.F. (1987) *The Functional Anatomy of the Vertebrates. An Evolutionary Perspective.* Saunders College Publishing, Philadelphia, PA.

cells and immune complexes (aggregations of antigen and antibody) around the body. In contrast, the lymphocytes of non-mammalian vertebrates seem not to use this system to the same extent.

3.5 Primary lymphoid tissues

As mentioned above, lymphoid stem cells emerge from haemopoietic tissue to continue their differentiation into mature T or B lymphocytes at other anatomical sites. These sites are classified as **primary** and **secondary lymphoid tissues**, according to the kind of differentiation that occurs within (Figure 3.11). Lymphoid stem cells that emerge from the haemopoietic tissues do not express antigen receptors on their plasma membranes, because the genes that encode the receptors are in the non-functional, germline configuration. To become functional the genes must first undergo rearrangement. This can only occur within the specialized cellular microenvironment provided by primary lymphoid tissues. A useful definition of a primary lymphoid tissue therefore, irrespective of species, is a tissue in which a differentiating lymphoid cell rearranges its antigen receptor genes.

3.5.1 The thymus is the primary lymphoid tissue for T lymphocytes

A true thymus gland has not been found in lampreys or hagfish, although larval lampreys have diffuse collections of lymphoid cells located near the gills that may fulfil the role of a rudimentary thymus. A true thymus is present in all the jawed vertebrates. In each case its function is the same – to provide the environment in which lymphoid stem cells differentiate into

Figure 3.11 Relationships between mammalian haemopoietic tissues, and primary and secondary lymphoid tissues in the production of lymphocytes. Committed lymphoid stem cells emerge from haemopoietic tissues and continue development in primary lymphoid tissues, where antigen receptor genes are rearranged. Mature, but naïve lymphocytes enter a pool of trafficking lymphocytes that circulate through secondary lymphatic tissues. In non-mammalian vertebrates the role of the lymphatic system is less important. The function of secondary lymphoid tissues is to trap antigen from the blood or lymph for activation of trafficking lymphocytes. See Figure 3.8 for the identities of primary and secondary lymphoid tissues in different animals.

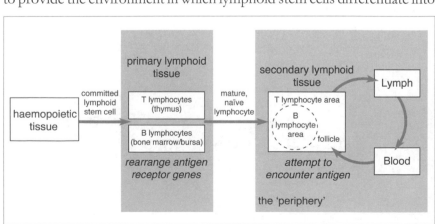

mature T lymphocytes. Lymphoid stem cells are released into the blood by the haemopoietic tissues, which are then attracted to the thymus by hormones it releases. Once inside the thymus these stem cells proliferate and differentiate. The progeny of the stem cells are called **thymocytes**.

In all jawed vertebrates, the thymus is at its most active during the juvenile stages in the life of the animal. Studies of newborn mammals, chicks and amphibian tadpoles have shown that surgical removal of the thymus (**thymectomy**) during this stage blocks the development of T lymphocytes. This abolishes the capacity of the animal to mount T lymphocyte-mediated responses upon reaching maturity. As a general rule the thymus begins to decline in size and activity (or **involute**) around the time an animal reaches sexual maturity, although it never disappears completely. Superimposed on this long-term trend are fluctuations in activity that depend on 'stress' factors such as disease and seasonal changes (see Chapter 7).

The vertebrate thymus has a complex embryological origin. It first develops during embryogenesis from a thickening of endodermal epithelial cells lining the **pharyngeal pouches** (Figure 3.12). These pouches develop on either side of the **pharynx** – the anterior chamber behind the mouth that is shared by the digestive tract and the respiratory tube. In fishes, the pharyngeal pouches fuse with clefts forming on the outside of the embryo to produce gill slits. In mammals, the pouches initially appear very similar to those of the fish embryo, but instead of producing gills they give rise to different structures, including the middle ear and Eustachian tubes, the thyroid and parathyroid glands, tonsils and thymus.

During development of the mammalian thymus, epithelial cells in the ectoderm are stimulated to proliferate in response to signals from the inwardly growing epithelial cells in the pharyngeal pouch. Ectodermal and endodermal cells merge to form a thymic rudiment. Mesodermal cells, including macrophages and dendritic cells, are also recruited, although the exact contribution that each makes may vary between species.

The anatomical location of the fully developed thymus also varies between species. In fishes, amphibians and reptiles, the thymus originates from the dorsal surface of the pouches. Hence the mature gland is located dorsally (Figure 3.8). In fishes only, the pair of thymi remain attached to the dorsal epithelium. In contrast, the thymus of birds and mammals originates from ventral pharyngeal epithelium, giving rise to a ventrally located gland. In all vertebrates other than the fishes, the thymic rudiments separate from the epithelium and migrate downwards and eventually fuse into a single gland. In frogs, the thymus is a pair of compact oval bodies although the glands are more diffuse in other amphibians. In reptiles and birds, epithelial cells detach during the downward migration, producing a characteristic pair

thymus *n.* a primary lymphoid tissue that hosts the maturation of T lymphocytes from lymphoid stem cells.

thymocyte *n.* an immature T lymphocyte within the thymus.

thymectomy *n.* surgical operation to remove the thymus.

pharynx *n.* in the vertebrates the tube behind the mouth shared by the alimentary and respiratory tubes; the throat.

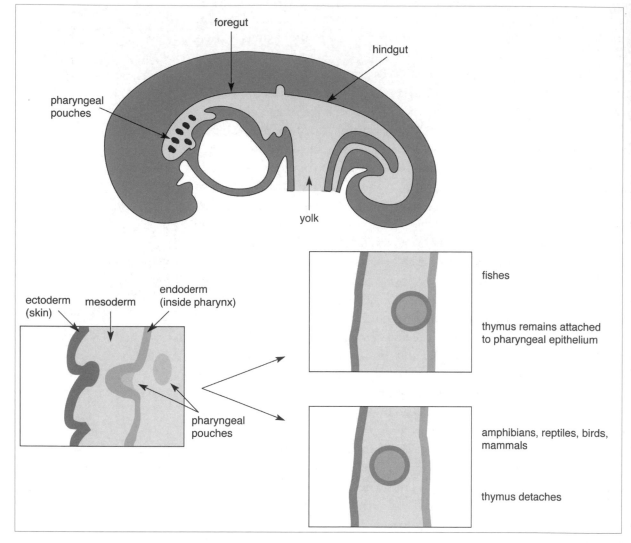

Figure 3.12 Generation of the thymus from pharyngeal pouches. Above is a diagrammatic representation of a mammalian embryo. The pharyngeal pouches are located on each side of the pharynx. Two thymic rudiments are formed from one or more pouches, depending on species of vertebrate. In fish they retain contact with the pharyngeal epithelium whereas in tetrapods the thymic rudiments detach and migrate posteriorly along the line of the gut before fusing (below). In mammals the thymic rudiment is formed by an anlage with epithelial cells of the ectoderm.

of strands (in the reptile) or chains of thymic glands in the bird (Figure 3.8). In humans and most other mammals, the thymus is located in the upper chest above the heart.

The internal organization of the thymus gland is essentially the same for all the jawed vertebrates. The mammalian thymus is surrounded by a capsule of connective tissue, from which are derived septa that separate the gland into lobules (Figure 3.13). An outer zone called the **cortex** contains very densely packed thymocytes. The epithelial cells, which are connected to each other at junctions called **desmosomes**, are pushed apart by the thymocytes. This gives the epithelial component of the cortex a delicate net-like structure and a spongy appearance. In the inner zone, the **medulla**, the thymocytes are less densely packed, and smaller numbers of macro-

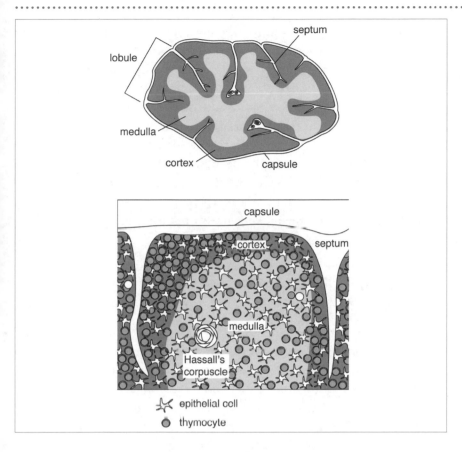

Figure 3.13 Cellular architecture of the thymus. The thymus is a lobed gland divided into distinct lobules. The epithelial stroma is spongy with the voids between the cells packed with thymocytes.

phages and interdigitating dendritic cells are found mostly in this part. Also scattered within the medulla are distinctive balls of flattened epithelial cells called Hassall's corpuscles, whose function is a mystery. The thymus is supplied by several blood vessels, and lymphatic fluid is carried away by efferent lymphatic vessels.

3.5.2 The primary lymphoid tissues for B lymphocytes: the bursa of Fabricius in birds and bone marrow of other tetrapods

In contrast to T lymphocytes, the anatomical location of primary lymphoid tissue for B lymphocyte development is not a distinct site but shared with the haemopoietic tissue. In fishes and urodele amphibians, B lymphocytes reach maturity in the spleen and kidney, whereas in other vertebrates this occurs mainly in the bone marrow. It is thought that in mammals, B lymphocytes may also undergo final stages of differentiation into a mature B lymphocyte after leaving the bone marrow *en route* for the secondary lymphoid tissues.

Birds are unusual as they have evolved a unique organ solely for B lymphocyte maturation called the **bursa of Fabricius** (Figure 3.8). Indeed

cortex *n.* the outer zone of an organ or gland, such as the **thymus** or **lymph node**.

medulla *n.* the central part of an organ or tissue such as the **thymus** or **lymph node**.

bursectomy surgical operation in birds to remove the **bursa of Fabricius**.

bursa of Fabricius *n.* a primary lymphoid tissue for B lymphocytes found only in birds; a pouch of lymphoid tissue connected to the dorsal surface of the cloaca.

it was the observation that surgical removal of the bursa from chickens (**bursectomy**) abolished antibody responses without affecting cell-mediated responses such as graft rejection that first revealed the division of lymphocytes into functional subpopulations. This division was soon thereafter confirmed in mice and other mammals.

The bursa is formed from a dorsal invagination of the cloaca. In the chicken embryo, lymphoid stem cells produced in the yolk sac populate the rudimentary bursa early in embryogenesis. Like the thymus, the bursa gradually involutes around the onset of sexual maturity. No equivalent of the bursa has been found in any other vertebrate, so it would seem that birds evolved this organ after their divergence from the reptiles.

3.6 Secondary lymphoid tissues

After differentiation and expression of the antigen receptor, the mature but naïve (antigen-inexperienced) T or B lymphocyte leaves the primary lymphoid tissue and enters a pool of lymphocytes that constantly recirculate around the body between the blood and the secondary lymphoid tissues (Figure 3.11). Every vertebrate has secondary lymphoid tissues that serve to trap antigens for lymphocytes. Unlike some primary lymphoid tissues, secondary lymphoid tissues appear late in ontogeny. Moreover they do not involute but remain throughout adult life.

ontogeny *n.* the history of the development of a cell or organism.

Antigen trapping in secondary lymphoid tissues is achieved in two ways: either by passive filtering of antigens, immune complexes and other particulate material that is carried in the blood or the lymph, or by the active immigration of phagocytes or dendritic cells carrying antigen acquired in the peripheral tissues. As we learned in Chapter 2, the actual number of lymphocytes carrying the appropriate receptor for a particular antigen is very low. Therefore, the encounter between antigen and antigen-specific lymphocyte would be very inefficient if left entirely to random encounter. By providing a site where antigens are deposited and through which lymphocytes can circulate, secondary lymphoid tissues help to increase the likelihood of this encounter occurring.

The second major function of secondary lymphoid tissue is to provide the specialized cellular microenvironment that is essential for the activation of naïve lymphocytes. The stringency for activation of naïve lymphocytes is high and it can only be met by this cellular microenvironment. High stringency for activation is a safety mechanism to help control the potentially damaging effects that lymphocyte effector mechanisms could inflict

Tolerance: Chapter 9

on the body if they were activated in error.

If a lymphocyte encounters antigen and becomes activated within the secondary lymphoid tissue, the lymphocyte begins to proliferate. The daughter cells undergo terminal stages of differentiation into effector and memory lymphocytes. For most lymphocytes in the population, however, this final antigen-dependent differentiation step never occurs and they remain as mature but naïve circulating cells. The life span of a lymphocyte is controversial, but a long life span may be dependent on activation by antigen.

3.6.1 Secondary lymphoid tissue exists in varying degrees of organization

During the course of vertebrate evolution there has been a progressive increase in the complexity of lymphoid tissues. A diagrammatic representation of this progression is given in Figure 3.14. The simplest tissues are **diffuse lymphoid tissues**. These consist of loose aggregations of lymphocytes and phagocytes that coalesce within connective tissues, usually near to sites where pathogens may gain entry, namely respiratory surfaces (the gills or lungs), lining of the gut, and urogenital and reproductive tracts. In agnathans and other fishes, diffuse tissues are the only kind of lymphoid tissue found.

Tetrapods have more complex tissues in addition to the diffuse lymphoid tissues retained along the usual portals of entry for microorganisms. An intermediate level of organization is the **follicle**, a discrete cluster of lymphoid cells. These first appear in the evolutionary sequence in anuran amphibians (frogs and toads). Follicles appear and disappear within diffuse lymphoid tissue and lymphatic systems according to need. In times of infections, solitary follicles may be more numerous. Follicles contain naïve B lymphocytes, a few helper T lymphocytes and specialized accessory cells for trapping antigen. Although B lymphocytes are capable of engaging free antigen with their surface immunoglobulin molecules, the immobilization and concentration of antigen on the surface of accessory cells enhances its ability to activate B lymphocytes.

A follicle starts life with a homogeneous appearance and little evidence of internal organization. This is called a **primary lymphoid follicle**. However, after B lymphocytes are activated and begin to proliferate, it evolves into a **secondary lymphoid follicle**. This has a distinctive **germinal centre** that arises in the centre of the follicle in which the proliferating B lymphocytes are located.

The highest level of organization is found in **multifollicular secondary lymphoid tissues**. As the name suggests, these consist of multiple primary

Anura *n.* one of three orders of extant amphibians containing the frogs and toads (see **Apoda** and **Urodela**).

lymphoid follicle *n.* a cluster of antigen-presenting cells and lymphocytes within secondary lymphoid tissue where antigen-stimulated B lymphocyte activation occurs (**primary lymphoid follicle** and **secondary lymphoid follicle**).

primary lymphoid follicle an undifferentiated follicle of B lymphocytes and accessory cells within secondary lymphoid tissue, prior to B lymphocyte activation and development into a **secondary lymphoid follicle**.

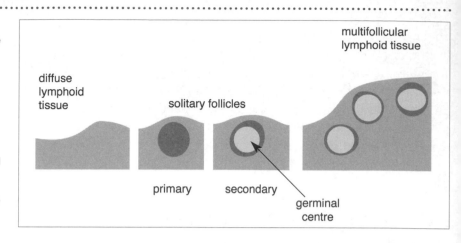

Figure 3.14 Highly schematic representation of the architecture of secondary lymphoid tissues. Diffuse lymphoid tissues are amorphous aggregations of T and B lymphocytes and phagocytes. These are located around areas where pathogens may gain entry to the body, particularly the gut and the respiratory organs (gills and/or lungs). Single follicles are formed mainly within diffuse lymphoid tissues and particularly in areas of local infections. B lymphocytes are concentrated in follicles whereas T lymphocytes are found mostly in the interfollicular diffuse lymphoid tissue. Activation of the B lymphocytes by antigen induces their proliferation and differentiation into plasma cells. This occurs in the centre of the follicle called a germinal centre. Multifollicular secondary lymphoid tissues include the lymph nodes, tonsils, Peyer's patches, spleen and appendix.

---

secondary lymphoid follicle an active lymphoid follicle with a **germinal centre** in which activated B lymphocytes are proliferating and differentiating into plasma cells.

germinal centre a focus of actively proliferating and differentiating B lymphocytes in the centre of follicles in lymph nodes and spleen.

spleen *n.* a multifunctional organ in vertebrates that is the site of haemopoiesis (blood cell production), and removal of effete erythrocytes, and acts as a secondary lymphoid tissue for collecting blood-borne antigens.

---

and secondary follicles within the same structure. Mammals have the widest variety of multifollicular tissues, including the **spleen**, **lymph nodes**, **Peyer's patches**, **tonsils** and **appendix.** The number of follicles within these structures may vary from a few, such as a small lymph node, to several thousand, such as in the lymphoid tissues of the spleen.

In general, secondary lymphoid tissues come in two varieties: those that specialize in trapping antigens carried in the blood, and those have evolved to trap antigen carried in the lymph.

3.6.2 Trapping antigen from the blood

When particulate materials such as bacteria gain entry to the blood they are quickly removed from circulation by antigen-trapping tissue. For antigens in the blood, trapping is usually performed by lymphoid tissues intermingled with haemopoietic tissues. In fishes and amphibians there is considerable functional overlap of haemopoietic and secondary lymphoid tissue. In fishes, trapping of blood-borne antigen is performed by the kidney and spleen, and by sessile phagocytes associated with the gills. In amphibians, reptiles, birds and mammals, blood-borne antigens are trapped mainly in the spleen. The first true spleen is seen in teleost fishes, although the haemopoietic and lymphoid areas of the fish spleen are intermingled. In tetrapods, the spleen continues to retain the dual functions of haemopoietic and lymphoid tissue, and the tissues that perform these functions are organized into discrete areas called **red pulp** and **white pulp**, respectively.

The spleen

The vertebrate spleen is a large multifunctional organ that develops during embryogenesis from an invagination of the gut and becomes detached to become a separate organ. The spleen is a major haemopoietic tissue in the majority of jawed vertebrates. It also removes effete (worn out) erythrocytes and other debris from the blood. As such it provides an important location for collecting and presenting blood-borne antigens to lymphocytes.

The majority of splenic tissue is red pulp, which is rich in erythrocytes and macrophages. Blood enters via a splenic artery which then branches into several arterioles and percolates through the red pulp through a network of **blood sinusoids**. Red pulp is the site where aged erythrocytes that have lost their ability to deform in the narrow capillaries are removed from circulation by macrophages.

Collected around the arterioles are areas of secondary lymphoid tissue. To the naked eye, white pulp can be seen in a sectioned spleen as pale spots. Each spot consists of diffuse lymphoid tissues and follicles. In some mammals such as the mouse, the arterioles are entirely sheathed in lymphoid tissue, called the **periarteriolar lymphoid sheath** (Figure 3.15). B lymphocytes and antigen-trapping cells are found predominantly in the follicles, whereas T lymphocytes and interdigitating dendritic cells are located in the interfollicular spaces and the periarteriolar sheath. Between the areas of red and white pulp is the **marginal zone** where lymphocytes move from the arterioles into the white pulp. Although the spleen does not have afferent lymphatic vessels, lymph is drained from a single efferent lymphatic vessel. This provides an exit route for the lymphocytes that entered with the blood.

3.6.3 Tissues that trap antigens from lymph

Trapping of antigen from lymph is performed by dedicated lymphoid tissues (i.e. not combined with haemopoietic tissue, as in the spleen) that are scattered around the lymphatic system.

Lymph nodes

Lymph nodes are encapsulated multifollicular structures. True lymph nodes do not occur in vertebrates that evolved before the marsupial mammals. In the monotreme, *Tachyglossus* (the echidna), nodules with a single germinal centre are found throughout the lymphatics suspended in the fluid by connections to the walls of the lymphatic vessel. However, these nodules are small and more closely resemble solitary follicles rather than true, multifollicular lymph nodes (Figure 3.16). A few aquatic birds have independently evolved multifollicular lymph nodes. However, the architecture of these is considerably simpler than that of the eutherian counterpart (Figure 3.16). In marsupials and eutherians, true lymph nodes are clustered throughout the body, particularly at junctions between lymphatic vessels. Many are strategically placed near to the portals of entry for microbes,

red pulp *n.* area of the spleen rich in erythrocytes and macrophages where effete erythrocytes are destroyed.

white pulp *n.* lymphoid tissue surrounding the arterioles in the spleen.

periarteriolar lymphoid sheath the T lymphocyte-rich **white pulp** of the **spleen** that surrounds the arteries.

marginal zone area of the **spleen** between the red and white pulp that is rich in lymphoid follicles.

lymph node *n.* a small secondary lymphoid organ that collects antigens drained by the lymph; contains antigen-presenting cells and trafficking lymphocytes.

Peyer's patch *n.* a cluster of lymphoid follicles that are found in the inner lining of the small intestine in mammals; secondary lymphoid tissue where antigens from the contents of the digestive tract collect.

Figure 3.15 Structure of a spleen of a eutherian mammal (mouse).

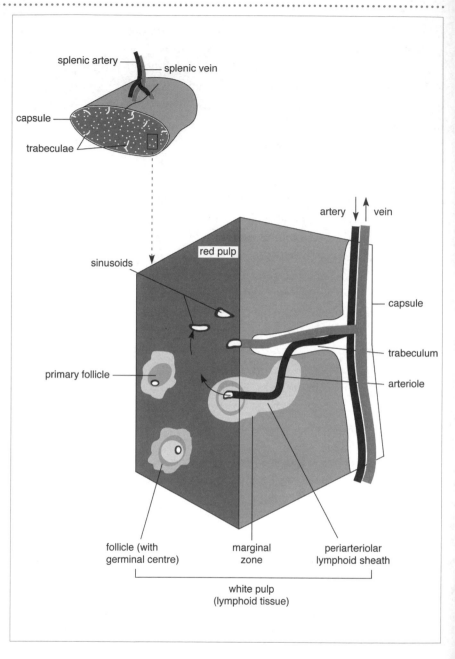

follicular dendritic cell *n.* antigen-presenting cell that resides in lymphoid follices in secondary lymphoid tissue; has characteristic long processes that retain antigen.

particularly draining the lower mesenteries of the gut, the neck, armpits and the groin and around the aorta (Figure 3.8).

The structure of a mammalian lymph node is shown in Figure 3.17. The node is organized into a dense outermost **cortex**, and a less dense central **medulla**. Lymphoid follicles are located in the outer margin of the cortex. Inside the follicles are **follicular dendritic cells** which have delicate arms (dendrites) that trap and hold antigen and immune complexes on their

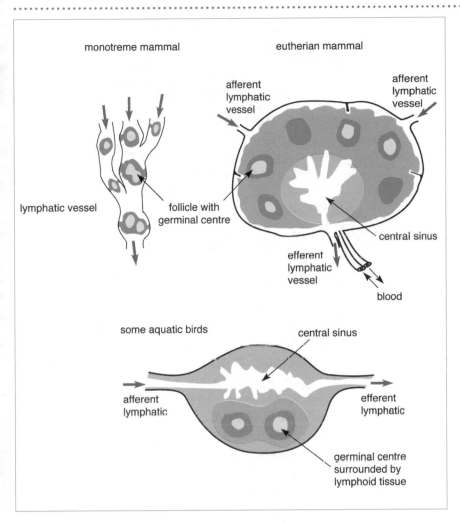

Figure 3.16 Architecture of lymph nodes from monotreme and eutherian mammals, and from a bird. Redrawn and modified from Manning, M.J. and Turner, R.J. (1976) *Comparative Immunobiology.* Blackie & Son, London.

surfaces for B lymphocytes. Lymph enters a lymph node from multiple converging afferent lymphatic vessels. This percolates through the tissues of the node and collects in sinuses in the medulla and leaves the node in a single efferent lymphatic vessel. Each lymph node is also supplied by a single artery from which lymphocytes can gain entry. This splits into several smaller arteries that penetrate deep into the cortex, which then branch into arterioles around each follicle. Postcapillary venules leave each follicle and then converge into a single vein that carries blood out of the lymph node.

Naïve T and B lymphocytes leave the blood and enter the tissues of lymph nodes by adhering to specialized endothelial cells in the venules called **high endothelial venule** cells (Figure 3.3). These cells have an unusual and characteristic 'plump' (or high) morphology, and are the specific entry points for naïve T and B lymphocytes into all secondary

Figure 3.17 Structure of a typical eutherian multifollicular lymph node.

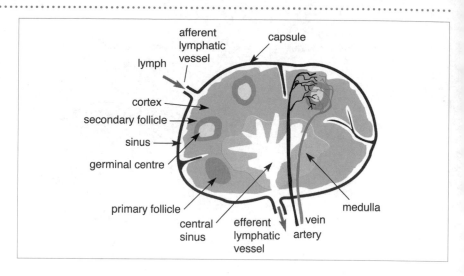

lymphoid tissues (except the spleen, where lymphocyte migration occurs in the blood sinusoids in the marginal zones). Normal endothelial cells can also acquire the 'high' endothelial cell morphology at sites of chronic inflammation in response to inflammatory cytokines. This allows activated lymphocytes to be recruited into the tissue. These 'inducible' high endothelial cells gradually revert to a normal, flattened appearance in the absence of antigen and become less adhesive for lymphocytes.

T and B lymphocytes circulate through different areas of the lymph node. T lymphocytes squeeze through the diffuse lymphoid tissue between the follicles and in the inner cortex near to the medulla, a region called the **paracortex**. The paracortex is rich in **interdigitating dendritic cells** which, as mentioned, carry antigen collected in non-lymphoid tissues back to draining secondary lymphoid tissues for presentation to naïve T lymphocytes.

Naïve B lymphocytes move predominantly through the outer margin of the cortex where they may encounter antigen percolating through with the lymph. Activation by antigen causes B lymphocytes to move into follicles. Inside, follicular dendritic cells and small numbers of helper T lymphocytes together provide the necessary cellular microenvironment for the differentiation of B lymphocytes into memory B lymphocytes. Antibody-secreting plasma cells are found mainly in the medulla, ideally positioned to discharge antibody into the outgoing lymphatic fluid.

Other multifollicular secondary lymphoid tissues

Mammals also have other multifollicular secondary lymphoid tissues (Figure 3.18). **Tonsils** are found in the pharynx, and like the thymus, are derived from the pharyngeal epithelium. These tissues collect lymph draining from

high endothelial venule an area of capillary venule consisting of plump endothelial cells where leucocytes can pass through into the tissues (see **extravasation**).

interdigitating dendritic cell antigen-presenting cell of lymphoid tissue that presents antigen to T lymphocytes.

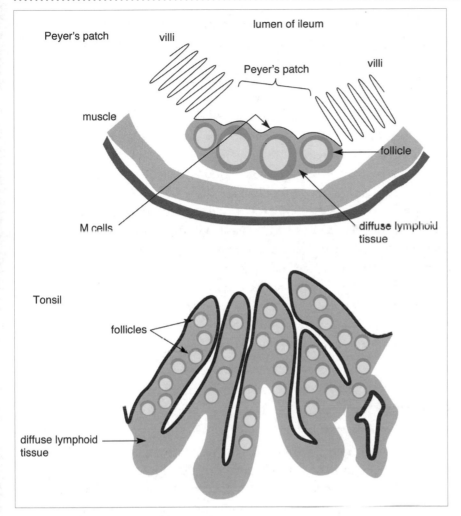

Figure 3.18 Organization of other mammalian multifollicular secondary lymphoid tissues. Peyer's patches in the ileum and tonsil found in the pharynx.

the head and neck. Analogous multifollicular structures, called **Peyer's patches,** are found in the mammalian ileum. Each patch consists of 20–30 follicles embedded in the villi. The surface of each follicle, which protrudes into the lumen of the gut, is covered by a specialized epithelium of **M cells**. These take up material from the gut lumen and transport it into the follicle. Peyer's patches are responsible for exposing the immune system to antigens ingested in the food.

The **appendix**, a vermiform (worm-like) appendage of the mammalian large intestine, is also a multifollicular secondary lymphoid tissue that contains several hundred lymphoid follicles. Together the Peyer's patches, the appendix, and the diffuse lymphoid tissues around the gut form an interconnected lymphoid tissue collectively termed **gut associated lymphoid tissue** (or **GALT**). Rabbits and rats also have diffuse lymphoid tissues and solitary follicles located near to the branch points of the airways

gut associated lymphoid tissue (GALT) a system of secondary lymphoid tissues draining the gastrointestinal tract of mammals, consisting of Peyer's patches, the appendix, and the diffuse lymphoid tissues.

in the lungs, collectively termed **bronchus associated lymphoid tissues** (or **BALT**), although the follicles seem to be absent from other mammals that have been studied.

3.6.4 Metamorphosis in amphibians

Amphibians represent an interesting group of vertebrates because a significant proportion of development occurs as a free-living larva. Lymphoid tissues therefore develop early in embryogenesis of amphibians, and although tadpoles have relatively few lymphocytes, they are immunologically competent. The major lymphoid organs of *Xenopus* tadpoles are the thymus and spleen, with diffuse lymphoid tissues associated with the kidney, liver and lining the pharynx next to the gills. In all amphibians, the metamorphosis from the larva to adult is associated with a profound change in body plan. In *Xenopus* the spleen and thymus are retained during this transformation, as are lymphoid tissues in the kidney and liver. However, gills and the associated pharyngeal lymphoid tissues are lost, whereas haemopoietic bone marrow and diffuse gut and skin associated lymphoid tissues appear after metamorphosis.

Summary

- Primary lymphoid tissues host the development of lymphocyte progenitors, characterized by antigen receptor gene rearrangements. Differentiation in primary lymphoid tissues is antigen independent. T lymphocytes differentiate in the thymus; B lymphocytes in haemopoietic tissues or in the case of birds, the bursa of Fabricius.

- Secondary lymphoid tissues are aggregations of lymphocytes and antigen-presenting accessory cells, and are the sites through which lymphocytes circulate. These tissues have become progressively more complex from fishes to mammals.

- The lymphatic system of tetrapods drains extracellular fluid and carries immune complexes and dendritic cells to secondary lymphoid tissues. Blood-borne antigens are trapped by different secondary lymphoid tissue, usually the spleen.

- Secondary lymphoid tissues serve to display antigen to T and B lymphocytes, and provide the required cellular microenvironment for their antigen-dependent activation and differentiation.

• Splenic secondary lymphoid tissue is phylogenetically ancient, and probably predates the evolution of the lymphatic system and its associated secondary lymphoid tissues.

3.7 Lymphocyte traffic

The patterns of lymphocyte movement around the body have been studied mostly in mammals. Although lymphocytes are a predominant component of the white blood cells, very little time is spent in the blood or lymphatic systems. In sheep for example, at any one time only a tenth of the total lymphocyte population is in the blood or lymph (2 and 8%, respectively) with the remainder trafficking through lymphoid and non-lymphoid tissues.

We have seen that naïve lymphocytes enter the secondary lymphoid tissues in the blood (Figure 3.11). While some lymphocytes leave the lymphoid tissue directly with the outgoing blood, the majority leave the blood circulation and migrate into the lymphoid tissue at the high endothelial venule cells. Once inside, lymphocytes may or may not encounter their specific antigen. Those lymphocytes that are not activated leave with the efferent lymphatic fluid and are returned to the blood, although many die within the lymphoid tissue.

Upon activation by antigen, the circulatory pathway of a lymphocyte changes. B lymphocytes enter the follicles where they proliferate and undergo terminal differentiation. During clonal expansion, some become plasma cells and move to the medulla, whereas others become memory cells. Effector and memory T lymphocytes leave via the lymph and re-enter the blood as before. Now, however, they have a different homing behaviour. Instead of trafficking through the secondary lymphoid tissues, activated T lymphocytes traffic between the blood and the tissues. It has been observed that they tend to return preferentially to the site in which the antigen gained entry. Activated lymphocytes are also particularly attracted to sites of local inflammation. As we saw earlier these cells can be seen to attach to the endothelial cells before migrating between the endothelial cells into the surrounding tissues.

3.8 Cell adhesion molecules

immunoglobulin gene **superfamily** a **gene family** whose members all possess one or more **immunoglobulin folds**.

integrin *n.* a member of a family of cell-surface adhesion molecules that bind mainly to extracellular matrix.

selectins *n. plu.* a family of carbohydrate-binding cell-surface adhesion molecules.

homodimer *n.* a protein consisting of two identical polypeptide chains.

These migratory behaviours of leucocytes, and indeed all cell-to-cell interactions, are controlled by cell adhesion molecules. Cell adhesion molecules are cell-membrane proteins that bind specifically to appropriate counter-structures on other cells. These interactions play a pivotal role in embryonic development and the maintenance of tissue organization of multicellular animals. The particular adhesion molecules a cell displays determine the anatomical sites in the body it occupies. Reciprocally, signals transmitted into cells through their adhesion molecules can influence the expression of the genes of other cell adhesion molecules.

The levels of expression of adhesion molecules on leucocytes are influenced by stimuli received from other cells. Lymphocytes, for example, undergo many changes in the density, affinity and particular combination of adhesion molecules expressed before and after activation by antigen. Similarly the expression of adhesion molecules on epithelial cells changes in response to local inflammation. It is through these dynamic molecular interactions that the trafficking behaviour of leucocytes is controlled.

There are at least four major families of cell adhesion molecules of importance in the immune system, including the members of the **immunoglobulin gene superfamily**, the **integrins**, the **selectins** and the **vascular addressins**. The following is an overview of their structure; the roles of individual molecules will be discussed where relevant in subsequent chapters. A brief description of certain molecules can also be found in Appendix A.

3.8.1 The immunoglobulin superfamily

This family, which contains over 100 genetically related proteins, is so called because the first member described was the antigen receptor of B lymphocytes (immunoglobulin). Members of the immunoglobulin superfamily are involved mainly in the formation of transient contacts between cells, particularly in the immune and nervous systems. Representatives of the superfamily can be found throughout the animal kingdom, indicating this is phylogenetically ancient family.

Evolution of the immunoglobulin superfamily: Section 7.2

The common denominator of members of the superfamily is possession of one or more immunoglobulin folds, shown in the previous chapter (Figure 2.5). Diagrammatic representations of some immunoglobulin superfamily members that are important in leucocyte adhesion are shown in Figure 3.19.

The majority of members are anchored to the membrane by a hydropho-

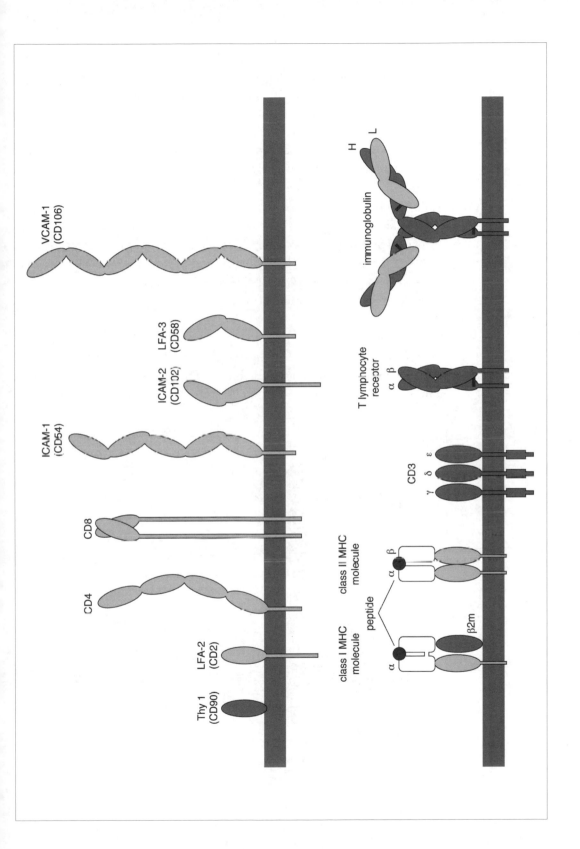

Figure 3.19 Representatives of the immunoglobulin superfamily (see Appendix 1 for brief descriptions). CD, cluster of differentiation; ICAM, intercellular adhesion molecule; LFA, leucocyte function associated antigen; MHC, major histocompatibility complex; VCAM, vascular cell adhesion molecule.

bic transmembrane domain. A few, including Thy-1 (CD90), a molecule of unknown function found only on T lymphocytes and on some cells in the central nervous system, are anchored to the lipid bilayer by a phosphatidyl inositol linkage. Notice that some members consist of a single domain. In addition to Thy-1, these include β2-microglobulin (β2m), which is the light chain of class I major histocompatibility molecules, and the CD3 polypeptides which are part of the T lymphocyte receptor complex. These will be encountered in later chapters.

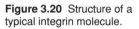

The structure of antigen receptors: Chapter 4

The majority of members of the superfamily are constructed of multiple domains linked in chains. More complex arrangements are created by the formation of disulphide bond-linked homodimers or heterodimers, such as the α and β chains of the T lymphocyte antigen receptor. The B lymphocyte receptor is the most complex member of all, comprising two heterodimers joined by interchain disulphide bonds.

3.8.2 Integrins

The integrins are a distinct family of cell-surface molecules with representatives found throughout the vertebrates. These molecules are involved in numerous cell-to-cell interactions. They also mediate the adhesion of leu-

Figure 3.20 Structure of a typical integrin molecule.

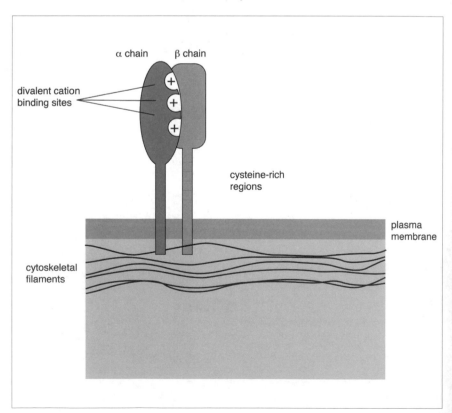

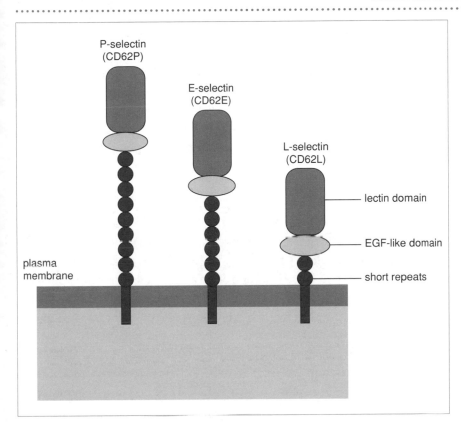

Figure 3.21 The selectins. EGF, epidermal growth factor.

cocytes to fibrous extracellular matrix proteins such as collagen, laminin and fibronectin, which provide the strength and elasticity of connective tissue. The intracellular domains of integrins are connected to the cytoskeleton which provides both the framework and the motive power required to change shape. These molecules thus 'integrate' the extracellular matrix with the cytoskeleton and enable tissue-infiltrating leucocytes to migrate to sites of inflammation.

All integrins are heterodimers comprising an α chain and a β chain of approximately 120 and 83 kDa, respectively (Figure 3.20). The α chains have three or four divalent cation (Ca^{2+} or Mg^{2+}) binding sites which must be bound to enable the integrin molecules to bind to their counter-receptors. For this reason, many cellular interactions require divalent cations to be present in the extracellular fluid.

Integrins can be classified according to the type of β chain they have. Most integrins have either the β_1 (CD29) or β_2 (CD18) chain. For example, there are three different integrins with the β_2 chain. These are leucocyte function associated antigen-1 (**LFA-1**), and the complement receptors 3 and 4 (**CR3** and **CR4**). Each is a heterodimer consisting of the common β_2 chain with a different α chain (see Appendix A for more details).

Complement receptors: see Section 1.4.4

LFA-1 is a particularly important adhesion molecule on the surface of leucocytes. It has two counter-receptors, the intercellular adhesion molecules, ICAM-1 and ICAM-2, which as we saw in Figure 3.18, are members of the immunoglobulin superfamily. The interaction between LFA-1 and ICAM-1 or -2 contributes to a variety of leucocyte activities that involve their contact with other cells, including the lysis of target cells by cytotoxic T lymphocytes and NK cells, delivery of 'help' by helper T lymphocytes to B lymphocytes, and the firm adhesion of neutrophils to vascular endothelial cells.

3.8.3 Selectins

Animal lectins: Table 2.1.

Selectins are adhesion molecules found on leucocytes and endothelial cells that are involved in the early stages of extravasation of lymphocytes and neutrophils into the tissues. The molecules are so called because they have the properties of lectins, which are proteins that bind to sugars. Selectins mediate their binding activity by binding to specific saccharide groups on cell-surface glycoproteins.

vascular endothelium
n. a layer of endothelial cells that form the blood vessels.

Three different selectins have been found in humans, called L-selectin, P-selectin and E-selectin (CD62; Figure 3.21). These are involved mainly in the attachment of neutrophils to vascular endothelium during inflammation and the attachment of lymphocytes to high endothelial venules. L-selectin is found on lymphocytes and neutrophils, whereas E-selectin is induced on vascular endothelial cells in response to inflammatory mediators. P-selectin is stored in granules in platelets and vascular endothelial cells and is released in response to products generated by the clotting cascade.

3.8.4 Vascular addressins and lymphocyte homing

These are a family of glycoproteins found on high endothelial venule cells in various secondary lymphoid tissues. Their carbohydrate groups show considerable structural diversity, and different vascular addressins are found in different secondary lymphoid tissues. These molecules are thus responsible for the selective or non-random trafficking of lymphocyte subpopulations through particular secondary lymphoid tissues (called homing). For example, homing to mucosal high endothelial venules is directed by the vascular addressin **MAdCAM-1** (mucosal addressin cell adhesion molecule-1) and to lymph nodes by GlyCAM (glycosylation-dependent cell adhesion molecule). These are recognized by L-selectin on the T lymphocytes. Other addressins bearing galactosyl and mannosyl groups interact with the **homing associated cell adhesion molecule** (CD44) on leucocytes. The assorted ligands of vascular addressins are collectively termed **homing receptors**.

3.8.5 Extravasation of neutrophils and lymphocytes

Inflammatory cytokines that are produced during inflammation in the tissues activate local vascular endothelial cells. This induces the surface expression of selectins and ICAM-1, and makes them more adhesive for passing leucocytes. The first to infiltrate the tissues are neutrophils. These are captured from the blood flow by P selectin on the activated endothelium which binds to carbohydrates on the neutrophil. This initial interaction is relatively weak and is easily broken by shear forces of the blood, although others are quickly reformed. Breaking and reforming adhesions causes the cell to roll along the endothelium for a few seconds, but this is quickly arrested as firmer contacts are made (Figure 3.4). E selectin on the endothelial cells is involved more in the rolling phase, although there is some overlap in the contributions made by both types of selectins. Firmer adhesions are then mediated, predominantly by the integrin LFA-1 on the leucocyte binding to ICAM on the epithelium. Finally, the leucocyte migrates between the junctions of the endothelial cells and into the tissue, although the molecular interactions involved in this stage are poorly understood.

The extravasation of lymphocytes at high endothelial venule cells, either in secondary lymphoid tissues or in inflamed tissue, is a similar multistep process. Initial capture is mediated by homing receptors on the lymphocyte and the tissue-specific vascular addressins. This is consolidated by the integrins, including LFA-1, binding to adhesion molecules on the high endothelial venule cell, prior to extravasation.

Summary

- Mature lymphocytes leave the primary lymphoid tissues and enter a circulating pool. This pool uses both the cardiovascular and lymphatic circulatory systems to move the cells between tissues where they perform their immune functions.

- Lymphocytes move out of the blood at high endothelial venules in secondary lymphoid tissues or in inflamed tissues.

- The circulatory pathways of naïve and activated/memory lymphocytes differ due to changes in the arrays of surface adhesion molecules that are expressed on the lymphocyte and on activated endothelium. These molecules are classified into different families according to their structure: integrins, selectins and members of the immunoglobulin super-family.

The future

- Much of the physiology of vertebrate immune systems is comparatively well understood. The frontiers lie at the molecular level.

- Much progress has been made in recent years in understanding the molecular interactions that underlie leucocyte adhesion and traffic in mammals. Major areas of uncertainty still surround the definition of memory lymphocytes, particularly at the molecular level.

- Comparatively less is known of the molecules involved in non-mammalian vertebrates, although as more mammalian molecules are identified their homologies in other species can be identified.

Further reading

Books
Burkitt, H.G., Young, B. and Heath, J.W. (1993) *Wheater's Functional Histology*. Churchill Livingstone, Edinburgh.

Manning, M.J. and Turner, R.J. (1976) *Comparative Immunobiology*. Blackie, Edinburgh.

Ratcliffe, N.A. and Rowley, A.F. (eds) (1981) *Invertebrate Blood Cells*. Academic Press, London.

Rowley, A.F. and Ratcliffe, N.A. (eds) (1988) *Vertebrate Blood Cells*. Cambridge University Press, Cambridge.

Satchell, G.H. (1991) *Physiology and Form of Fish Circulation*. Cambridge University Press, Cambridge.

Turner, R.J. (ed.) (1994) *Immunology. A Comparative Approach*. J. Wiley, London.

Review articles
Girard, J.-P. and Springer, T.A. (1995) High endothelial venules (HEVs): specialized endothelium for lymphocyte migration. *Immunology Today*, **16**(9), 449–57.

Haas, W., *et al.* (1993) Gamma delta cells. *Annual Review of Immunology,* **11**, 637–86.

Imhof, B.A. and Dunon, D. (1995) Leucocyte migration and adhesion. *Advances in Immunology,* **58**, 345–416.

Springer, T.A. (1990) Adhesion receptors of the immune system. *Nature,* **346**, 425–34.

Turner, M.L. (1992) Cell adhesion molecules: a unifying approach to topographic biology. *Biological Review,* **67**, 359–77.

Young, A.J., Hay, J.B. and Mackay, C.R. (1993) Lymphocyte recirculation and life span *in vivo. Current Topics in Microbiology and Immunology,* **184**, 161–71.

Review questions

Fill in the blanks

Vertebrates have three categories of blood cells: erythrocytes (or _____[1] cells) leucocytes (or _____[2] cells), and _____[3]. The leucocytes can be classified according to the granularity of the cytoplasm into _____[4] and _____[5] cells. Cells that contain cytoplasmic granules can be subclassified into the _____,[6] _____[7] and _____[8]. The other major lineage consists of the _____[9] and the _____[10]. All blood cells develop during _____[11] from a single _____[12] stem cell. The different types of fully differentiated cells are often described simply as lymphoid (which are mostly _____[13]) and non-lymphoid or _____[14] cells. Leucocytes differ from other blood cells as their functions take place in the _____[15] rather than the circulation. These leave the blood and enter inflamed tissue by a process called _____[16]. Mature T lymphocytes can be distinguished according to function into _____[17] T lymphocytes and _____[18] T lymphocytes. This function correlates with the phenotypic expression of one or other of the coreceptor molecules, _____[19] or _____[20], respectively. Another category of cell that circulates is the dendritic cell. These are potent _____ _____[21] cells that have the unique capacity to activate _____[22] T lymphocytes. Unlike other leucocytes, lymphocytes can leave the blood and enter _____[23] lymphoid tissues by adhering to _____ _____ _____[24] cells. Normal endothelium can assume similar properties in response to cytokines released during _____[25].

Short answer questions

1. True or false; if you think the answer is false, give your reasons.

    (a)   Multifollicular lymph nodes are found only in mammals.

    (b)   Fishes have a true lymphatic system.

    (c)   The vertebrate thymus is a haemopoietic tissue.

    (d)   Follicular dendritic cells and interdigitating dendritic cells are different forms of the same cell.

2. What are the functions of primary and secondary lymphoid tissues?

3. How does the recirculation pathway of naïve T lymphocytes differ from that of activated T lymphocytes?

4. Describe what happens to antigen that gains entry to the blood, the connective tissues or the gastrointestinal tract of mammals.

5. The nude mouse has a recessive gene that causes many skin defects. Homozygous (*nu/nu*) mice lack hair follicles, and during embryogenesis, epithelial cells fail to proliferate around the thymic rudiment formed by the pharyngeal pouches. What effect do you think this has on the immune system of the nude mouse?

6. What are the functions of the vertebrate spleen?

7. Which pairs of adhesion molecules are involved in the following cell-to-cell interactions?

    (a)   The initial adhesion and rolling of neutrophils to inflamed vascular epithelium.

    (b)   The firm adhesion of neutrophils to inflamed vascular endothelium prior to extravasation.

    (c)   The homing of lymphocytes to mucosal lymphoid tissues.

The antigen receptors

4

In this chapter

- How the antigen receptors of T and B lymphocytes were first characterized.
- Antigen receptor genes, rearrangement and the generation of diversity.
- How diversity translates into structure.
- The distinguishing features of B lymphocytes – release of antibody, class switching, and somatic mutation.
- How antibodies trigger effector mechanisms.

Introduction

The antigen receptors of B lymphocytes (called immunoglobulins) and of T lymphocytes are membrane-bound molecules that, together with a complex of other membrane proteins, relay signals to the interior of the lymphocyte caused by the extracellular binding of antigen. This triggers lymphocyte activation. Activated B lymphocytes begin to manufacture soluble immunoglobulins (called antibodies) which diffuse away from the cell. Although lacking any signalling function, antibodies retain their antigen-binding property and function instead to opsonize antigens or mark them for the deposition of complement. Here we will focus on this antigen-binding property of T and B receptors.

Antigen receptors are the key to the vertebrate adaptive immune system. They underlie the ability of the system to respond to any foreign antigen and its ability to retain a memory of previous exposures. These remarkable properties all stem from the way the antigen receptor genes are assembled from smaller gene segments by a DNA recombination process called antigen receptor gene **rearrangement**. Most of what we know of this process has come from the study of mammals, and of mouse and man in

Signal transduction by antigen receptors: Section 8.1.2

particular. Using the mammals as a model we will examine the genetics and structure of antigen receptors and how antibodies trigger effector functions. Other vertebrates will be compared with the mammals in Chapter 7, when we examine evolutionary aspects of these important molecules.

4.1 Historical overview

Antibodies have been the object of study for biochemists and crystallographers since the 1950s, and long before the existence of the T lymphocyte receptor was appreciated. Early biochemical studies revealed that antibody molecules are very heterogeneous. Efforts to understand the genetics behind this molecular diversity broke new technical grounds, and forged much of the molecular biology revolution of the early 1980s. It was during this time that the gene encoding the T lymphocyte antigen receptor was first cloned, and, in contrast to antibodies, long before much was known of its structure. In contrast to immunoglobulin, the T lymphocyte receptor exists only as a membrane-bound molecule and it is not released in a soluble form. Nevertheless, these two kinds of antigen receptor share many important structural and genetic features.

4.1.1 Immunoglobulins

The starting material in the biochemical characterization of antibodies was serum – the cell-free fluid component of blood that remains after it has been allowed to clot. Mammalian serum normally contains several hundred different proteins. Because different proteins have different net electrical charges imparted by their amino acid side chains, mixtures of proteins can be separated from each other in an electric field (a technique called **electrophoresis**). Positively charged proteins migrate to the cathode, negatively charged proteins to the anode, both at a speed that is proportional to the amount of charge. By this technique, serum proteins resolve into the **albumins** and three **globulin** fractions, called α, β and γ.

Figure 4.1 shows an electrophoretic pattern of mammalian serum before and after the animal was immunized against a foreign protein antigen. After immunization, the size of the γ-globulin fraction increases. If this is the first time the animal has encountered this particular antigen, the increase in the γ-globulin fraction occurs gradually, over a period of several days. As we will see later (Section 4.5.5), the response is much more rapid if the antigen has been encountered before.

It was noticed that if this serum was mixed with the original antigen, a

electrophoresis *n.* a technique for the separation of proteins or fragments of DNA or RNA on the basis of charge or molecular mass by migration through a gel matrix in an electric field.

albumins *n. plu.* soluble proteins of serum, milk, synovial fluid and other mammalian fluid secretions remaining after insoluble protein (**globulins**) have been precipitated.

globulin *n.* the serum protein fraction that is precipitated by the addition of salt.

precipitate formed and the size of the γ-globulin fraction returned to normal. Because of this, the antigen-specific serum produced by immunization was called **antiserum**, and the proteins responsible for precipitating the antigen were called antibodies (or immunoglobulins). Subsequently, it was found that immunoglobulins could be extracted from the γ-globulin fraction by subjecting it to ultracentrifugation. A band of low molecular weight proteins consisting of immunoglobulins was produced, which were named immunoglobulin G (or IgG).

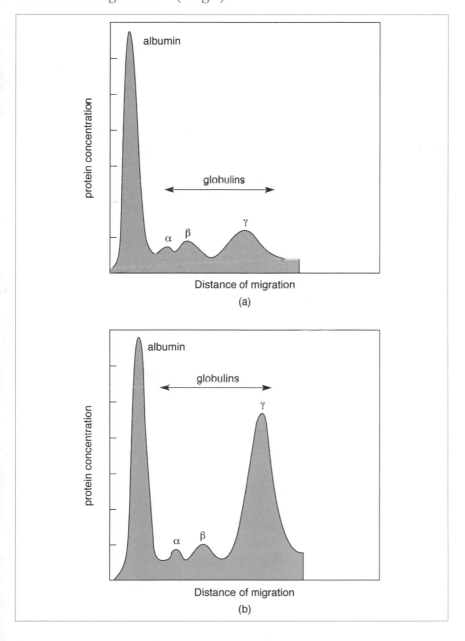

Figure 4.1 Electrophoresis of rabbit serum proteins into fractions before (top) and after (below) immunization against ovalbumin from chicken egg white. Adapted from the work of Tiselius, A. and Kabat, E.A. (1939) *Journal of Experimental Medicine*, **69**, 119–31.

papain *n.* an endopeptidase.

Fab fragment a crystallizable fragment of immunoglobulin consisting of one light chain and a similarly sized piece of the heavy chain; produced by mild digestion with **papain**.

Fc fragment a crystallizable fragment of immunoglobulin consisting of the C-terminal halves of both heavy chains; produced by mild digestion with **papain**.

Figure 4.2 Determination of the basic structure of IgG by enzymatic digestion. Compare with Figure 2.4.

The availability of purified IgG in the 1950s and 1960s enabled biochemists to determine the basic structure of the molecule. One approach was to digest IgG with proteinases (Figure 4.2). After digestion with papain, two types of fragment were generated. Two-thirds of the mixture consisted of fragments with a molecular weight of 45 000. These retained the antigen-binding capacity of the immunoglobulin and were designated **Fab fragments** (for *f*ragment-*a*ntigen *b*inding). The remaining third consisted of fragments with a molecular weight of 50 000, and which did not bind the antigen. These fragments crystallized in the cold and were therefore called **Fc fragments** (for *f*ragment-*c*rystallizing). This property indicated the Fc fragments of IgG, unlike the Fab fragments, were homogeneous. Digestion with another enzyme, pepsin, generated a single 100 000 molecular weight fragment, termed F(ab')2. Not only did this bind antigen, but it retained the ability to precipitate it.

Meanwhile, in other studies, mercaptoethanol was used to reduce the disulphide bonds within IgG and dissociate polypeptide chains from each other. These studies revealed the molecule was constructed of two kinds of polypeptide chains, heavy (or H) chains with a molecular weight of

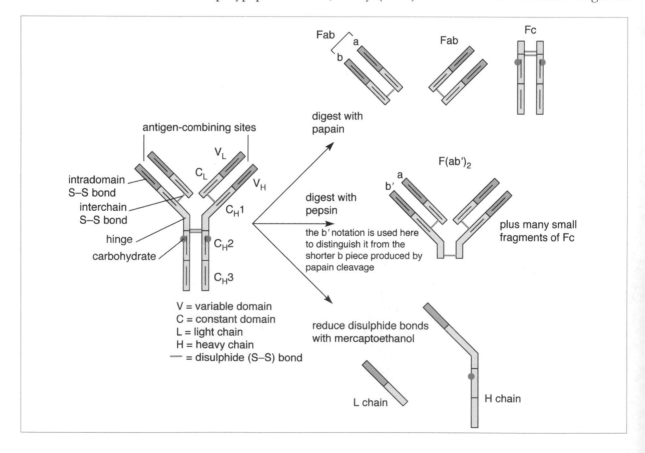

50 000 and light (or L) chains weighing 25 000. It was also noticed that antibodies raised in goats against rabbit Fab fragments precipitated heavy chains *and* light chains, whereas antibodies raised to Fc fragments only bound to heavy chains. Piecing together all these observations led to the model of IgG, shown in Figure 4.1, in which the molecule is constructed from two identical heavy chains and two identical light chains linked by disulphide bonds.

Electron microscopy, and later X-ray diffraction studies, established that the whole molecule is in the shape of a 'Y', with the Fc region being the stem, and the Fab regions, containing the antigen-combining sites, forming the arms.

Further progress in understanding the structure of the antigen-binding site required the amino acid sequence and three-dimensional structure to be determined. Purified immunoglobulins from serum are too heterogeneous for either amino acid sequencing or crystallization for X-ray diffraction studies. Indeed, this was the first real evidence for antigen receptor diversity. This problem was first overcome with help from human patients with **multiple myeloma.** This is a cancer that arises spontaneously in a clone of immunoglobulin-secreting plasma cells (**plasmacytomas**). Unlike healthy plasma cells, which produce antibodies for a few days and eventually die, a plasmacytoma continues to proliferate and produce antibody indefinitely. Patients with myeloma produce large amounts of clonally derived, and thus homogeneous, immunoglobulin molecules. In these patients, light chains are sometimes excreted with the urine (**Bence-Jones protein**). This provided the first source of homogeneous immunoglobulin light chain for sequencing and crystallography.

The first amino acid sequence data were obtained from the light chains of two myeloma patients. Comparison of the sequences revealed two distinct, but equally sized regions of 100–110 amino acids in length. The carboxyl-terminal half of the molecule had the same sequence in both proteins and was termed constant (C), whereas the amino-terminal half was very different and termed variable (V). As more sequences were compared, two types of constant region sequences emerged, called κ and λ.

Heavy chains were obtained for sequencing studies from plasmacytoma proteins. Although longer, heavy chains also featured a V region, which was approximately the same size as the V region of a light chain. The remainder of the heavy chain was much less variable, although five basic sequences emerged (μ, δ, γ, α and ε). These chains give rise to different **isotypes** (or **classes**) of immunoglobulins with different properties. These are named according to the type of heavy chain. Thus IgG contains γ chains, IgM has

myeloma *n.* a cancer of bone marrow cells; also used to describe the disease caused by a malignant clone of plasma cells (**plasmacytoma**) that produce homogeneous immunoglobulin.

plasmacytoma *n.* a cancer of plasma cells.

IgG immunoglobulin G; the predominant serum antibody produced in secondary responses in mammals.

IgM immunoglobulin M; found in all jawed vertebrates; exists as monomer in membrane of B lymphocytes; secreted forms polymeric (pentameric in mammals).

IgD immunoglobulin D; mammalian B lymphocyte receptor with IgM.

IgA immunoglobulin A; the predominant antibody found in extravascular secretions in mammals (mucus, tears, milk, colostrum); exists as monomer or dimer.

IgE immunoglobulin E; mammalian antibody involved in inflammatory responses via Fc receptors on mast cells and basophils.

μ, IgD has δ, IgA has α and IgE has ϵ. We will return to these different heavy chains in Section 4.5.1.

Attention then turned to the genetic mechanisms behind immunoglobulin diversity. It is estimated there are in the order of 10^8 different antigen specificities in a typical mammal such as the mouse. A vast amount of genetic information would be needed if each receptor was encoded by a separate gene. It was proposed, therefore, that immunoglobulins were encoded by two sets of gene segments, a large number of variable gene segments and a single constant segment for each class, and that a variable segment would somehow become joined to the constant segment to produce a functional gene. This model turned out to be correct, although the details of the rearrangement process were more complex than expected.

The first direct evidence in support of the gene rearrangement idea came from experiments with purified immunoglobulin κ light chain mRNA. We have encountered the principle of this experiment earlier, in the overview

BOX 4.1 Cloning new genes

In eukaryotes, a gene that encodes a particular protein represents a very small fraction of the genome. Gene cloning technology is the process by which a particular gene is isolated from the rest of the genome. This necessitates inserting the gene into a carrier DNA molecule, or **vector**, so that a single copy of the gene can be replicated to provide unlimited quantities. A vector DNA containing an insert of foreign DNA is **recombinant DNA**. Many vectors have been designed for propagation in the bacterium *Escherichia coli*. These include **plasmids**, **bacteriophage** λ and **cosmids**. The choice of vector depends on the size of the DNA fragment to be cloned.

Plasmids

Plasmids are naturally occurring circles of DNA that replicate independently within bacteria. Plasmids will accept fragments of foreign DNA of 5–8 kilobase pairs (kb) in size and several bacterial plasmids have been modified for gene cloning in the laboratory. All have several features in common:

- An **origin of replication** to allow the plasmid to replicate autonomously.

- A bacterial **promoter** which initiates transcription of the inserted gene into RNA.

- A **multiple cloning site** containing a cluster of several unique restriction enzyme sites, into which the foreign DNA can be inserted.

- An **antibiotic resistance gene**, such as the gene that encodes resistance to ampicillin. Bacteria containing recombinant plasmids are transformed into antibiotic-resistant cells, which allows them to be **selected** from non-recombinant cells in the presence of the antibiotic.

Plasmids are cloned by spreading out low numbers of transformed bacteria onto plates of agar. Each bacterium grows into a colony, or clone.

Bacteriophage λ

The vectors of choice for large fragments of DNA are based on the λ **phage,** which is a virus that infects *E. coli*. Foreign DNA fragments of up to 15 kb in size can be accommodated by λ phage genomes. Recombinant λ phage is used to infect *E. coli*, which can be cloned as before. The genome of λ phage is a linear double-stranded DNA molecule that has at each end short, single-stranded overhangs (or **cohesive ends**) that are complementary to each other. Inside the bacterial cell the cohesive ends join to circularize the molecule. In the **lysogenic phase** of the life cycle, the circularized genome inserts itself into the bacterial chromosome which is then replicated as the bacteria replicate. Recombinant phage DNA can also be obtained 'packaged' inside phage virus particles, which allows the DNA to be introduced into other bacterial cells by the normal infection process. In the **lytic phase** of the viral life cycle, the circularized genome replicates producing linear genomes, linked end to end by the cohesive ends, called **concatemers**. The junctions of the cohesive ends, called **cos sites,** are necessary for the packaging process. Meanwhile, viral genes are transcribed into the proteins which spontaneously assemble into new phages. The concatemers associate with assembling particles and single λ genomes are cleaved at the cos sites and packaged into each phage particle. Finally the bacterial cells then burst (**lyse**) and the phage particles escape.

Cosmids

If large genes are to be isolated, it is technically more convenient if longer genomic DNA fragments are cloned. **Cosmids,** a kind of hybrid between plasmids and the λ phage, are able to accommodate 45 kb of foreign DNA. Cosmids lack the genes required for host cell lysis, but retain the cohesive ends of λ phage required for particle assembly. Cosmids also contain plasmid-derived genes for DNA replication and antibiotic resistance. Cosmid DNA therefore replicates inside bacteria like a plasmid and is packaged into particles like the original λ phage, but without lysing the host bacterial cell.

Cutting and rejoining DNA

All gene cloning is made possible by **restriction endonucleases**. These are enzymes in bacteria that cleave foreign DNA while leaving their own genome intact. Different restriction endonucleases cleave at specific sites, or **restriction sites**, formed by unique combinations of nucleotides. Usually when double-stranded DNA has been cut with a restriction enzyme it produces compatible or cohesive ends which can be religated with a **ligase** enzyme. Importantly, a foreign piece of DNA bearing the appropriate cohesive end can also be ligated to produce recombinant DNA. In this way a cloned gene that has been excised using a particular restriction enzyme can be inserted and ligated into vector DNA with the appropriate restriction site (Figure 1). In practice the restriction sites used for insertion of foreign DNA into cloning vectors are

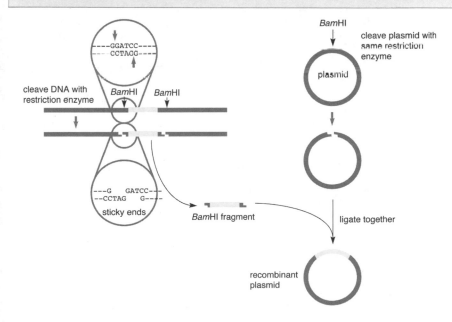

Figure 1 Cloning DNA.

BOX 4.1   *continued*

Figure 2 Producing DNA.

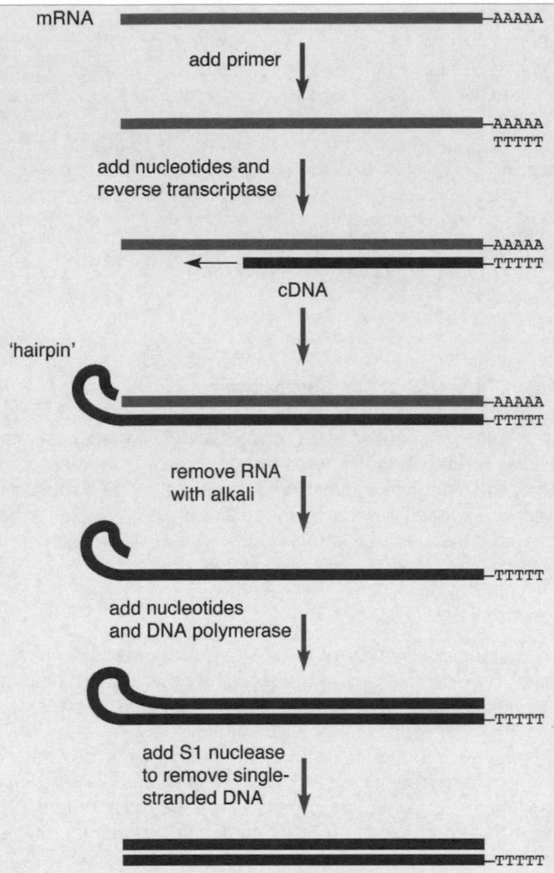

mRNA ▬▬▬▬▬—AAAAA

add primer ↓

▬▬▬▬▬—AAAAA
                 TTTTT

add nucleotides and
reverse transcriptase ↓

▬▬▬▬▬—AAAAA
 ←▬▬▬▬—TTTTT
cDNA

↓

'hairpin'
▬▬▬▬▬—AAAAA
▬▬▬▬▬—TTTTT

remove RNA
with alkali ↓

▬▬▬▬▬—TTTTT

add nucleotides
and DNA polymerase ↓

▬▬▬▬▬—TTTTT

add S1 nuclease
to remove single-
stranded DNA ↓

▬▬▬▬▬—TTTTT

unique within the vector. Some restriction enzymes produce blunt ends which can also be religated to other blunt-ended DNA. Although the efficiency is low it is a useful for recombining DNA with incompatible ends that can be blunt-ended by, for example, 'filling in' using DNA polymerase.

DNA libraries

The usual place to start when cloning a novel gene from a eukaryotic cell is to construct a **DNA library**. Two different approaches can be taken to construct a library.

- In the so-called 'shotgun' method, genomic DNA is purified from cells and cleaved into large fragments using a restriction enzyme. These are then ligated into suitable vector genomes, usually λ phage, to produce a library of recombinant vector clones containing different genomic DNA fragments.
- Alternatively, mRNA is purified from the eukaryotic cells and then reverse transcribed with **reverse transcriptase** to produce a hybrid molecule of RNA and complementary DNA or **cDNA**. The RNA is removed in alkali (DNA is insensitive to alkali) and then the single-stranded cDNA copied with DNA polymerase (Figure 2). The double-stranded molecules are then ligated into appropriately cut vectors.

The cDNA library differs in two important respects. Because mRNA has been processed to remove non-coding intron sequences (see Box 4.2), the cDNA molecules produced by reverse transcription lack the original arrangement of intron and exons seen in the genomic DNA. Also, none of the transcription regulatory sequences located outside the coding sequence are represented in cDNA. If the regulatory sequences are required, it is essential to analyse genomic DNA rather than cDNA. Second, the cDNA library is representative of only those genes that were actively transcribed at the time the mRNA was prepared. This has advantages if one is interested in cell or tissue-specific gene expression.

There are several methods available for screening genomic libraries for genes of interest. If a gene has already been cloned, it may itself be used as a probe to identify identical or homologous genes in other genomic or cDNA libraries (Figure 3). Screening of libraries can be performed by making replicas of colonies of recombinant bacteria on nitrocellulose membranes and lysing the cells to release the plasmids. Libraries in λ phage are screened by allowing recombinant phages to infect a confluent lawn of bacteria on a plate. After a few hours phages replicate and lyse the host cells, causing holes or **plaques** in the lawn of cells. The phages can be picked up on a nitrocellulose membrane in the same way. DNA is released by lysing the cells or phages and then denatured (separated into single strands of DNA) in alkali. DNA then binds to the membrane. This is then incubated with radioactively labelled, denatured DNA as the probe. The position of any colonies that contain the gene or a homologue (a similar sequence to which the probe also hybridizes) can be revealed by exposing the washed membrane to photographic film. Any bound probe will cause black spots to appear on the developed photographic emulsion.

An antibody may also be used as a probe. Indeed, many proteins have been characterized biochemically before their genes were cloned. Specific antibodies to the protein are made and then used to screen special, **expression libraries**, in which each cDNA species is expressed into protein. Any cloned cDNAs expressing protein recognized by the antibody can then be recovered. For example, antibodies were the first handles on the genes of the major histocompatibility complex (MHC; the subject of Chapter 6) and of immunoglobulins.

plate of bacteria
containing recombinant
plasmids or phages

Figure 3 Screening cDNA libraries.

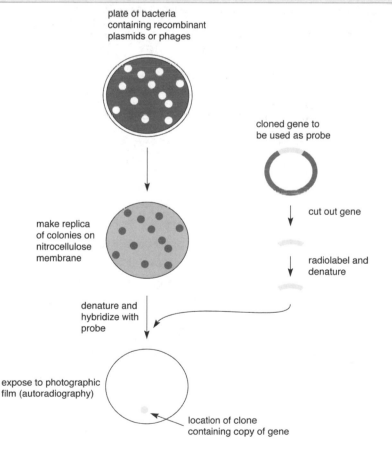

cloned gene to
be used as probe

make replica
of colonies on
nitrocellulose
membrane

cut out gene

radiolabel and
denature

denature and
hybridize with
probe

expose to photographic
film (autoradiography)

location of clone
containing copy of gene

BOX 4.1 *continued*

Very long stretches of genomic DNA are often cloned as separate, but overlapping, fragments. However, these can be pieced together by a technique called **chromosome walking**. One clone in a library is used to rescreen the library for another clone containing partially overlapping sequences by allowing them to hybridize. The second clone is then used to obtain a third, and so on. Clones spanning the entire human MHC have been obtained in this way.

Characterization of a gene

Once a gene has been cloned it can be used for a variety of other purposes. The gene may be expressed into protein in eukaryotic cells to study its function in the cell. DNA can be introduced into cultured eukaryotic cells (**transfection**) by a variety of means, including treatments to render the plasma membrane more permeable to the passage of DNA or via a recombinant virus vector. DNA can be introduced into large cells, such as the oocytes (eggs) of *Xenopus*, in fine needles (**microinjection**). The functions of proteins can also be studied in cell-free expression (*in vitro* **translation**) systems. A common *in vitro* translation method is the use of lysed rabbit reticulocytes, to which are added cloned mRNA. Because the genetic code is universal, the gene may be translated in eukaryotic and prokaryotic cells. Prokaryotes, such as the bacterium *E. coli*, can be grown in large numbers and large quantities of the gene product produced. Prokaryotes lack the post-translational mechanisms characteristic of eukaryotes, so studies of the function of eukaryotic gene products may require expression of the gene in eukaryotic cells.

Cloned genes can also be used as a source of probes. Single-stranded copies of the gene are complementary to the opposite strand and complementary to the mRNA that is transcribed from it. The cloned gene can thus be used as a source of a probe to study patterns of expression of the gene, in different tissues and in related species. Finally, the gene can be sequenced, thereby enabling the amino acid sequence of the gene product to be inferred. Sequence information from homologous genes in different species can be used to calculate phylogenetic relatedness.

Further reading

Old, R.W. and Primrose, S.B. (1994) *Principles of Gene Manipulation*, 5th edn. Blackwell Scientific Publications, Oxford.

Sambrook, J., Fritsch, E.F. and Maniatis, T. (1987) *Molecular Cloning. A Laboratory Manual*, 2nd edn. Cold Spring Harbor Laboratory Press, Cold Spring Harbor, NY.

Lodish, H., *et al* (1995) *Molecular Cell Biology,* 3rd edn. Scientific American Books/ W. H. Freeman, New York.

of antigen receptors in Chapter 2 (see Figure 2.6). DNA from mature B lymphocytes (actually mouse plasmacytoma cells), and from embryonic mouse cells, were digested with the restriction endonuclease, *Bam*HI, to produce a mixture of variously sized fragments. The fragments were then separated on slabs of agarose gel according to their size by electrophoresis. Each band on the gel was then probed with radioactively labelled whole κ chain mRNA. Two bands were recognized in the digested embryonic DNA, whereas only a single band was recognized in the digested plasmacytoma DNA. The data showed that the coding segments are some distance apart in the germline DNA but these are joined to form a contiguous stretch of DNA during lymphocyte differentiation. These rearranging segments coded for the constant and variable regions. This was demonstrated when the 3-end half of the κ chain mRNA (which codes only for the constant region) was used as a probe. While this probe still recognized a single band in the plasmacytoma DNA, it now only recognized a single band in the germline DNA.

The isolation and sequencing of immunoglobulin genes that ensued led much of the technological advances in molecular biology in the late 1970s and early 1980s. Once any gene has been cloned and sequenced, the amino acid sequence of the protein it encodes can be inferred. DNA sequencing is also simpler than protein sequencing, and the amino acid sequences of a multitude of immunoglobulins have since been obtained via the DNA sequencing route.

4.1.2 The T lymphocyte receptor

Unlike immunoglobulin, the antigen receptor of T lymphocytes is not secreted, and it is not readily cleaved off with enzymes. Therefore, attempts were made to produce T lymphocyte receptor-specific antibodies which could be used to extract the receptors from T lymphocytes using biochemical methods.

Antibodies as a tool: Section 5.1.5

Immunizing animals with mouse T lymphocytes to produce specific antibodies for these studies met with little success. This was because, like immunoglobulins, T lymphocyte receptors display enormous diversity. The problem was resolved when methods were developed for cloning mouse T lymphocytes and expanding the populations in the laboratory. Clones were then used to immunize rabbits to produce anti-T lymphocyte antibodies. Antibodies that distinguished one T lymphocyte clone from another recognized the part of the cell surface that was unique to that clone, i.e. the receptor itself. This allowed pure receptors to be extracted from the clone by biochemical means and studied. Such approaches revealed that the T lymphocyte receptor consisted of two, similarly sized, chains which were called α and β.

While these biochemical studies were in progress, attempts were also being made to clone the genes of the mouse receptor from cDNA libraries. There were several difficulties to be overcome. The early problems with producing antibodies against polyclonal T lymphocytes encouraged researchers to find a way of screening libraries without using antibodies. Also, unlike plasma cells, which synthesize large amounts of antibody and which have abundant immunoglobulin mRNA, T lymphocytes do not secrete their antigen receptors and the mRNA coding the receptor is less abundant and mixed with numerous other mRNA species.

The strategy that was developed and used for the mouse receptor is outlined in Figure 4.3. Firstly, only mRNA associated with rough endoplasmic reticulum of T lymphocytes was used for the library (see Box 4.2 *Export of proteins* later in this chapter). This step enriched the mRNAs for those coding proteins destined for export or for expression at the plasma membrane and eliminated many of the irrelevant mRNA molecules. The cDNAs obtained were then reverse transcribed into cDNA.

Approximately 98% of the 300 or so mRNA transcripts in T lymphocytes are also found in B lymphocytes. Therefore the library could be further enriched for T-specific cDNAs by hybridizing them with mRNA from B lymphocytes. Those cDNAs that remained unhybridized were then recovered. This process, which removed from the library any sequences that were common to B lymphocytes, is called **subtractive hybridization**. Finally, the T lymphocyte-specific cDNAs were screened. It was anticipated that, like immunoglobulin genes, T lymphocyte antigen receptor genes also underwent somatic rearrangement. Each cDNA was therefore screened to see if it hybridized to a rearranged T lymphocyte gene. To do this the cDNAs were hybridized with genomic DNA from T lymphocytes and against germline DNA as a control (from liver tissue), in much the same way as described earlier for B lymphocytes. Any cDNAs that hybridized differently to T cells than the control DNA contained a putative T cell receptor gene, and were sequenced.

Using this approach, a cDNA with a high degree of homology with immunoglobulin genes was identified, and shown to encode the mouse T lymphocyte receptor β chain. A few months later the α chain was cloned. The loci for α and β are distinct and located on different chromosomes. They are called TCRA and TCRB, respectively. Unexpectedly, two completely new T cell receptor loci were also discovered. The products of these two genes had hitherto escaped discovery by conventional biochemical approaches. These chains are termed γ and δ, and their loci are called TCRG and TCRD. T lymphocytes bearing the γδ type of receptor appear to have different properties to the αβ T lymphocytes.

γδ T lymphocytes: see Mononuclear cells, Section 3.2.2

Several dozen rearranged receptor genes have since been obtained from T lymphocyte clones and sequenced. This information reveals that T lymphocyte receptor chains are approximately the same length as the immunoglobulin L chain, and show a similar organization of V and C regions. Figure 4.4 shows the main structural features of the αβ and γδ T lymphocyte receptors, for comparison with immunoglobulin depicted earlier. Figure 4.5 shows the overall organization of T and B lymphocyte receptor chains for comparison.

The DNA sequences encoding T lymphocyte receptors and immunoglobulins are homologous. This indicates that the T lymphocyte receptor should have a similar three-dimensional structure as

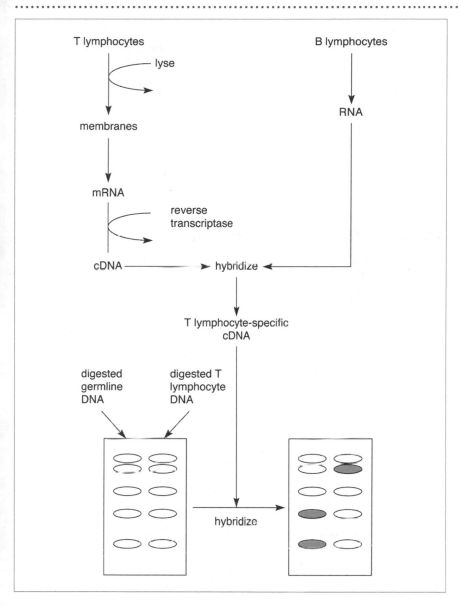

Figure 4.3 Strategy for cloning the T lymphocyte antigen receptor gene by subtractive hybridization. A cDNA library was made from mRNA associated with the membrane-rich fraction of lysed T lymphocytes. The cDNA coding genes also found in B lymphocytes were removed by subtractive hybridization with B lymphocyte RNA. The remaining T lymphocyte-specific cDNAs were tested for identity with a rearranging gene by hybridizing each against germline and T lymphocyte DNA. Any cDNA clone that hybridized to different fragments in the germline versus the T lymphocyte DNA was itself the product of a rearranging gene and was sequenced. Compare with Figure 2.6. Based on the experiments of Hedrick, S., *et al.* (1984) *Nature*, **308**, 149–153.

immunoglobulin. Answering this directly by crystallography of the T lymphocyte receptor is proving tricky, but at the time of writing, the structure of the β chain had just been resolved. The connection between the V and C domains seems to be more rigid than in immunoglobulin, although overall, the predicted structural similarities between immunoglobulins and T lymphocyte receptors based on sequence homology seem to be correct.

Figure 4.4 The main structural features of αβ and γδ T cell receptor heterodimers. V, variable region (shaded); C, constant region.

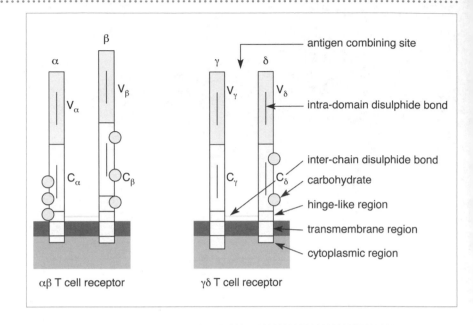

4.2 The structural basis of diversity

4.2.1 Hypervariable regions make contact with antigen

We encountered variability plots of immunoglobulins in Chapter 2, which show quite clearly the diversity of these molecules. (Recall that in immunological parlance, variability of antigen receptor molecules is described as 'diversity'.) If you re-examine the plot shown in Figure 2.5 of murine immunoglobulin V regions, you can see three well-defined 'hot-spots' of extraordinary variability. These regions are called **hypervariable regions**. Before the crystal structure of immunoglobulin had been resolved, it had been hypothesized that the hypervariable regions of the molecule were the parts that interacted with the antigen. This hypothesis was later confirmed using radioactively labelled antigens; by causing the antigens to bind irreversibly to specific immunoglobulins and then separating the immunoglobulins into individual chains, the position of the radiolabel was determined and found to be associated with the hypervariable regions.

The significance of the hypervariable regions in antigen binding becomes apparent when we locate them on the three-dimensional structure of a V domain, as shown in Figure 4.6. The hypervariable regions map to prominent loops located at the end of the molecule. Because they together determine which complementary surface the receptor binds, these loops are

hypervariable region one of three loops in the variable domain of each chain of an immunoglobulin or T lymphocyte receptor that form the antigen-combining site of the receptor.

referred to as **complementarity-determiningregions** (CDR1, CDR2 and CDR3). Thus the hypervariable regions, which are separated from each other in the primary amino acid sequence, come together to form a single antigen-binding site in the folded immunoglobulin molecule. In reality this is only half of the antigen-binding site. The hypervariable loops of the other V domain of the heterodimer lie alongside, creating an antigen-binding site of six loops – three from V_H and three from V_L. Variability plots of T cell receptor V regions show they too have several CDR-like regions. It is likely that the hypervariable regions of the T lymphocyte antigen receptor also interact with its ligand.

Recently, the crystal structure of antibody bound to antigen has been determined, enabling us to see the interface (Plate 1). The region on the surface of the antigen that is defined by the antigen-binding site is called the **epitope**. It has an irregular surface, approximately 20×30 Å, with protuberances and depressions. The corresponding surface of the antigen-binding site of the antibody is also an undulating surface, with a roughly complementary surface topography. All six hypervariable loops of the antigen-binding site are involved in the interaction.

As we will examine over the course of the next two chapters, the nature

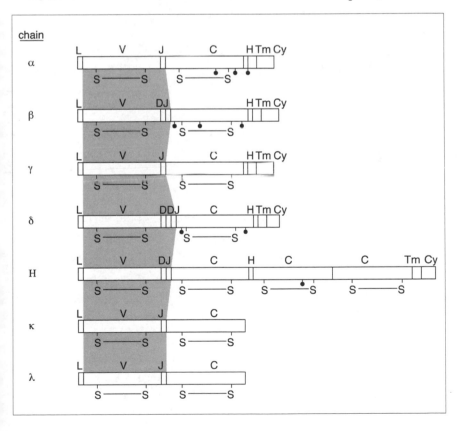

Figure 4.5 Organization of antigen receptor chains, with domains drawn to scale. L, leader sequence; V, D and J, sequences encoded by the V, D and J gene segments, and which together comprise the variable domain of the chain (shaded); C, constant domain; H, hinge region; Tm, transmembrane region; Cy, cytoplasmic region; black dots, carbohydrate; s–s, intradomain disulphide bond.

Figure 4.6 Ribbon diagram of the V domain of an immunoglobulin light chain. Hypervariable regions are in colour. Reprinted with permission from Edmundson, A.B. *et al.* (1975) *Biochemistry*, **14** (18), 3953–3961, copyright 1975 American Chemical Society.

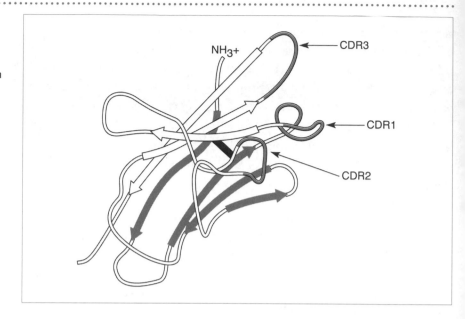

of the antigen recognized by the T lymphocyte receptor is very different to that of immunoglobulin. Rather than being a whole antigen as depicted in Plate 1, the ligand of the T lymphocyte receptor is a complex of a short peptide epitope derived from the antigen and a peptide-presenting molecule on the surface of another cell. Despite this difference, the surface of this ligand is also an undulating surface, of about 20×30 Å, with all six CDR-like loops of the T lymphocyte receptor thought to be involved in the interaction.

Summary

- Antigen receptors on B lymphocytes (immunoglobulins) exist in membrane-bound and soluble (antibody) forms. Exposure of an animal to antigen causes an increase in the concentration of antigen-specific antibodies in the serum.

- An immunoglobulin molecule is a pair of identical heterodimers each comprising a heavy (H) chain and a light (L) chain.

- Each chain has a variable domain (V_H or V_L), which are different between clones of B lymphocytes. The antigen-binding site is created by V_H and V_L. An immunoglobulin molecule therefore has two identical antigen-binding sites.

- Each chain also has one or more constant domains. Light chains have one (C_L) and heavy chains have three (C_H1, C_H2, C_H3) or more.

- Membrane-bound immunoglobulins are anchored into the plasma membrane by transmembrane domains, one located in each heavy chain.

- The T lymphocyte receptor is also a heterodimer, α and β or γ and δ. Like the immunoglobulin light chain, each chain has one V domain and one C domain. The antigen-binding site is created by $V\alpha V\beta$ or $V\gamma V\delta$. All four chains have transmembrane domains.

- Immunoglobulins and T lymphocyte receptors have short hyper-variable regions within their V domains. These coincide with the complementarity-determining regions – the external loops of the molecule that together create the antigen-binding site and which interact directly with antigen.

4.3 The genetic basis of diversity

The type of somatic rearrangement displayed by antigen receptor genes is a vertebrate 'invention'. In no other animals have such deliberate and cell-specific rearrangements of genes been found. As we will now see, the process enables a relatively small amount of genetic material to generate a vast amount of diversity in its polypeptide products.

4.3.1 Antigen receptor gene loci consist of multiple V(D)J gene segments

The arrangement of gene segments in the seven different antigen receptor loci of the mouse is illustrated in Figure 4.7. The organization is complex, although the families of segments all follow the general pattern of V-(D)-J-C. This has been emphasized in the diagram. Some families of segments within some loci have undergone duplication. For example, the DJC motif has been repeated in the TCRB locus, and in the λ locus there are repeats of a VJCJC motif. Notice that only immunoglobulin H chains, and T cell receptor β and δ chains, have D gene segments.

4.3.2 Rearrangement is an ordered sequence of events

We learned in the previous chapter that in mammals, antigen receptor gene rearrangements occur in primary lymphoid tissues. T lymphocytes

Figure 4.7 Schematic representation of antigen receptor loci of the mouse drawn to emphasize the repeating organization of families of gene elements (not to scale). Dotted lines through the D segment column indicate no D gene segments have been found at that locus. The number of gene segments is indicated, and non-functional pseudogenes are signified by 'ψ' and grey symbols. The TCRA and TCRD loci have been drawn on different strands of DNA for clarity. In reality they overlap; the V_α and V_δ segments are intermingled, and the remainder of the TCRD locus is located between the TCRA V and J segments.

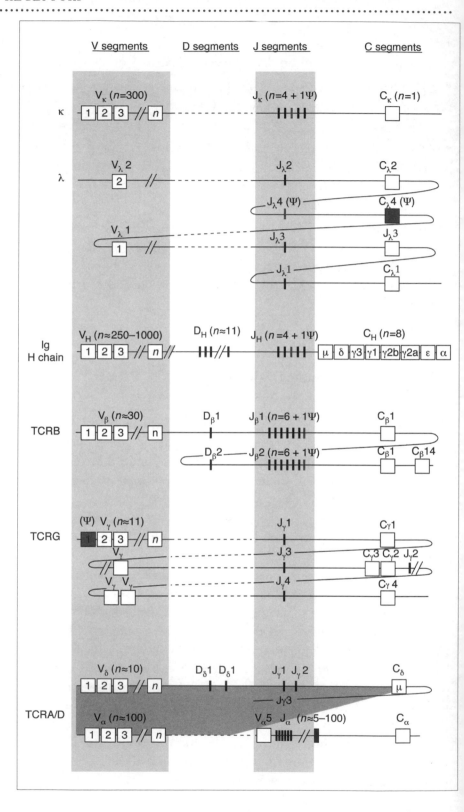

rearrange their receptor genes in the thymus and B lymphocytes in the bone marrow. In each, the first step in rearrangement is the joining of a D segment, if present, to a J segment, to form a DJ segment. This is followed by the joining of a V segment to the DJ to give a VDJ, yielding a functional, rearranged V gene. In loci that do not have D segments, the formation of a rearranged gene is a single-step joining of a V segment to a J segment.

The assembly of V(D)J gene segments is controlled by a system of motifs, called **recombination signal sequences**, in the DNA sequences flanking the gene segments. As shown in Figure 4.8, these are located downstream of V segments, upstream of J segments, and at both ends of D segments. During rearrangement, joining between segments occurs between two rearrangement signals on the same chromosome. There are two kinds of lymphocyte rearrangement signals. They consist of a distinctive motif of seven nucleotides followed by a spacer of either 12 or 23 nucleotides, followed by a nine nucleotide sequence. A signal with a 12 nucleotide spacer will only join to one with a 23 nucleotide spacer and *vice versa*. This prevents two V gene segments from being recombined together or two J segments being recombined.

DNA between rearrangement signals is looped out and enzymes remove the intervening DNA and religate the chromosome. These enzymes, called recombinases, are encoded by genes called RAG-1 and RAG-2. If these genes are artificially inserted into non-lymphoid cells, the cell rearranges its antigen receptor genes. Moreover, the same enzymes are used for the rearrangement of T lymphocyte receptor and immunoglobulin genes.

4.3.3 Successful rearrangement blocks further rearrangements

The majority of all rearrangements fail to produce a protein product. This is because the joining of gene segments is not precise and nucleotides can be lost in the process. This may cause a **frameshift mutation**. Because the mRNA transcript of a gene is 'read' or translated in codons of three nucleotides, two of every three such mutations will place the rearranged gene segments in different reading frames. This will fail to produce a full-length protein product. Such rearrangements are termed **non-productive**.

Antigen receptor genes, like other autosomal genes, have two alleles – one on the paternal chromosome and one at the same position (locus) on the homologous maternal chromosome. Rearrangement occurs on both chromosomes simultaneously, until either all the available gene segments have been exhausted, in which case the lymphocyte dies, or a rearrangement is productive. By an unknown mechanism, a productive rearrangement on

RAG-1 and **-2 (recombinase associated genes)** genes whose products are required for rearrangement of T and B lymphocyte antigen receptor genes; so far found only in vertebrates.

frameshift mutation deletion or insertion of nucleotides into a coding sequence of DNA that causes a change in the reading frame.

Figure 4.8 Recombination signal sequences associated with antigen receptor gene segments.

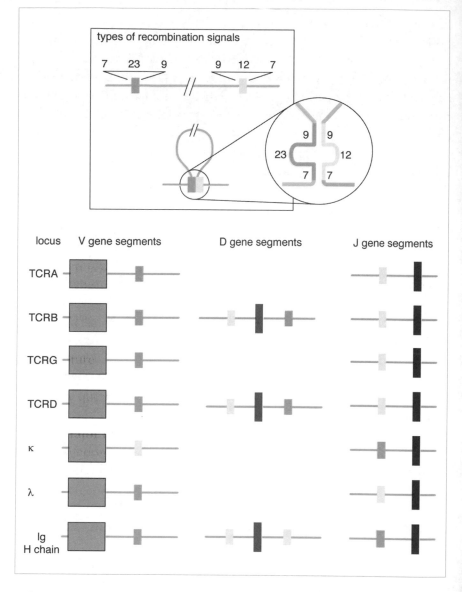

allelic exclusion a mechanism that allows only one of a pair of allelic forms of a gene to be expressed; although T and B lymphocytes rearrange both copes of a receptor gene, allelic exclusion ensures only one is expressed.

either chromosome blocks further rearrangement on both chromosomes. This phenomenon is termed **allelic exclusion**.

Rearrangements occur in an ordered sequence, with a productive rearrangement at one locus shutting down further rearrangement at that locus and initiating rearrangement at the next locus in the sequence (Figure 4.9). In B lymphocytes, the H chain locus rearranges first. The appearance of H chain polypeptides blocks further rearrangements at the H locus and initiates rearrangement of the κ locus. If κ fails, then λ rearrangements occur.

Mature T lymphocytes express either αβ or γδ receptors. These two

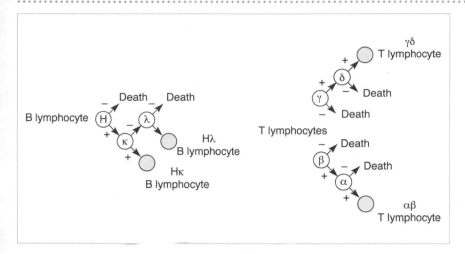

Figure 4.9 The sequence of antigen receptor gene rearrangements: +, productive rearrangement; –, non-productive rearrangement.

phenotypes are thought to arise from the same differentiating stem cell in the thymus. It is not clear, though, whether the two lineages split before gene rearrangements or whether the outcome of rearrangements determines the final phenotype. It is known that γδ cells appear first in the mouse thymus. One model suggests that γ rearranges first, and if this rearrangement is productive, the T lymphocyte rearranges δ to produce a γδ receptor. If, on the other hand, it is non-productive, the cell attempts to rearrange the β locus. If β is successful, α then rearranges to produce an αβ receptor. The other model suggests the rearrangements occur after thymocytes commit to one or the other lineages, as shown in Figure 4.9.

4.3.4 Diversity is generated from the number of gene segments

There are several sources of antigen receptor diversity. A main source is the number of gene segments available for rearrangement. Table 4.1 lists the number of gene segments in different antigen receptor loci of the mouse. The V segments are the most numerous at every locus. Because the antigen-binding site is formed by a heterodimer, the number of different combinations possible is the product of the number of V gene segments at the two loci. Thus, there are up to 300 000 immunoglobulin combinations, 2500 αβ T lymphocyte receptor combinations and 110 γδ T lymphocyte receptor combinations. This kind of diversity is called **combinational diversity**.

A major generator of diversity is the process of rearrangement itself, which occurs before the polypeptide chains associate into heterodimers. This diversity comes from four sources. Firstly, there is the incorporation of J (and D) segments. For example, there are $30 \times 2 \times 12 = 720$ possible VDJ combinations within the TCRB locus. Secondly, there is imprecise

combinational diversity *n.* diversity of antigen receptor V domains that arises from the different ways that single V, D and J gene segments can be combined (see **junctional diversity**).

Table 4.1 Potential diversity of immunoglobulin and T lymphocyte receptors of the mouse

| | Immunoglobulin | | TCR α:β | | TCR γ:δ | |
	H	κ	α	β	γ	δ
V segments	250–1000	300	100	30	11	10
V combinations	75 000–300 000		3000		110	
D segments	11	0	0	2	0	2
J segments	4	4	50	12	4	2
Ds read in all frames	rarely	—	—	often	—	often
N-region addition	V–D, D–J	none	V–J	V–D, D–J	V–J	V–D1, D1–D2, D2–J
Junctional combinations	$\approx 10^{11}$		$\approx 10^{15}$		$\approx 10^{18}$	

Ninety-five percent of murine antibodies contain the κ chain; λ has been omitted from this table. Based on Davis, M.M. and Bjorkman, P.J. (1988) *Nature*, **334**, 395–402.

joining of V, D and J segments. This may cause nucleotides to be lost and frame-shifts at the junctions. Thirdly, insertion of nucleotides at the junctions by DNA repair enzymes results in up to six extra nucleotides being incorporated. As a result of this process, which is called **N** (new) **region addition**, variable numbers of amino acids, not encoded in the germline, can be incorporated at the junction. Fourth, the D segments can in some cases be translated in all three reading frames. Thus a single D segment can give rise to three different amino acid sequences, depending on the reading frame created by the random incorporation of nucleotides in the preceding N region. The relative contributions of these four mechanisms differs between loci. The latter two mechanisms, for example, are used more by T lymphocytes than B lymphocytes. Since these mechanisms are focused at the junction of the V(D)J gene segments, the diversity they generate is collectively termed **junctional diversity.**

The potential number of junctional combinations of mouse lymphocytes is huge (Table 4.1). These figures are amplified by the combined effects of combinatorial and junctional diversity. Although the potential repertoires differ slightly in size, they all out-number the total number of lymphocytes produced during the lifetime of the animal. Consequently, the antigen-binding site expressed on each new T or B lymphocyte is unlikely to be found on any other lymphocyte, which ensures that at least one or more lymphocytes in the repertoire will be able to recognize any given antigen that is encountered.

junctional diversity *n.* diversity of antigen receptor V domains that arises at the junction of V(D)J gene segments, in part as a result of the number of D segments but also because of the loss or addition of variable numbers of nucleotides at the junction(s). See also **combinational diversity**.

4.3.5 The third hypervariable region (CDR3) is a product of junctional diversity

We saw earlier that the hypervariable regions of immunoglobulins map to the prominent CDR loops. Figure 4.10 shows these hypervariable regions

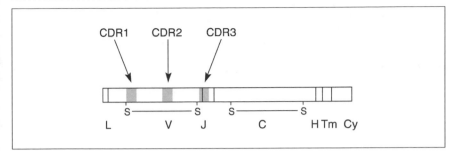

Figure 4.10 Location of hypervariable regions in antigen receptor chains. Shown is a T lymphocyte receptor α chain. Notice that the CDR3 region maps to the junction of the V and J gene segments, and is therefore the product of junctional diversity.

superimposed onto a linear diagram of an antigen receptor chain. It can be seen that the CDR3 region is constructed from the junctional region of the molecule where the V and J gene segments are joined by rearrangement. Thus the variability of the CDR3 loops is generated by the mechanisms of junctional diversity. Notice that the CDR1 and CDR2 loops lie wholly within the region of the molecule contributed by the V gene segment. Therefore, the CDR1 and CDR2 loops are significantly less variable than CDR3.

4.3.6 Birds and some mammals generate immunoglobulin diversity by gene conversion

In most of the vertebrates that have been studied, including teleost fishes, *Xenopus*, reptiles and most of the mammals, immunoglobulin diversity is generated by gene rearrangement. In the chicken, however, the loci that encode the immunoglobulin H and L chains contain only a single functional V and J gene segment. Rearrangement is a simpler process than in mammals and diversity is instead generated by the replacement of short sequences of the functional V gene segment with regions copied from a reservoir of non-functional V gene segments, or **V pseudogenes** (Ψ**V**), located nearby. Upstream of the chicken V_L gene segment are around 25 ΨV gene segments, and upstream of the V_H segment are around 90 ΨV gene segments. Rabbits, like other mammals, have several hundred V_H gene segments. Some are functional and some are pseudogenes. However, unlike in other mammals, only one functional V gene segment is used during rearrangement and diversity is created by a similar mechanism to that used in the chicken.

It is thought that birds and rabbits copy sections of the pseudogenes into the functional V gene by **gene conversion** during proliferation of the immature B lymphocyte. A B lymphocyte in the chicken may accumulate as many as gene conversion events as it matures in the bursa of Fabricius. Often the sections inserted overlap, which makes it very difficult to determine precisely how the mosaic originated.

pseudogene *n.* a non-functional gene derived from a once functional gene by the accumulation of mutations.

gene conversion a phenomenon caused by the repair of **heteroduplex DNA** in which the incorrect strand is used as the synthesis template, leading to an apparent 'conversion' of one allele to another.

Recombination and gene conversion: Box 2.2

4.4 Translation into protein

While V(D)J rearrangements are unique to lymphocytes, the subsequent process of transcription into RNA and translation into protein are the same as other eukaryotic genes. Box 4.2 gives a generalized overview of the transcription of genes into mRNA and translation into protein for expression at the cell surface.

BOX 4.2 Export of proteins

The genomes of eukaryotes consist of expanses of apparently non-functional DNA within which are discrete patches of protein-encoding DNA – the genes. A eukaryotic gene consist of several distinct sequence elements. The polypeptide-coding sequence is broken up into segments called **exons**, which are separated from each other by non-coding **introns** (Figure 1). The number of exons can vary between different genes and each usually corresponds to a functional region or **domain** of the final polypeptide product. Upstream of the polypeptide-coding region of the gene are non-coding sequences of DNA to which enzymes and regulatory proteins bind and modulate the expression of the gene (see below).

The process begins in the nucleus with **transcription** – the synthesis of a strand of RNA using the gene as a template. The enzyme RNA polymerase II copies the sequence of nucleotides in the antisense strand of the DNA into a complementary copy of RNA. In most eukaryotic genes, RNA polymerase II interacts with a sequence within the **promoter** region called the **TATA box**, so called because of a distinctive motif of thymines (T) and adenines (A). Recognition of the TATA box is assisted by a complex of other proteins, collectively called **transcription factors**. Transcription actually starts approximately 30 bases downstream of the TATA box, at the RNA **initiation site**. Efficient transcription of most eukaryotic genes also requires various other regulatory sequences. **Enhancers** elevate the level of transcription and can be located close to or far away from the initiation site. In several genes enhancers are located within the coding sequence. During transcription, a methylated structure called a **cap** is added to the 5′ end of the growing RNA molecule.

Transcription proceeds until a **transcription termination region** is encountered which stops transcription. The **primary RNA transcript** (or **pre-mRNA**) is then processed into **mRNA** (messenger RNA) by removing the introns and religating the exons into a continuous coding sequence – a process called **splicing**. The 3′ end of the molecule is cleaved at the **polyadenylation site** and several (as many as 200) adenylate residues are immediately added. This assists passage of the mRNA through the nuclear pores and into the cytoplasm, where it will be translated into polypeptide by ribosomes. The coding region of the mRNA is defined by the **translation initiation** (or 'start') **codon** located downstream of the transcription initiation site, and the

Figure 1 Sequence of events in the transcription of a typical eukaryotic gene into mRNA and translation into polypeptide.

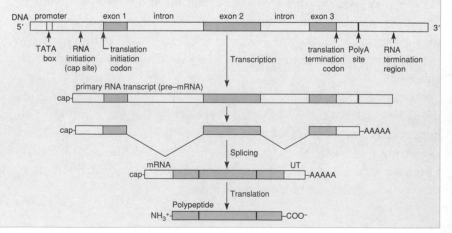

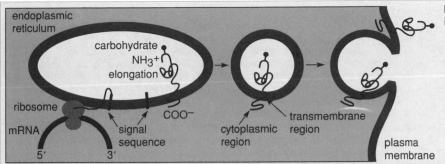

Figure 2 Translation of mRNA into protein in the cytoplasm.

translation termination codon. The coding region is flanked on either side by non-coding sequences – the 5′ cap and the 3′ untranslated region (represented by UT in Figure 1).

The next stage is the **translation** of mRNA into a polypeptide chain of amino acids by the ribosome (Figure 2). Ribosomes catalyse the formation of peptide bonds between amino acids donated by **tRNA** (transfer RNA) molecules. Each amino acid is carried by a tRNA molecule bearing a particular nucleotide triplet, or **anti-codon**, which associates with the appropriate complementary codon in the mRNA. The translation start codon is the site at which translation of the coding sequence commences.

Figure 2 shows a diagram of the export process through the various intracellular compartments; a life-like drawing of a cell is also shown in Figure 3 for comparison. In proteins destined for export, a **signal** (or **leader**) **sequence** is incorporated into the polypeptide. This is a hydrophobic stretch of amino acids that attaches the ribosome to the cytosolic face of the **endoplasmic reticulum** (ER). Figure 2 shows the signal

Figure 3 Drawing of a generalized animal cell showing major organelles and compartments.

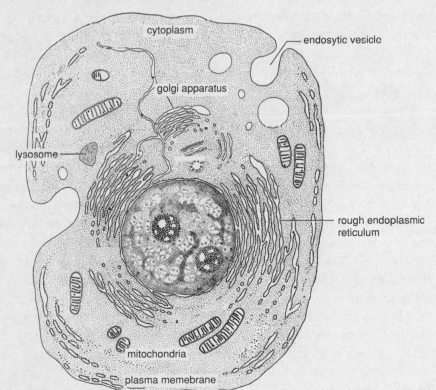

BOX 4.2 *continued*

sequence causing the growing polypeptide to cross through the membrane and into the lumen of the ER. The remainder of the polypeptide is then extruded into the lumen as it is translated.

The transmembrane region of the polypeptide cannot tolerate the aqueous environment either side of the membrane and remains within it, usually adopting an α-helical conformation and thereby anchoring the polypeptide into the membrane. Translation stops at a translation termination signal, leaving the cytoplasmic domain of the molecule protruding into the cytoplasm.

Within the ER, the newly synthesized polypeptide folds into the most stable conformation, a process often assisted by other polypeptides called **chaperones**, and disulphide bonds are formed between cysteine residues to stabilize the structure. **Post-translational modification** of the polypeptide, including the attachment of carbohydrate (**glycosylation**) and removal of the signal sequence, occurs in the ER.

Once properly folded, vesicles of membrane containing the protein pinch off from the ER membrane and fuse with the nearest face of a stack of membrane-enclosed compartments called the **Golgi apparatus**. The Golgi is where additional glycosylation occurs and where sorting of proteins destined for different intra- and extracellular compartments occurs. Membrane-bound and exported proteins are transported through the Golgi and leave within a transport vesicle bound for the plasma membrane. Finally, fusion with the plasma membrane releases the protein into the cell surface.

exon *n.* block of DNA that forms part of the amino acid-encoding part of a eukaryotic gene, and separated from the next exon by a non-coding sequence of DNA called an **intron**.

splice *v.* the processing of pre-mRNA by the removal of introns and the joining of the exons into a contiguous sequence.

Upon completion of the rearrangement of V(D)J gene segments, the whole locus is transcribed within the nucleus into a primary RNA transcript. Shown in Figure 4.11 is the sequence of events in the transcription and translation of immunoglobulin H chain. Upstream of the recombined V(D)J gene is an exon coding for a leader sequence, whereas downstream, separated by a long intron, are one or more **constant** (or **C**) **gene segments**. In the λ and H chain loci of immunoglobulins, and the β and δ loci of T lymphocyte receptors, there are several C gene segments (see Figure 4.7), although only the two C gene segments nearest to VDJ are shown in Figure 4.11.

The primary transcript is then processed into mRNA (see Box 4.2 for details) prior to export to the cytoplasm. In the immunoglobulin H chain transcript, as well as in transcripts of other antigen receptor genes, the intron between the leader sequence and the rearranged V gene is removed by splicing during RNA processing. When the mRNA is translated, the leader sequence becomes the first few amino acids of the newly translated polypeptide. This inserts itself into the lumen of the endoplasmic reticulum and causes the remainder of the polypeptide to follow it in. Similarly, the intron between V(D)J and the nearest downstream C gene segment is spliced out, as are the introns within the C gene segment itself. Thus in most cases, the segment nearest downstream from the recombined V(D)J segment is usually translated, although in the case of immunoglobulin H locus the nearest (Cμ) and second nearest (Cδ) C gene segments are translated (see below). Newly synthesized antigen receptor chains then assemble in the endoplasmic reticulum, held together by disulphide bonds, as depicted in Figure 2.4.

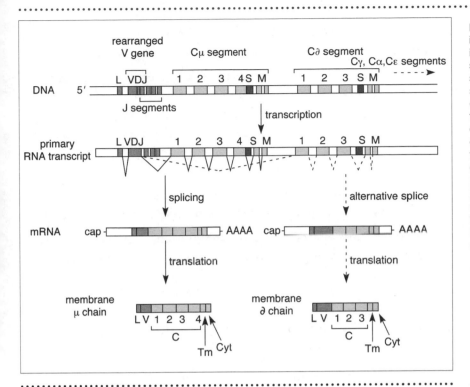

Figure 4.11 Expression of immunoglobulin μ and δ chains in the H chain locus. Coding sequences of DNA (exons) are shaded. S, secretory coding sequence; M, two 'membrane' exons encoding the transmembrane (TM) and cytoplasmic (Cyt) regions of the chain. The rearranged V gene is transcribed with the C_μ *and* C_δ gene segments into a very long primary transcript. Expression of μ or δ chains is achieved by alternative splicing of this transcript during processing into mRNA, as shown by the solid and dashed 'V's.

Summary

- In mammals, immunoglobulin light chains are encoded by genes at different loci, κ or λ. Heavy chains are encoded by genes at a single locus.

- T lymphocyte receptor chains α, β, γ and δ are encoded by different genes, at loci termed TCRA, TCRB, TCRG and TCRD, respectively.

- Each of these seven loci contains clusters of multiple gene segments called variable (V) and joining (J) gene segments. In addition, the immunoglobulin H, TCRB and TCRD loci also have a cluster of diversity (D) gene segments situated between the V and J clusters.

- During lymphocyte development, the DNA at each locus undergoes rearrangement to create a single V(D)J gene combination. This encodes the V domain of the chain.

- The C domain(s) of the chain is encoded by a C gene located downstream of the recombined V(D)J.

- Rearrangements in both types of lymphocytes are often non-productive and are repeated until a polypeptide chain is produced. This halts further rearrangement of the locus, a process called allelic exclusion.

- Each successful rearrangement is capable of creating a receptor that is unlike others in the repertoire. This is caused by the particular combination of gene segments used in the rearranged gene (combinational diversity) and the loss or addition of nucleotides at the junctions between the gene segments that occurs during the rearrangement process (junctional diversity).

- The region of the V domain encoded at the junction of the V(D)J gene segments is the third hypervariable loop (CDR3) in the binding site.

- Although birds and rabbits have multiple gene segments, diversity is generated mainly by gene conversion rather than rearrangement.

4.5 Unique features of immunoglobulin

So far we have examined the structural and genetic features common to T and B lymphocyte antigen receptors. After successful development, mature but naïve lymphocytes then enter the peripheral circulation.

Contact with antigen in secondary lymphoid tissues stimulates the mature lymphocyte to undergo a terminal, **antigen-dependent** differentiation into effector cells. At this point, several changes occur in the expression of immunoglobulin genes that do not occur in T lymphocytes.

class switching change from one class (or isotype) of antibody to another by the rearrangement of C region genes.

somatic mutation mutation of DNA in somatic cells and not inherited.

- Immunoglobulin genes undergo further rearrangement, causing the rearranged V(D)J to be relocated next to a different constant (C) gene segment. This is called **class** (or **isotype**) **switching**.

- Immunoglobulin genes change from producing membrane-bound immunoglobulin to producing soluble immunoglobulin, or antibody.

- Immunoglobulin genes undergo **somatic mutation** within the rearranged V(D)J region as the clone of B lymphocyte expands. This changes the affinity of the receptor for the antigen.

4.5.1 Immunoglobulin classes are defined by the C domains of the heavy chain

Several antigen receptor loci have more than one C gene segment. The mouse, for example, has eight C gene segments in its immunoglobulin heavy chain locus. The λ locus and TCRG have four each, and TCRB has two. Only the C gene segments of the immunoglobulin heavy chain have any influence on the function of the receptor.

In eutherian mammals, five major types of C gene segments can be distinguished: C_μ, C_δ, C_γ, C_ε and C_α. The rearranged V gene of the heavy chain can be transcribed with any one of the C gene segments, yielding different heavy chain polypeptides that are called μ, δ, γ, ε or α. These assemble with κ or λ light chains to produce the assembled immunoglobulin molecule. As mentioned earlier the heavy chain polypeptide determines the **class** of the immunoglobulin (μ for IgM, δ for IgD, γ for IgG, ε for IgE and α for IgA). Some C gene segments have closely related sequences and the immunoglobulins they encode are **subclasses**. For example, there are several human C_γ isotypes, called $C_\gamma1$, $C_\gamma2$, $C_\gamma3$ and $C_\gamma4$. These give rise to the IgG1, IgG2, IgG3 and IgG4 subclasses of human IgG. These different classes and subclasses of immunoglobulin are shown in Figure 4.12. Marsupials and monotremes have fewer classes than eutherian mammals, and in birds, reptiles and amphibians the predominant classes are IgM and IgY.

Evolution of the immunoglobulin superfamily: Section 7.2

4.5.2 IgM and IgD are expressed constitutively on B lymphocytes

On mature mammalian B lymphocytes, surface IgM is produced first, but this is gradually replaced by surface IgD. Frequently, both classes can be detected on the cell surface during the transition period. The basis of this co-expression lies in the way the μ and δ heavy chain genes are expressed. B lymphocytes transcribe their rearranged V_H gene with the two (sometimes three) C_H gene segments located immediately downstream, to produce an exceptionally long primary RNA transcript. In Figure 4.11, the details of the C_μ and C_δ gene segments have been expanded to show the arrangement of the introns and exons. The first three or four exons (depending on isotype) encode the immunoglobulin folds of the C domains. These are exons 1–4 of C_μ and 1–3 of C_δ.

Initially, V_H is translated into polypeptide with the C_μ gene, yielding μ chains, and ultimately surface IgM. However, V_H can instead be translated with C_δ. This is achieved during splicing of the primary RNA transcript. By splicing out the entire C_μ gene segment, the V_H can be brought alongside C_δ instead. This is called **differential** or **alternative splicing**, a process that is widely used in eukaryotic genes for altering the properties of their products.

alternative (or **differential**) **splicing** removal of the introns in different ways from the same primary RNA transcript (or pre-mRNA) to produce different translated polypeptides; the process by which mammalian and avian immunoglobulin genes produce membrane-bound or soluble H chains.

Figure 4.12 Five major immunoglobulin H chain isotypes of mammals. Isotype (or class) is determined by the particular C$_H$ gene segment translated. The basic immunoglobulin molecule is produced by the association of two *identical* heavy/light chain heterodimers to produce a four chained structure. Intradomain disulphide bonds have been omitted for clarity. Different classes have different functions (see text for details).

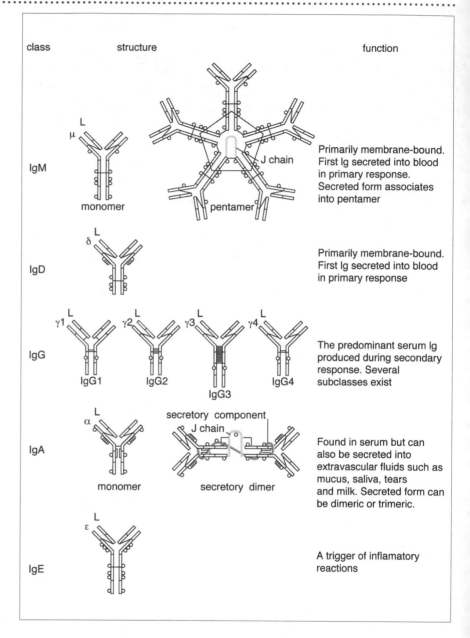

class   structure   function

IgM   monomer   pentamer   J chain   Primarily membrane-bound. First Ig secreted into blood in primary response. Secreted form associates into pentamer

IgD   Primarily membrane-bound. First Ig secreted into blood in primary response

IgG   IgG1   IgG2   IgG3   IgG4   The predominant serum Ig produced during secondary response. Several subclasses exist

IgA   monomer   secretory dimer secretory component J chain   Found in serum but can also be secreted into extravascular fluids such as mucus, saliva, tears and milk. Secreted form can be dimeric or trimeric.

IgE   A trigger of inflamatory reactions

4.5.3 Immunoglobulins are produced in soluble form after B lymphocyte activation by antigen

One of the consequences of activation of a B lymphocyte by antigen is a shift from the production of membrane-bound immunoglobulin to the production of soluble immunoglobulin (antibodies). The membrane-bound immunoglobulin is a receptor that, when bound by specific antigen, activates the B lymphocyte and the change-over to the production of soluble receptor. In reality, the shift to antibody production is a complex event requiring several other signals (discussed in Chapter 8). However, there is an absolute dependency for stimulation through the membrane-bound receptor, which ensures that only B lymphocytes that recognize the antigen are activated, and that the antibodies produced are of the required specificity.

The difference between the membrane-bound and soluble forms of a mammalian immunoglobulin lies in their heavy chains, shown in Figure 4.13. The membrane-bound form has transmembrane and cytoplasmic domains, whereas these domains are absent when the antibody is produced in soluble form. Antibody has instead a small hydrophilic region, encoded by the **secretory coding sequence** (see Figure 4.12).

This change is achieved by differential splicing of the pre-mRNA of the heavy chain gene. As an example, Figure 4.14 shows the differential splicing pathways that yield the membrane-bound or soluble forms of the µ chain. One mRNA, the result of splicing out both the secretory sequence and the intron between the exons encoding the transmembrane and cytoplasmic domains, results in a membrane-bound IgM. The other mRNA species, produced by leaving these introns in place, causes the transmembrane and cytoplasmic domains to be removed during processing, resulting in a soluble IgM.

A similar process also applies to δ, γ, ε or α chains. Moreover, alternative splicing enables a B cell to produce membrane-bound and secreted immunoglobulin simultaneously. Immunoglobulin light chains, which do not exist in membrane-bound forms, have secretory sequences but are devoid of exons encoding transmembrane and cytoplasmic domains. The reverse situation is found in T cell receptors, which exist only as membrane-bound forms.

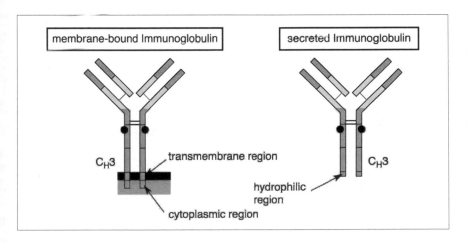

Figure 4.13 Membrane-bound and soluble forms of the same immunoglobulin. The membrane-bound form has a hydrophobic transmembrane region and a short cytoplasmic region, whereas the secreted form has instead a short hydrophilic region.

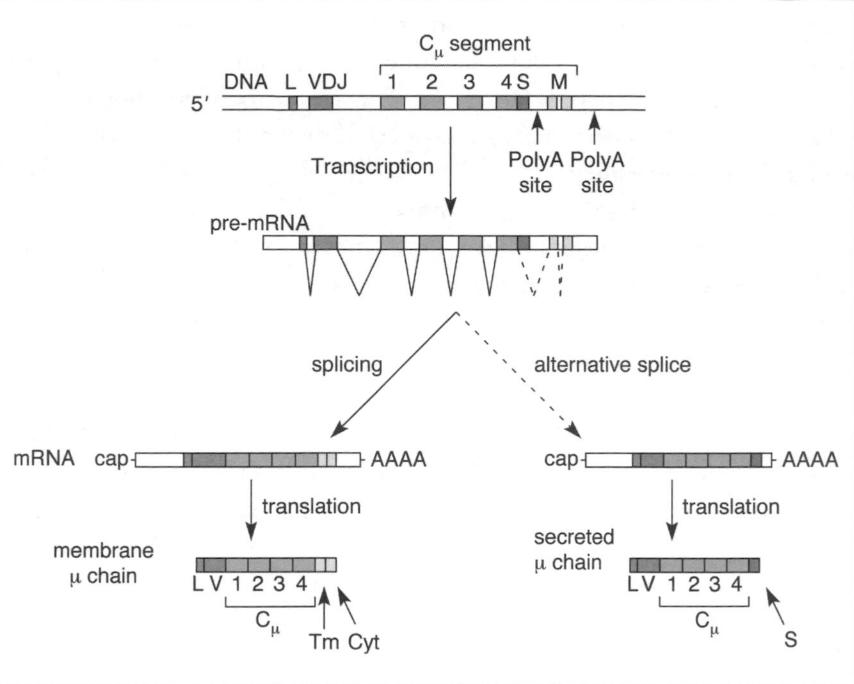

Figure 4.14 Soluble and membrane-bound immunoglobulin heavy chains are produced by differential splicing of the same primary RNA transcript. The example shown is of the C_μ segment. Polyadenylation sites (see Box 4.1 for explanation) are located on either side of the transmembrane and cytoplasmic domain sequences (M). The splice indicated by the dotted lines removes the secreted (S) sequence and poly-A site, to yield a polypeptide with transmembrane and cytoplasmic regions (bottom left). These anchor the chain in the plasma membrane of the B lymphocyte. Alternatively, leaving the S sequence in place allows polyadenylation at the site preceding the M exons and results in a soluble polypeptide (bottom right).

IgY immunoglobulin Y; the predominant antibody found in birds, reptiles and amphibians.

IgY(ΔFc) lower molecular weight species of IgY lacking the third and fourth C domains that predominates in secondary responses.

4.5.4 Alternative splicing is also used by some birds to produce IgY without an Fc region

As we will see in Chapter 7, the predominant soluble immunoglobulin found in birds, reptiles and amphibians is termed IgY. The heavy chain of this immunoglobulin, called upsilon (υ), has one variable domain and four constant domains. As in mammals, this immunoglobulin can exist in membrane-bound and soluble forms, and studies in ducks have confirmed that the soluble form, like its mammalian counterparts, is also produced by alternative splicing of the pre-mRNA of the IgY heavy chain (Figure 4.15).

In a few species, including turtles and Anseriform birds (waterfowl), a second lower molecular weight species gradually predominates as the antibody response matures. In the duck, it has been shown that this is also achieved by alternative splicing of the pre-mRNA of the IgY heavy chain. These birds are unique as they have a small terminal exon located between the second and third C region exons. If this is not spliced out the product lacks the third and fourth C domains, thereby producing an IgY molecule devoid of an Fc region. This molecule has been called **IgY(ΔFc)**. As we will see below, the Fc region of mammalian antibodies is critical for eliciting effector functions, either via receptors for Fc on phagocytes, or by triggering complement. It is unclear at present what function in birds the truncated IgY performs.

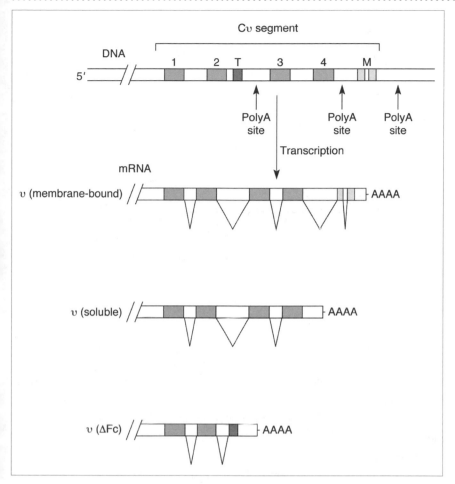

Figure 4.15 Alternative splicing of the duck IgY heavy chain gene is used to produce membrane-bound and two different forms of soluble IgY. Adapted from Magor, K.E., *et al.* (1994) *Journal of Immunology,* **153**, 5549–55.

4.5.5 Class switching occurs after B lymphocyte activation

During the course of an antibody response to antigen, the class of antibodies found in the blood gradually changes. This can be demonstrated experimentally if an animal such as a rabbit is immunized against a protein antigen and then given a secondary immunization several weeks later (Figure 4.16). The **primary response** after the first immunization is characterized by a gradual and transient appearance of antibody in the blood, reaching a peak after about two weeks. The predominant antibody classes of a primary response are IgM and IgG. Several pathogenic microorganisms have exploited this lag phase as a way of evading elimination by the immune system.

Antigenic variation: Box 5.2

A second immunization given later restimulates the previously activated clones of B lymphocytes that remain in the circulation as **memory cells**. The result of activating memory cells is a much more vigorous **secondary response** from the clone that peaks more quickly (usually in less than 1 week) and results in much higher titre of antibody in the blood that remains

Figure 4.16 Class switching during an antibody response to antigen.

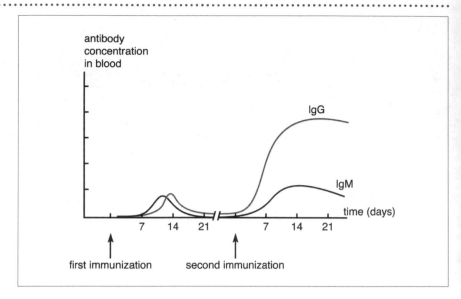

elevated for months. In the secondary response of mammals, IgG is the predominant class seen.

This change in the class of the antibody produced changes the effector functions elicited by the antibody, while retaining its antigen specificity. Mammalian B lymphocytes achieve this by changing the C_H gene segment transcribed with the V_H sequence, from μ or δ to any of the other classes. This process, called **class switching**, is accomplished by DNA rearrangements that are not dissimilar to the rearrangements that created the original VDJC sequence.

Switching is mediated by **switch sequences** located at the 5′ end of each C_H gene segment (except $C\delta$). The process is illustrated in Figure 4.17. During switching, the rearranged V_H gene, along with an upstream promoter and downstream enhancer (see Box 4.2 for explanation), are moved alongside a different C_H gene segment further downstream. The intervening DNA is probably removed by a looping-out and excision process. The switch sequences are different to the recombination signals found flanking V, D and J gene segments, and the precise details of how switch sequences work are not known. Notice that there is no switch sequence between C_μ and C_δ. As we learned above, this is because the change from IgM to IgD expression is achieved by a differential splicing.

4.5.6 Somatic mutation occurs in immunoglobulin V genes after B lymphocyte activation

In addition to secretion and class switching, mammalian B lymphocytes have a third distinguishing feature that sets immunoglobulins apart from T lymphocyte receptors. During proliferation in the germinal centre of a

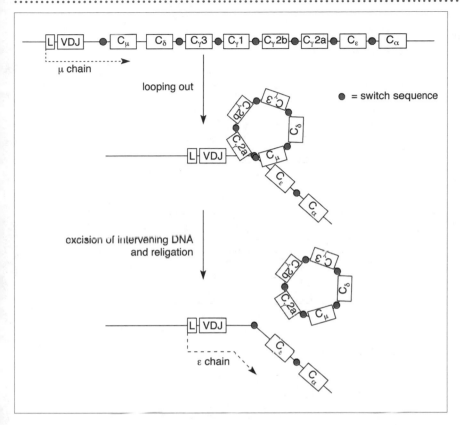

Figure 4.17 Class switching of immunoglobulin by rearrangement of the heavy chain locus. In this example a switch from C_μ to C_ε (IgM to IgE) is made by the joining of the switch signals at the 5' ends of C_μ and C_ε and looping out the intervening DNA. This is then excised and the DNA religated. This is only one model of switching, since switching back can occur.

lymphoid follicle, multiple base pair substitutions (**point mutations**) accumulate in the V genes of both the immunoglobulin heavy and light chains. This is illustrated by the experiment described in Figure 4.19. Mice were immunized against a simple antigen once, twice or three times. The second and third 'booster' immunizations restimulates the clones of residual memory cells that have been activated previously by the primary immunization. The aim is to cause the same antigen-specific clones to undergo repeated cycles of proliferation over the course of the experiment. B lymphocyte clones producing antibodies to the antigen were obtained at different times during the experiment and their V genes sequenced. It can be seen from Figure 4.18 that as the response matures, the V genes accumulate progressively more point mutations.

In other types of cells, point mutations also occur during DNA replication, but these are corrected by proof-reading enzymes. It has been estimated that during replication of an 'average' eukaryotic gene, a point mutation arises once in every 10^7 nucleotides (i.e. a mutation rate of 10^{-7} per base per cell division). Mutation within immunoglobulin V genes is much higher, with a rate of 10^{-3} to 10^{-4} per base per cell division. It appears that in B lymphocytes this is tolerated or encouraged, although the precise

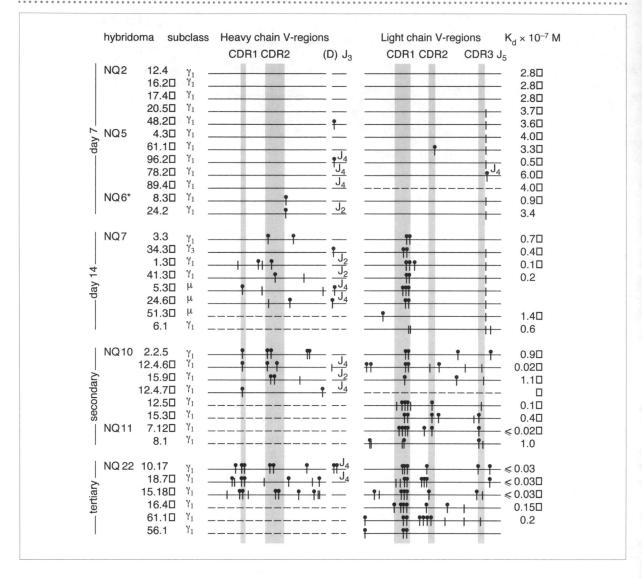

Figure 4.18 Somatic mutation *in vivo*. In this experiment, different mice were immunized with an antigen, 2-phenyl-5-oxalone (phOx). B lymphocytes were obtained on day 7 and 14 during this primary response and fused with non-antibody-secreting B lymphocyte myeloma cells to produce immortal hybridomas. These were then separated into different clones and allowed to grow. Clones producing anti-phOx antibodies were identified and the sequence of the antibody V region genes

mechanisms which promote somatic mutation in B lymphocytes are not known. In contrast, T lymphocytes do not undergo somatic mutation after activation.

Somatic mutation of V genes has a remarkable effect on the quality of the mammalian immune response. While most of the mutations that accumulate in the expanding clone of B lymphocytes are deleterious and reduce the affinity of the antibody for the antigen, some cells will acquire a higher affinity receptor than the original. If we measure the affinity of antibodies produced by the clonal population at different times during its expansion, we see that the overall affinity gradually *increases* with time. This is because cells in the clone with the higher affinity receptors receive positive signals continue to produce antibodies, whereas those with low affinity receptors die. This is called **affinity maturation**

To obtain positive signals, proliferating B lymphocytes must receive periodic restimulation. Using their surface immunoglobulins, individual B lymphocytes make repeated contacts with germinal centre follicular dendritic cells bearing the specific antigen, and in so doing solicit the additional signals required for their continued survival and proliferation. Those B lymphocytes in the expanding clone that acquire deleterious mutations, which reduce the affinity of its surface immunoglobulin, fail to receive the positive signals and die by apoptosis.

Although mutations are an important part of the affinity maturation of a response, we also have to invoke a mechanism to suppress further mutation in the high affinity B cells, otherwise a good antibody could be ruined by a deleterious mutation.

> **affinity maturation** the gradual increase in overall binding strength (affinity) of antibodies produced during the course of an immune response to antigen.

Apoptosis: Box 9.2

4.6 How antibodies trigger effector functions

Notwithstanding the effects of somatic mutation, antibodies retain the same antigen specificity as their membrane-bound predecessors, and they combine with the antigen in precisely the same manner. However, while its *function* (which is to bind antigen) is unchanged, the *consequences* of binding to its antigen are very different. The membrane-bound immunoglobulin is part of a signalling complex. When antigen is bound by the extracellular antigen-binding sites the immunoglobulin relays signals, via associated signalling proteins, to the nucleus inside the B lymphocyte causing cellular activation. In contrast, binding of the same antigen by soluble immunoglobulin produced by the same clone instead elicits an effector function that ultimately leads to the removal of the antigen.

4.6.1 Effector function is determined by the Fc region of antibody

Antibodies do not possess any intrinsic cytotoxic or cytopathic activities. Instead, their main role is to mark antigen for elimination by effector mechanisms. As we saw in the first chapter these mechanisms include agglutination and opsonization for ingestion by phagocytic cells, triggering of the complement cascade by the classical pathway, and antibody-dependent cell-mediated cytotoxicity. The antibody molecule is a versatile adapter molecule, the V domains interacting with antigen and the C domains of the heavy chains mediating the other properties.

Figure 4.18 *continued opposite page*
then determined. This was repeated on animals 3 days after receiving a second immunization to elicit a secondary response and on animals 3 days after receiving a third immunization over 1 year later. The horizontal lines represent the heavy and light chain V regions clones producing anti-phOx antibodies. Unbroken lines indicate nucleic acid sequence identity. Broken lines indicate sequences from other germline genes. Vertical bars show nucleotide substitutions, bars with black circles indicates also a predicted amino acid substitution. On the right is shown the affinity of the interaction between the antibody secreted by the hybridoma and phOx. See Box 5.1 for an explanation of the dissociation constant, K_d. Redrawn with permission from Berek, C. and Milstein, C. (1987) *Immunological Review*, **96**, 23–41, Munksgaard International Publishers Ltd, Copenhagen, Denmark.

At the simplest level, a useful protective property of antibody arises from its agglutinating properties. An antibody molecule, by virtue of being constructed from two identical heavy/light heterodimers, has two antigen-combining sites with the same antigen specificity. This enables it to cross-link two antigens simultaneously. This effect is amplified in IgM, which exists as a pentamer in most mammals. In the presence of multiple antibodies, soluble antigens can be precipitated out of solution. Similarly, larger particulate antigens, such as bacteria, viruses and other microorganisms, can become agglutinated. Antibodies are particularly efficient in clumping particulate material on which there are multiple copies of a single type of antigen, such as the capsid of a virus. Aggregates of antigen and antibody are called **immune complexes** and their formation helps phagocytes remove antigen more efficiently. In addition, antibodies can prevent infection of a cell by potential intracellular parasites, such as viruses, by binding to the surface of the parasite and blocking any interaction with its cell-surface receptor.

immune complex *n.*
aggregations of antigen
and antibody.

Opsonization

Phagocytosis: Section
1.4.1

One of the most important functions of antibodies is as opsonins. Macrophages and other phagocytic cells bear on their surfaces a variety of **Fc receptors** which bind to the Fc regions of antibodies within immune complexes (Figure 4.19). The most abundant are receptors for the Fc regions of IgG molecules. These are called FcRγI, FcRγII and FcRγIII and are found mainly on macrophages and monocytes. There are also receptors for IgE on mast cells and for IgA on macrophages and monocytes. Table 4.2 lists these receptors found in humans. With the exception of the FcRεII receptor for IgE (which is a selectin), Fc receptors are members of the immunoglobulin gene superfamily.

Immunoglobulin
superfamily: Section 3.8.1
Selectins: Section 3.8.3

Antibody-dependent cell-mediated cytotoxicity (ADCC)

Several types of cytotoxic cells with innate recognition capabilities also have

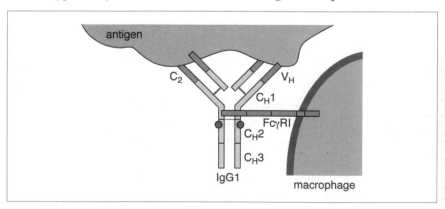

Figure 4.19 Opsonization by antibody. Schematic view of the interaction of an Fc receptor in the plasma membrane of a macrophage with the C$_H$1 domain of an antibody in an immune complex. FcγRI (or CD64) shown here is a high affinity receptor for the IgG1 subclass. It is also a member of the immunoglobulin superfamily with three extracellular immunoglobulin fold domains, a transmembrane domain and a cytoplasmic domain.

Table 4.2 Fc receptors

Receptor[a]	Cellular expression	Ligand
FcγRI (CD64)	monocytes, macrophages	
	IFN-γ induces on neutrophils	IgG2a
FcγRII (CD32)	B lymphocytes, neutrophils, monocytes	IgG1, IgG2b
FcγRIII (CD16)	macrophages, NK cells, neutrophils	IgG3
FcεRI	mast cells, basophils	IgE
FcεRII (CD23)	macrophages, eosinophils, B lymphocytes	IgE
FcαR	monocytes/macrophages, granulocytes	polymeric IgA1, IgA2

[a]The Greek letter denotes the class of antibody the receptor binds and the R stands for 'receptor'.

Fc receptors. This enables them to direct their cytotoxic activities to target cells that have been tagged by antibodies. We saw in Chapter 1 that natural killer (NK) cells have an innate 'NK receptor' able to detect structures peculiar to tumour cells or cells infected with viruses. Some NK cells express FcRγIII which allows them also to kill foreign cells recognized by antibodies. Killer (or K) cells, which are mononuclear cells that bear Fc receptors, are defined as cytotoxic cells that kill only by antibody-dependent cell-mediated cytotoxicity. Macrophages, monocytes and neutrophils are also able to mediate killing in this way.

Mast cell degranulation
FcRεI is a high affinity receptor for the Fc region of IgE, and is found on the surfaces of mast cells and basophils. Binding of IgE–antigen complexes to FcRεI stimulates a rapid degranulation of mast cells and the release of several preformed vasoactive mediators, principally histamine and serotonin. These are inflammatory mediators and cause local vasodilation, smooth muscle contraction and increased vascular permeability to infiltrating leucocytes.

Activation of complement
As we saw in the first chapter, complement is an effector mechanism of around 30 serum and membrane-bound components, that is triggered directly or indirectly by microorganisms. The indirect route is by antigen-bound antibody via the **classical pathway.**

This pathway is triggered by the first component, C1, binding to the Fc regions of antibodies within immune complexes. C1, shown in Figure 4.20, is a complex of four subunits – C1q, C1r, C1s and C1INH (for 'inhibitor'). The C1r and C1s subunits are inactive proteinases, whereas C1q mediates binding to Fc.

The efficiency of binding to Fc is dependent upon the isotype of the antibody. Only IgM, IgG1, and IgG3 bind C1q efficiently. The C1q binding

Figure 4.20 The complement C1 complex. C1INH inhibits the enzymatic activity of C1 and also prevents the C1 complex from binding to antibody by stearic hindrance.

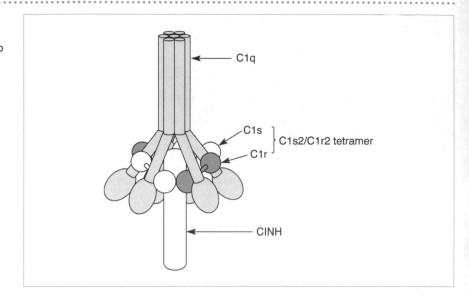

sites are located in the Cγ2 domain of IgG and the Cμ3 domain of IgM. Since each Fc region contains two H chains, there are two such sites on the Fc region of an antibody. Other Ig subclasses bind C1q very weakly, and IgA, IgE and IgD fail to bind C1q altogether.

The C1INH component is a proteinase inhibitor that inhibits the activity of spontaneously activated C1. This is achieved by binding to the active site of the C1r enzyme. It also prevents the interaction of the C1 complex with an Fc region by stearic hindrance. For full activation of the C1s and C1r

Figure 4.21 Activation of the classical complement pathway by an immune complex. Shown are two IgG molecules bound by their antigen-combining sites to the surface of a foreign antigen. The C1q components of the C1 complex bind with high affinity to immunoglobulin CH2 domains of the immune complex. In turn, a conformational change in C1 triggers its activation.

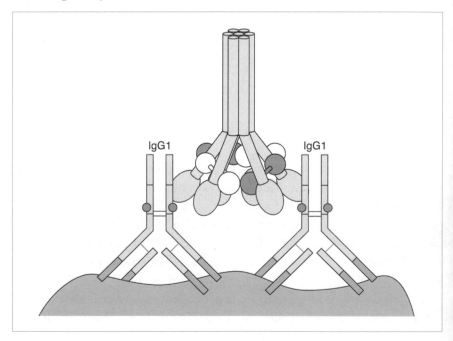

enzymatic subunits, a conformational change must occur within the C1 complex. This only occurs when more than two Fc subunits are engaged by C1q (Figure 4.21). A single IgM pentamer, or two or more densely packed IgG molecules, can satisfy this requirement. Thus, immune complexes, and antibodies bound to the surface of foreign cells and viruses, are particularly efficient activators of the classical complement pathway. However, neither free IgM nor IgG can trigger the cascade. Pentameric IgM has multiple binding sites, but these are only exposed when it is distorted by binding to antigen. Serum IgG never reaches the density required to provide the close proximity of Fc regions required for bridging by the C1q subunits.

After this conformational change within the C1 complex, C1INH is displaced and C1r molecules cleave each other and the C1s molecules. The C1 complex is now activated. The subsequent sequence of events, from the cleavage of C4 onwards has been discussed in detail previously.

The mammalian complement cascade: Box 1.2

Secretion into extravascular fluids

Most vertebrates have evolved ways of transferring low molecular weight immunoglobulin isotypes from the blood into extravascular fluids. These can be found in mucus lining gastrointestinal and genital tracts, where they assist mechanical and chemical barriers in preventing the entry of micro-organisms. Antibodies are also transferred from mothers to developing embryos. This provides the embryo with **passive immunity**. Birds transfer immunoglobulin into egg white during oogenesis, whereas some mammalian embryos receive maternal immunoglobulin *in utero*. Mammals also continue to provide passive immunity to their young after they are born by immunoglobulins that are secreted into breast milk. These adaptations are discussed in Chapter 7.

Transfer of immunoglobulin into extravascular fluids is understood best in mammals. In humans, the major class of immunoglobulin found in exocrine secretions (tears, saliva, milk, colostrum and mucus of the alimentary, urogenital and respiratory tracts) is IgA. This is transported from the serum through the vascular epithelial cells of exocrine tissue or of the gut by a process called **transcytosis** (Figure 4.22). IgA undergoes distinctive structural modifications before passage into such fluids. In the blood, IgA is linked into polymers by a short, 18 amino acid-long polypeptide called the **J chain,** secreted by the B lymphocyte. (An identical J chain is found in the IgM pentamer.) Dimers are the predominant polymer, although monomers and polymers of up to five IgA molecules are found in the blood.

For transport through the vascular epithelium, a receptor called the **poly-Ig receptor** on the vascular epithelial cells captures polymeric IgA which is then endocytosed. IgA dimers are transported through the cell

passive immunity immunity acquired by the transfer of preformed antibodies, as in maternal antibodies passed on to newborn mammals in milk.

exocrine *n.* glands that secrete fluids into ducts.

transcytosis *n.* the transport of material through a layer of epithelial cells by endocytosis.

J chain *n.* a polypeptide chain that joins IgA or IgM into multimers.

Figure 4.22 Transcytosis of IgA through vascular endothelium.

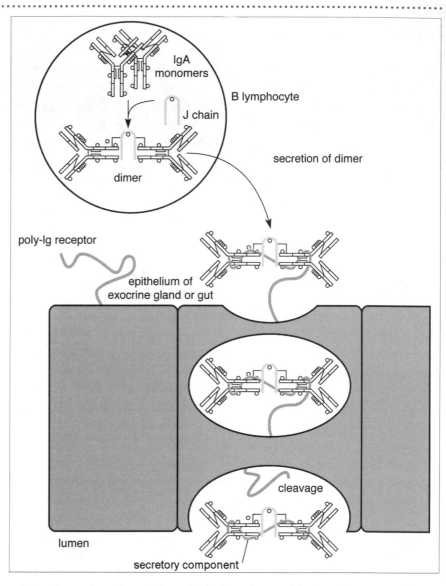

secretory component *n.* a fragment of the **poly-Ig receptor** that remains with secreted IgA and IgM.

within the endocytic vesicle, which then fuses with the plasma membrane on the luminal face of the cell. Proteolysis of the poly-Ig receptor then releases the IgA dimer into the extravascular fluid. A fragment of the poly-Ig receptor, now called the **secretory component**, remains attached to the secreted dimer. The secretory component enhances the resistance of the dimer to proteolytic enzymes and may help prolong the activity of IgA in hostile environments such as the gut.

In suckling mice and rats, the uptake of maternal IgG provided in milk is also mediated by transcytosis. Antibodies are captured by **neonatal Fc receptors** found on the intestinal epithelial cells which transport maternal IgG from inside the gut into the blood.

Summary

- Isotypes (or classes) of immunoglobulin are defined by the C domains of the heavy chain. The best characterized are from eutherian mammals. These have C_μ, C_δ, C_γ, C_ϵ and C_α which define IgM, IgD, IgG, IgE and IgA. Each class participates in different kinds of effector functions.

- IgM (and IgD in mammals) are constitutively expressed as membrane-bound receptors on mature but naïve B lymphocytes. Encounter of a B lymphocyte with specific antigen triggers clonal proliferation and differentiation.

- Newly dividing cells undergo immunoglobulin class switching. This is caused by DNA rearrangement within the C genes of the heavy chain locus, by a process not unlike that which creates the VDJ gene.

- Differentiated B lymphocytes (plasma cells) also release soluble antigen receptors (antibody) into the blood. This is achieved by removing the transmembrane and cytoplasmic domains of the heavy chain by alternate splicing of the RNA transcript of the H chain gene.

- Newly dividing mammalian B lymphocytes acquire mutations within the rearranged V gene. B lymphocytes bearing higher affinity receptors receive stimulation for growth. This is reflected as a gradual increase in affinity of serum antibodies (affinity 'maturation').

- The biological functions of different classes of antibody are determined by C domains of the H chains. Different classes interact differentially with a variety of Fc receptors on host cells such as phagocytes and cytotoxic cells, or to trigger the complement cascade by binding C1q.

The future

- The T lymphocyte receptor has been partially crystallized. The full crystal structure will allow us to understand the interaction with the peptide–MHC complex.

- Several gaps exist in our understanding of antigen receptor phylogeny and evolution. Inroads are being made into mechanisms of antigen receptor diversity in several other mammals other than man and mouse, and some non-mammalian vertebrates.

phylogeny *n.* the evolutionary line of descent of a species.

Further reading

Immunoglobulins

Amzel, L.M. and Poljack, R.J. (1986) Three-dimensional structure of immunoglobulins. *Annual Review of Biochemistry,* **48**, 961–98.

Hulett, M.D. and Hogarth, P.M. (1994) Molecular basis of Fc receptor function. *Advances in Immunology,* **57**, 1–127.

Loffman, R.L., Lebman, D.A. and Rothman, P. (1993) Mechanism and regulation of immunoglobulin isotype switching. *Advances in Immunology,* **54**, 229–70.

McCormack, W.T., Tjoelker, L.W. and Thompson, C.B. (1993) Immunoglobulin gene diversification by gene conversion. *Progress in Nucleic Acids Research,* **45**, 27.

Reynaud, C.-A. *et al.* (1994) Formation of the chicken B-cell repertoire: ontogenesis, regulation of immunoglobulin gene rearrangement, and diversification by gene conversion. *Advances in Immunology,* **57**, 353–78.

Sim, R.B. and Reid, K.B.M. (1991) C1: molecular interactions with activating systems. *Immunology Today,* **12**(9), 307–11.

Warr, G.W., Magor, K.E. and Higgins, D.A. (1995) IgY: clues to the origins of modern antibodies. *Immunology Today,* **16**(8), 392–8.

T lymphocyte receptors

Davis, M.M. and Bjorkman, P.J. (1988) T-cell antigen receptor genes and T-cell recognition. *Nature,* **334**, 395–402.

Raulet, D.H. (1989) The structure function and molecular genetics of the γδ T cell receptor. *Annual Review of Immunology,* **7**, 175–208.

Review questions

Fill in the blanks

Immunoglobulins are found in the _____[1] fraction of mammalian serum. Enzymatic cleavage of antibodies with _____[2] yields two kinds of fragments; the antigen-binding parts of the molecule are called _____[3] fragments. The other kind of fragments are the _____[4] fragments, so called because in the original studies these fragments could be made to crystallize. The antigen-binding fragments did not crystallize because they are _____ [5]. This problem was resolved when it was discovered that some humans with _____ [6] excreted immunoglobulin _____ [7] chains in their urine. Sequencing of these chains from different patients revealed two equally sized parts to the molecule: a _____ [8] domain and a _____ [9] domain. To understand the reason for this, experiments were conducted with purified λ light chain mRNA. DNA from B lymphocytes and from embryonic cells was digested with a _____ [10] and the fragments probed with the λ chain mRNA. The mRNA hybridized with two fragments of digested _____ [11] DNA but only one fragment of digested _____ [12] DNA. This showed immunoglobulin genes in the _____ [13] configuration underwent _____ [14] in B lymphocytes. The T lymphocyte receptor was characterized by cloning its gene first. First, a cDNA library was obtained from a _____ [15] of T lymphocytes, which was then enriched for T lymphocyte-specific cDNAs using a novel technique called _____ [16]. Each was then used as a probe to see which hybridized with rearranging DNA. In this way the mouse _____ [17] chain was cloned.

> **cDNA library** a set of complementary DNA (cDNA) clones derived by reverse transcription of different mRNAs in a cell.

Short answer questions

1. Draw a schematic diagram of a membrane IgM molecule and of an $\alpha\beta$ T lymphocyte receptor. Include the extracellular, transmembrane and cytoplasmic domains. Identify the antigen-combining sites and the disulphide bonds, and annotate your diagram with the names of each chain and domain.
2. What are the differences between junctional and combinational diversity? What other mechanisms have been found to generate diversity in antigen receptors?
3. What is allelic exclusion, and what effect does this have on the antigen receptor of a mature lymphocyte?
4. Lymphocytes are activated when their antigen receptor binds antigen. Describe three changes that occur in the expression of immunoglobulins that do not occur in T lymphocyte receptors.
5. The Fc region of immunoglobulins are recognized by Fc receptors on cell surfaces. Name five different cellular phenomena in which such an interaction is involved.

5 Antigen recognition

In this chapter

- The nature and differences between epitopes recognized by B and T lymphocyte antigen receptors.
- How and where epitopes are located within antigens.
- Immunoassays for antibody–antigen interactions and measurement of T lymphocyte responses.

Introduction

In this chapter we turn to the ligands of the T and B lymphocyte antigen receptors. Of the two kinds of receptor, recognition by immunoglobulin is the more conventional ligand–receptor interaction (Figure 5.1). It is not dissimilar to the interactions between other proteins, such as protein hormones and their receptors, or enzymes and their substrates.

Antigen recognition by T lymphocytes is a more unusual interaction, since the ligand of the T lymphocyte receptor is formed by the association of two distinct components. One is a peptide derived from the antigen, and the other is a peptide-presenting molecule on the surface of another cell. These peptide-presenting molecules, which are encoded by the major histocompatibility complex (MHC), are the subject of Chapter 6.

5.1 The ligand of immunoglobulin

The most informative way of examining the interaction between two macromolecules, such as an antigen receptor and its antigen, is by X-ray crystallography. X-rays are passed through the crystal at different angles and the diffraction patterns that are generated can be used to determine the three-dimensional arrangement of atoms within the molecules. Good resolution is dependent on the homogeneity of the molecules in the crystal. A heterogeneous protein may, at best, produce 'blurred' diffraction patterns

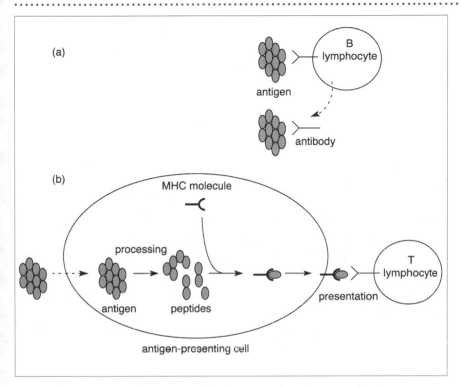

Figure 5.1 Schematic representation of antigen recognition by T and B lymphocyte receptors. (a) Immunoglobulin (as antibody or membrane bound) binds to free, whole antigen. (b) In contrast, before the T lymphocyte receptor can bind antigen, the antigen must be processed into peptides by enzymatic degradation within an antigen-presenting cell. The peptides are then exported to the cell surface by newly synthesized MHC molecules. The dashed line indicates that the antigen can originate outside the cell or from within.

and low resolution of the molecule or, at worst, will fail to crystallize.

We learned in the previous chapter that the first source of homogeneous immunoglobulin was the myeloma proteins (or Bence-Jones proteins) produced in humans with multiple myeloma. This led to the first structural analyses and X-ray crystallographic studies of immunoglobulin molecules. Although much was learned about the structure of immunoglobulin, the antigen recognized by these antibodies is not known. Therefore the nature of the interaction between a myeloma protein and its specific antigen could not be determined.

This problem was resolved by the development, in the mid-1970s, of a technique to propagate single clones of B lymphocytes with a defined antigen specificity. These **monoclonal antibodies** provided unlimited supplies of antibody specific for a single antigenic determinant. We will learn more about this technique later. The breakthrough then made it possible to co-crystallize an antibody together with its specific protein antigen.

Monoclonal antibodies: see Section 5.1.5

monoclonal antibody *n.* antibodies produced by a single clone of B lymphocytes and having the same antigen specificity; produced in the laboratory by the immortalization and cloning of plasma cells.

5.1.1 The epitope is complementary to the antigen-combining site

To date, several monoclonal antibodies and their protein antigens, including hen egg lysozyme and neuraminidase of influenza virus, have been co-crystallized and analysed by X-ray diffraction. In the previous chapter we saw (Plate 1) the three-dimensional structure of complex of hen egg lysozyme and a Fab fragment of an anti-lysozyme monoclonal antibody (called Hy HEL-5). The images of the two molecules have been separated by the computer to reveal the interfaces. These surfaces are complementary, such that wherever there is a protrusion on one surface there is a corresponding depression on the other. This feature is typical of the other antigen–antibody interactions examined. Sometimes the interaction is actually so close as to exclude water molecules. The area on the surface of the antigen that is defined by the antigen-combining site is called the **epitope** and it has roughly the same dimensions as the antigen-combining site itself.

Several epitopes defined by different monoclonal antibodies to lysozyme have been analysed by crystallographic studies. Three of these, depicted in Plate 2, together cover about 40% of the available surface area of lysozyme. Notice that the boundaries of two of these epitopes partially overlap. In reality, the surface of any protein antigen can be regarded as an overlapping patchwork of many epitopes, with all parts of the molecule accessible to antibodies representing potential epitopes.

5.1.2 Weak non-covalent bonds hold the molecules together

Because of the complementarity, the two opposing surfaces of epitope and antigen-combining site are very close to each other, and several non-covalent bonds – hydrogen bonds, van der Waals' interactions, salt bridges and hydrophobic bonds – can spontaneously form between them.

BOX 5.1 Interactions between macromolecules

Covalent bonds are stable, strong bonds that link together the units that make up macromolecules. Thus, the amino acids of polypeptide chains are linked by an amide linkage called a peptide bond, the nucleotides of nucleic acid chains are linked by phosphodiester bonds and the sugars of polysaccharides are linked by glycosidic links. These covalent bonds maintain the precise sequence of the units the macromolecule requires for its particular function and are normally only broken by the activity of specific enzymes.

However, this chain must also fold into a particular three-dimensional shape in order to carry out its function. This is achieved by much weaker **non-covalent** bonds that require approximately 1/100th the energy to break them. Non-covalent bonds form between units of the same chain and hold it in the conformation required for functional activity. Although in theory a given macromolecule can form many different weak intramolecular bonds and consequently many different possible conformations, only one conformation is adopted. This is usually the conformation involving the most bonds, which is consequently the most stable. Non-covalent bonds also form between two interacting macromolecules, for example between two strands of complementary DNA, or between interacting proteins, such as protein hormones and receptors, enzymes and substrates, and antigens and antigen receptors.

In its folded conformation, the surface of a given macromolecule has a characteristic landscape of protuberances and depressions created by the electron clouds around each of its atoms. Owing to random diffusion within aqueous environments, macromolecules continually encounter other macromolecules, forming and breaking non-covalent bonds. The stability of the interaction is dictated by the surface topography of the two interacting molecules. Two interacting macromolecules have surface regions that are complementary in shape to each other. This close association allows the formation of multiple non-covalent bonds between the two macromolecules at the interface. The greater the area of interaction, and the closer the matching of the interacting surfaces, the higher the forces of interaction. Thus while each individual non-covalent bond is relatively weak, the multiple bonds that form between two complementary molecules can lead to very high binding energy (i.e. more energy is required to break the interaction). On the other hand, the greater the overlap of electron clouds, the greater the repulsive forces that reduce the stability of any non-covalent associations.

Non-covalent bonds are usually classified into three types:

- **Ionic** (or **electrostatic**) **bonds.**

- **Hydrogen bonds.**

- **van der Waals' bonds.**

Ionic bonds form between surfaces of opposite charge. In non-aqueous environments these are particularly strong bonds, accounting for example for the strength of crystal lattices in minerals. In the presence of water, these charged groups interact via interposed water molecules and are consequently much weaker.

Hydrogen bonds form between hydrogen atoms and two other electronegative atoms such as oxygen and nitrogen. In water, macromolecules are dissolved by hydrogen bonding to water molecules; these must be broken before hydrogen bonds can reform between two macromolecules.

Van der Waals' forces occur at very close ranges between two atoms. Fluctuations in the electrical charge within electron clouds can lead to attractive or repulsive forces between atoms, depending on the distance between them. At very close distances atoms repel each other, whereas at a particular optimal distance, which is different for each element, a weak attraction is created. Van der Waals' interactions become very important when the surfaces of two macromolecules share large areas of complementary surface.

Figure 1 Non-covalent interactions.

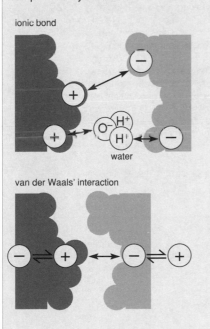

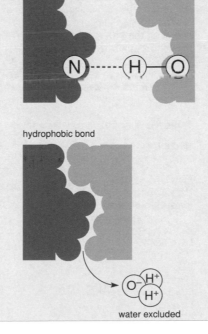

BOX 5.1 *continued*

All of these bonds are strengthened by a fourth kind of weak bond, which is created by the behaviour of hydrophobic subunits in aqueous environments. These tend to be pushed together to minimize the instability they cause in the network of hydrogen-bonded water molecules. Water is squeezed out, resulting in a very close association called a **hydrophobic bond**. See Figure 1.

Measuring the interaction

Random thermal motion both brings molecules together and pulls them apart. The stronger the attraction between them, the more energy is required to break them apart and consequently the longer the molecules remain together before dissociating. Low energy bonds are readily broken and the association time is briefer.

The strength of bonding between two macromolecules is represented by the **affinity** (or **equilibrium**) **constant** (K), which is expressed as litres per mole. Consider the binding of two macromolecules A and B to form a complex AB. At equilibrium the rate of association equals that of dissociation. This is dependent on the concentrations of free A and B, and on the strength of the interaction. The affinity constant for this reaction, K, is calculated from the formula [AB]/[A][B]. The stronger the binding, the larger the value of K. Frequently the inverse of the affinity constant is used to describe the strength of an interaction; this is called the **dissociation constant** (K_d) (See Table 1). In biological systems the dissociation constant between macromolecules ranges from 10^{-3} (low affinity) to 10^{-10} (high affinity) moles per litre. Some examples are shown in the table. Mathematically, an unbreakable association can only be formed with an infinitely high binding energy. In reality this is not possible and even the most favourable interactions eventually break. Similarly, bonds can occur between unfavourable surfaces.

Antigen–antibody interactions

The binding of antibody to antigen is reversible and dependent on non-covalent interactions. The majority of these interactions are between the surface of the antigen and the six complementarity-determining regions (CDRs) that form the antigen-combining site of the antibody. There are several qualities of the interaction.

The **specificity** of antibodies is their most striking and useful property. An antibody raised to a particular antigen will bind to it, and only weakly, or not at all, to other antigens. Because antibodies bind to relatively small areas on the surface of antigen it is possible that an antibody may **cross-react** with a similar surface topography on a different molecule (non-specific binding).

The specificity of an antibody is a relative measure and cannot be quantified. However, the **affinity** is a quantitative measurement of the strength of the interaction between antigen and antibody. For antibodies, the range of affinities varies enormously, with dissociation constants ranging from 10^{-2} to over 10^{-12} M. A polyclonal antiserum raised in response to a particular antigen will contain antibodies with different affinities. As we saw in the previous chapter, the average affinity of the antibodies gradually increases during the course of an immune response (affinity maturation). The affinity

Table 1   Dissociation constants of some receptor–ligand interactions

Receptor	Ligand	K_d (M)
PDGF receptor	PDGF	3×10^{-11}
High affinity IL-2 receptor	IL-2	10^{-10}
Immunoglobulin (after affinity maturation)	protein antigen	$10^{-8} - 10^{-10}$
IL-1 receptor	IL-1	3×10^{-10}
Low affinity IL-2 receptor	IL-2	4×10^{-9}
P-selectin	saccharide	7×10^{-8}
CD2	LFA-3	4×10^{-7}
High affinity LFA-1	ICAM-1	4×10^{-7}
T lymphocyte receptor	peptide–MHC	10^{-5}
Low affinity LFA-1	ICAM-1	10^{-4}

and specificity of an antibody for antigen are not synonymous. Sometimes, antibodies with a high affinity for a particular epitope may also form bonds with energetically unfavourable surfaces. Similarly, some antibodies with a low affinity may be exquisitely specific for a particular antigen.

A measure of the overall stability of the interaction between antigen and antibody is **avidity.** This takes into consideration several parameters that influence the strength of the interaction, including the affinity and the valency of the antibody, and the spatial arrangements of the epitopes on the antigen. (The valency is related to the number of binding sites on the antibody molecule – IgG has two sites whereas IgM has 10.) High avidity interactions occur with antigens with multiple or repeating epitopes on their surfaces. Separate antigens can be cross-linked by antibodies acting like molecular bridges. In this way a lattice structure of many antigen and antibody molecules can be produced. The overall rate of dissociation is very slow and these complexes are relatively stable. Avidity is measured as an antibody **titre**, which is the lowest dilution of antibody or serum that still gives a visible reaction with antigen.

The contribution by the immunoglobulin to the non-covalent bonds is made almost entirely by amino acids located in the six complementarity-determining regions (or 'hypervariable' loops) although some bonds are also made from interposed framework amino acids. Individually, non-covalent bonds are relatively weak. However, as with other interactions between macromolecules, a snugly fitting interface ensures multiple non-covalent bonds can form, which together create a strong attractive force between the two surfaces.

5.1.3 Epitopes recognized by antibodies are discontinuous

The three-dimensional structure of a protein antigen is created by folding of the polypeptide chain in a particular way. Because of this, the amino acids contained within an epitope recognized by an antibody are often located in different parts of the primary amino acid sequence of the antigen (Figure 5.2), juxtaposed in the folded molecule. Epitopes recognized by antibodies are thus described as **discontinuous** and **conformation dependent.** Changing the conformation of an antigen, such as by denaturation with heat, abolishes binding of the specific antibodies. (Although, conversely, denaturation will *create* different epitopes that will be recognizable by *other* clones in the repertoire of B lymphocytes.)

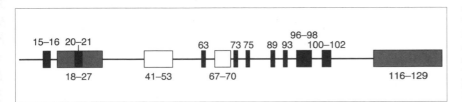

Figure 5.2 Epitopes recognized by antibodies are discontinuous. The location of the amino acids within the three epitopes depicted in Plate 2 on a linear trace of the antigen lysozyme. Notice in each case that the amino acids that form an epitope are located in different regions of the molecule. Antibodies: filled box, HyHEL-10; shaded box, D1.3; open box, HyHEL-5.

5.1.4 Natural sequence variants of antigens also show epitopes are discontinuous

Crystallography of antigen–antibody complexes is the most direct means of determining the structure of an epitope. However, only a few antigens have been studied in this way. Another less direct approach to map epitopes exploits the availability of antigens that have slightly different sequences in different animals. These model antigens include the lysozymes of different bird species, the myoglobins of different mammals and the haemagglutinin molecules on the surface of different strains of the influenza virus. By examining the reactivities of a particular antibody for different variants of its antigen, and then comparing the sequences of antigen variants that bind with those that fail to bind, it is possible to identify some of the amino acids that are part of the epitope.

For example, the anti-chicken egg lysozyme antibody D1.3 (see Plate 2) fails to bind to lysozymes from partridge, Japanese quail, turkey, pheasant and guinea fowl. These lysozymes differ from chicken lysozyme by an amino acid substitution at position 121. This particular residue is part of the epitope recognized by D1.3 and a substitution at this position for another amino acid sufficiently alters the surface topography of the epitope to reduce the binding energy to undetectable levels. By studying a sufficiently large number of variants a picture of the epitope can be assembled. Also, if the three-dimensional structure of the antigen has been determined by crystallography, the relevant amino acids can be mapped onto the structure and the rough area of the epitope seen.

The haemagglutinin molecule of influenza virus is another good example of the principle. This molecule is found on the surface envelope of the virus. Its function is to mediate the fusion of the envelope of the virus with the plasma membrane of a host cell in order to gain entry. Antibodies to this structure are generated during an infection and these provide the host with good protective immunity to subsequent infections by the same strain. However, permanent immunity to influenza viruses is particularly difficult to establish because new strains with different haemagglutinin molecules are constantly emerging (see Box 5.2).

In the laboratory, mutants that escape recognition can also be encouraged to emerge by allowing the virus to replicate *in vitro* in the presence of neutralizing antibodies. When the haemagglutinin genes of different escape mutants are then sequenced and the amino acid sequences deduced, it is possible to identify the amino acid substitutions that have arisen. Plate 3 shows where these substitutions are located on the surface of the haemagglutinin molecule. Notice they are found to cluster into four discrete epitopes. As with the lysozyme model, the neutralizing epitopes on

haemagglutinin *n.* a substance that agglutinates erythrocytes; an envelope protein of influenza virus.

As we saw in the last chapter (Section 4.5.5), the first exposure to a particular antigen elicits a gradual and transient appearance of serum antibodies, called the primary response. This precedes the maturation of the response, during which is seen antibody class switching, affinity maturation and a sustained elevation of antibodies in the blood. It is during the initial lag phase of an immune response that a host is particularly vulnerable to the disease.

Several pathogens survive in hosts by virtue of their ability to vary the structure of their surface antigens and thereby keep ahead of the developing secondary response. For simple microorganisms, such as bacteria and viruses, this is achieved by high rates of mutation that occur during replication. New strains of bacteria and viruses are constantly emerging in this way. New strains that differ in their antigenic structure may escape recognition by the memory lymphocytes established by previous infections. Other, more complex microorganisms, such as protozoa and eukaryotic worms, have evolved more deliberate means, involving the recombination of genetic information during particular stages of their life cycle in their host, to enable them to constantly side-step a developing immune response.

Influenza virus

The best example of this process in a virus comes from influenza (Figure 1). These viruses infect the respiratory epithelium of mammals and birds. A primary infection is cleared mainly by the actions of cytotoxic T lymphocytes, although antibodies are also produced against surface structures of the virus. These are the viral haemagglutinin and neuraminidase molecules that are embedded in the lipid envelope surrounding the virus, and which are used by the virus for attachment to the surface of the target cell and entry into the cytosol.

Immunity to future infection by the same strain is mediated by antibodies against the haemagglutinin molecules (and to a lesser extent the neuraminidase). This immunity is established by clones of memory B lymphocytes that remain after the primary infection. Upon restimulation from a subsequent infection, they rapidly manufacture anti-haemagglutinin antibodies. These both block viruses from binding to its receptor on respiratory epithelium and also opsonize the virus particles for removal by phagocytes.

However, influenza and other viruses with RNA genomes have very high mutation rates. That of influenza is 10 000 times greater than that of an average eukaryotic gene. As a consequence the haemagglutinin molecule gradually accumulates two or three amino acid substitutions a year. This process is called **antigenic drift**. Alterations within

BOX 5.2 Antigenic variation

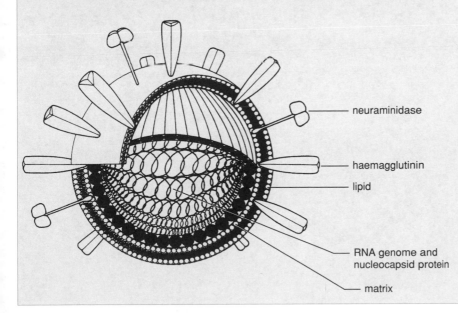

Figure 1 Influenza virus showing the haemagglutinin and neuraminidase molecules protruding from the lipid envelope. Reproduced from Collier, L. and Oxford, J. (1993) *Human Virology*, by permission of Oxford University Press.

neuraminidase

haemagglutinin

lipid

RNA genome and nucleocapsid protein

matrix

BOX 5.2 *continued*

epitopes recognized by neutralizing antibodies may allow such a mutant to escape recognition by the memory cell and elicit another primary infection (see Plate 3 in the text). Disease caused by influenza thus occurs in waves or epidemics, each caused by a novel antigenic form that escapes neutralization by memory B lymphocytes. Mammalian influenza may also exchange entire regions of genetic information with other human influenza or related avian influenza strains when both co-exist within the same host. This causes dramatic and stepwise changes in the surface antigenic structures. This process, called **antigenic shift**, also contributes to the emergence of novel strains and evasion of pre-existing immunity.

Gram-negative bacteria

Bacteria also display this form of escape. Gram-negative bacteria have complex cell walls surrounded by a lipid membrane bearing chains of lipopolysaccharide (LPS) constructed of lipid and carbohydrate (see Figure 1 in Box 2.1).

The lipid moiety is buried in the outer membrane and anchors a chain of sugar residues called the **core polysaccharide**. Attached to the core polysaccharide is a shorter polysaccharide called the **O-side chain.** The O-side chains, although readily accessible to antibodies, are very variable within the same species of bacteria, giving rise to different strains. Antibodies elicited to one strain will tend to provide poor protection against other strains. Other variable structures are the short, hair-like processes of Gram-negative bacteria called **fimbriae** (or **pili**) made of a protein called **pilin.** These are the targets of antibodies. Strains of human pathogens such as *Neisseria gonorrhoeae* and *Pseudomonas aeruginosa* are able to evade opsonization or complement fixation by antibodies by changing the structure of their pilins.

Protozoan parasites

True antigenic variation as an escape strategy involves the deliberate changing of antigenic structures during existence within the same host. The best known example is the **variable surface glycoprotein** that covers the entire surface of the protozoan parasite *Trypanosoma brucei*. As we learned in Box 2.2, the parasite has several copies of the surface glycoprotein gene, each encoding a slightly different version.

Figure 2 The life cycle of *Plasmodium*. Sporozoites are injected into a blood vessel by a mosquito. Sporozoites enter liver hepatocytes, where they replicate to produce many merozoites. These are released into the blood, where they penetrate erythrocytes, in which further replication takes place. This cycle is repeated as the merozoites are released from ruptured erythrocytes. Male and female stages are also produced in the blood which are taken up by mosquitoes when they feed. The male and female gametes combine in the mosquito gut to produce a zygote and then an oocyst in the gut wall. Further multiplication produces more sporozoites that enter the salivary glands ready for transmission to a human host. Adapted from Cox, F.E.G. (1992) *Nature*, **360**, 417–18.

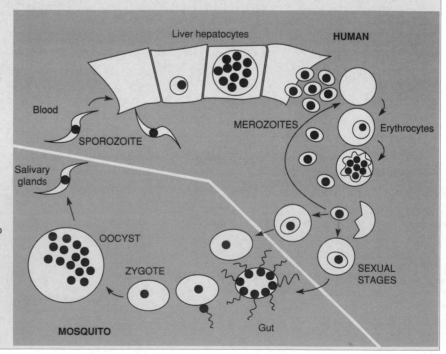

During cell division, gene conversion may occur in some of the trypanosomes, causing a copy to be duplicated into the actively transcribed locus, causing a switch in the gene expressed there. In this way, some of the progeny are surrounded by a different surface glycoprotein. Switching occurs at low frequency and the majority of the parasites are gradually eliminated as the immune response matures. However, some minor variants bearing a switched surface glycoprotein survive and soon dominate. Gradually these too are eliminated, only to be replaced by more variants. The disease – African trypanosomiasis (or sleeping sickness) – therefore occurs in weekly waves of blood-borne parasites (**parasitaemia**).

A similar effect is achieved by protozoan parasites that undergo cycles of morphological changes during their life cycle. The parasite that causes malaria in humans, *Plasmodium falciparum*, is a good example of this (Figure 2). Each stage has a distinct set of antigens to which the human host responds.

influenza haemagglutinin are located on the surface of the antigen, and each comprises discontinuous amino acids. Naturally, spontaneous substitutions also occur inside the haemagglutinin molecule, but since these are inaccessible to antibodies they contribute very little to escape from recognition by antibody.

The discontinuity of amino acids within epitopes is not unique to antigen–antibody interactions. Two macromolecules that need to interact have surfaces that are complementary in shape to each other. Amino acids at the surface of the interfaces make contact with the opposing interface. Because of the globular nature of many such proteins, is not unusual to find that critical contact residues that make up these sites are juxtaposed only in the folded molecule. Examples include the active site of an enzyme, the DNA-binding site of a transcription factor, the binding site of a viral surface protein for its receptor on a cell membrane, or the complementarity-determining regions of an antigen receptor.

An immunoglobulin has the capacity to bind specifically to a particular antigen but with lower binding energy to unrelated antigens, even those with very similar conformations. Clearly, complete denaturation of a protein will abolish binding by one particular antibody. Binding can also be abolished by far more subtle changes to the conformation of the epitope, such as by substitution of a single amino acid. This ability to discriminate between epitopes with very similar topographies is typical of antibodies and has meant that antibodies elicited in animals to particular antigens have proved to be remarkably useful experimental probes with applications in virtually every branch of biological and medical research.

5.1.5 Antibodies as a tool

The exquisite specificity of an antibody binding to its antigen, and the ease with which antibodies specific for a defined antigen can be produced, purified, stored and transported, have been exploited to yield a variety of useful diagnostic techniques, collectively called **immunoassays**. The

Figure 5.3 Principles of immunoassays.

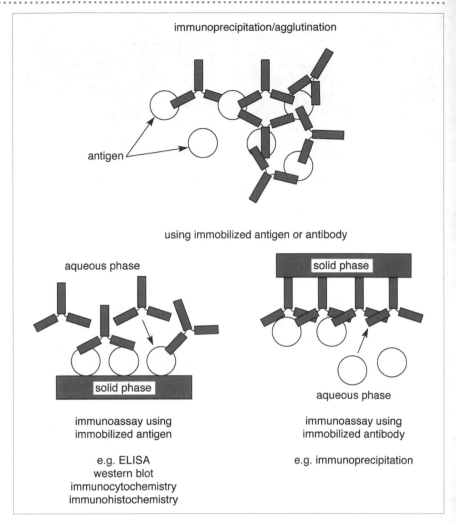

immunoprecipitation/agglutination

antigen

using immobilized antigen or antibody

aqueous phase

solid phase

solid phase

aqueous phase

immunoassay using
immobilized antigen

immunoassay using
immobilized antibody

e.g. ELISA
western blot
immunocytochemistry
immunohistochemistry

e.g. immunoprecipitation

details and uses of immunoassays are very diverse, although the basic principle of all of them is similar: provided one component of the antigen–antibody interaction is available, it can be used to specifically detect the presence of the other within an undefined or complex mixture.

The interaction is made to occur wholly in a liquid or gel medium, or more commonly with either the antigen or the antibody immobilized, as represented in Figure 5.3. Since one of the components is defined, there are essentially two categories of immunoassay, represented in Figure 5.4: those where a defined antigen is used to detect the presence of its antibody (antigen-based) and those where an antibody of defined specificity is used to detect the presence of its antigen (antibody-based).

The applications of antibodies are numerous. Detection of antibodies to defined antigens or pathogens is used as an indicator of a current or previous history of infection with the pathogen. For example, the current test for

HIV infection relies on the detection of antibodies to the virus, rather than the virus itself. Antibody-based assays, when used at their simplest, are used for detecting the presence of a particular antigen in biological material. For example, human pregnancy tests are based on the detection of human chorio-gonadotrophic (HCG) hormone in urine using HCG-specific antibodies. Antibody-based assays can also be used to characterize the antigen (determine its molecular weight, quantity, purity, conformation, relatedness to other antigens and so on) or to physically extract the antigen from complex mixtures of molecules.

Seeing the interaction

A requirement for all immunoassays is that the interaction between antigen and antibody is made visible in some way. There are essentially two 'read-out' systems.

One exploits the way in which an antibody, by virtue of having two identical antigen-combining sites, can cross-link antigens (Figure 5.3). If the ratio of antigen and antibody concentrations is optimal, extensive three-dimensional networks called **immune complexes** are formed. These are readily visible: small soluble antigens produce cloudy precipitates, whereas larger antigens, such as cells or antigen-coated beads, form clumps (**immunoprecipitation** and **agglutination**, respectively).

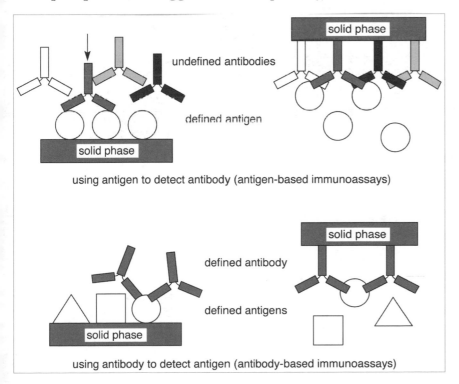

using antigen to detect antibody (antigen-based immunoassays)

using antibody to detect antigen (antibody-based immunoassays)

Figure 5.4 Many immunoassays feature immobilization of antibody or antigen.

Table 5.1 Some commonly used immunoassays and immunodetection techniques (see Appendix B for details)

Assay	Nature of antigen	Method of visualization	Major applications
ELISA (enzyme-linked immunosorbent assay)	Usually purified protein samples; first adsorbed onto plastic multiwell plates.	Primary antibody, followed by secondary antibody coupled to chromogenic enzyme.	If the antigen is defined, can be used to estimate the titres of specific antibody in serum by using serial dilutions.
Immunocytochemistry	Cell-surface antigens; cells are first allowed to adhere to glass surface (usually a microscope slide).	Ditto.	Used for counting cell phenotypes in heterogeneous cell populations (usually by light microscopy).
Immunohistochemistry	Whole tissue; sections are first cut from embedded tissue on a microtome.	Ditto.	Used to visualize the architecture of antigens in tissue.
Western blotting	Usually complex protein mixtures (cell lysates, etc.); first resolved into individual proteins by electrophoresis, and then blotted onto nitrocellulose membranes.	Ditto; can use radiolabelled secondary antibody visualized by autoradiography.	If antibody is defined, can be used to detect the presence of particular antigen in complex mixtures; can also provide the molecular weight of the antigen.
Immunofluorescence	(1) Surface antigens on live cells in suspension. (2) Intracellular antigens in permeabilized, adherent cells. (3) Antigens in tissue sections.	Primary antibody, followed by secondary antibody coupled to fluorochrome.	Used for counting cell phenotypes in heterogeneous cell populations (usually by flow cytometry); also to visualize the distribution of antigens in cells (cytoskeletal filaments, etc.) or tissue sections.
Radioimmunoprecipitation assay (RIPA)	Usually complex protein mixtures (cell lysates, etc.); proteins are first radiolabelled and antigen is then captured using immobilized antibody ('immunoprecipitated'); immune complexes then resolved into separate polypeptides by electrophoresis.	Autoradiography	Like the Western blot, can be used to detect the presence and molecular weight of particular antigen in complex mixtures, with the advantage that the antigen is not denatured first.

The other way to visualize the antigen-antibody interaction is to immobilize one component of the interaction and incubate it with the other component in the fluid phase that has been 'tagged' to disclose its presence (Figure 5.5). Antibodies can be labelled with fluorescent dyes, chromogenic enzymes, or radioactive isotopes. These techniques are summarized in Table 5.1 and detailed in Appendix B.

Techniques using antibodies tagged with fluorochromes are types of immunofluorescence. Fluorochromes, such as fluorescein and rhodamine, emit a particular wavelength in the visible spectrum (i.e. of a particular colour) upon excitation with UV light. Antibodies conjugated to fluorochromes are widely used for detecting cell-surface or intracellular antigens, which are then visualized by fluorescence microscopy or quantitated in a flow cytometer (a machine that measures the level of fluorescence on individual cells in suspension stained by immunofluorescence). Chromogenic enzymes, such as peroxidase or phosphatase, convert a colourless substrate into a coloured compound. These are used in Western blotting, ELISAs and other immuno-detection methods in which the antigen is immobilized on a solid support. Radioactive isotopes are the third major type of label, and can be coupled to either antibody or antigens. In radio-immunoprecipitation assays, for example, cellular antigens are labelled. Alternatively, antibodies are radiolabelled and can be an alternative to enzyme-conjugated antibodies in Western blotting. Either way, the immune complexes are normally precipitated. Radioactivity can be measured directly, or after the proteins have been separated by electrophoresis, and the radioactive bands identified by exposing the gel to photographic film (autoradiography).

Antibodies, like any other protein, are themselves antigens and can be bound by another antibody. While in some antibody-based assays the antibodies can be tagged directly, it is more common to use a tagged **secondary antibody** that binds to the constant regions of the antigen-specific, primary antibody (Figure 5.5). The antigen is thus detected indirectly. Secondary antibodies have to be produced in a xenogeneic species (i.e. a different species to the source of the primary antibody) since animals generally fail to respond to their own proteins. Thus if the primary antibody has been produced in a rabbit, a labelled anti-rabbit antibody, produced perhaps in a goat, is then used to detect it. There are advantages to indirect detection. Since secondary antibodies bind to the constant regions of primary antibodies, they can be used with a variety of different primary antibodies. Moreover, several secondary antibodies can bind to a single primary antibody, which amplifies the signal.

A requirement in all immunoassays is the inclusion, where possible, of

flow cytometer *n.* a machine for quantitating the level of fluorescence on individual cells in suspension stained by immunofluorescence (see **fluorescence activated cell sorter**).

secondary antibody *n.* an antibody that is coupled to a fluorescent dye, radioisotope, or chromogenic enzyme, which is used in immunoassays to visualize the primary antibody bound to antigen.

xenogeneic *a.* two genetically dissimilar individuals of different species.

Figure 5.5 Direct and indirect visualization of antigen.

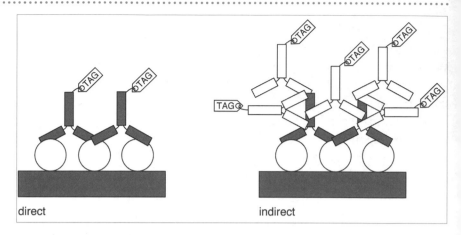

antigen-negative and antigen-positive controls to confirm that primary antibody binds specifically to the intended antigen. Ideally, the primary antibody should bind to the antigen-positive control and not the antigen-negative control. Similarly, the specificity of binding by the secondary antibody, if one is used, to the primary antibody, is also controlled. This is usually achieved by using the secondary in the absence of the primary antibody. Any binding seen here is considered to be 'background'.

Production of antiserum

Antibodies of a defined specificity are produced by immunizing mammals with the desired antigen (Figure 5.6). Rabbits, goat and swine are probably the most widely used animals. Antigens can be in the form of whole cells, or as proteins. Synthetic peptides, when coupled to a larger carrier protein, can also be used to elicit anti-peptide-specific antibodies. This is particularly useful if all that is known of a given protein is the sequence of its cloned gene, from which its amino acid sequence can be inferred.

Protein and peptide antigens are usually administered in an **adjuvant**. Adjuvants perform two functions. One is to retain the antigen at the site of injection and prevent it dissipating too rapidly. The other is to act as a local

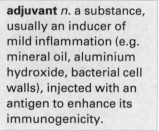

adjuvant *n.* a substance, usually an inducer of mild inflammation (e.g. mineral oil, aluminium hydroxide, bacterial cell walls), injected with an antigen to enhance its immunogenicity.

Figure 5.6 Production of polyclonal antiserum.

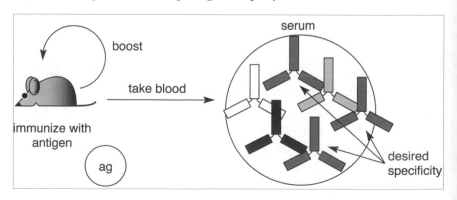

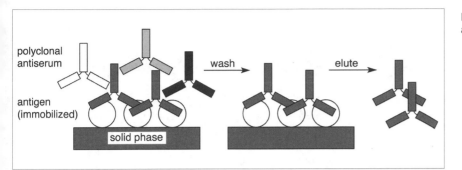

Figure 5.7 Affinity purification of antibodies.

irritant, causing an inflammatory infiltration of leucocytes and thereby promoting the exposure of the antigen to the immune system. High titres of serum antibodies are produced by periodic re-immunization ('boosting') with the antigen.

Despite high titres of the desired antibodies, antiserum also contains the residual background of other antibodies present before immunization. These may cause problems of high background in immunoassays. It is often desirable therefore to remove these from serum by **affinity purification** (Figure 5.7). Here, the antigen used for immunization is immobilized, usually on microscopic beads held in a column through which the immune serum is passed. Non-specific antibodies pass through whereas the antigen-specific antibodies are retained. After washing the beads, the antibody can be released from the column by raising the ionic concentration.

Antiserum is the product of a polyclonal B lymphocyte response, consisting of antibodies with varying affinities and with specificities directed towards different epitopes on the surface of the antigen. Often it is desirable to have antibody from a single B lymphocyte clone. This is achieved by cloning antibody-secreting cells *in vitro* and selecting the clone(s) producing antibody with the desired properties. These **monoclonal antibodies** are conventionally produced in mice or rats, illustrated in outline in Figure 5.8. The first step is to elicit a polyclonal response *in vivo* by immunization. Antibody-secreting plasma cells are then removed from the spleen and then immortalized by fusing them with myeloma cells, a cancer of non-immunoglobulin-secreting B lymphocytes. The fused hybrid cells or **hybridomas** have the dual properties of continuous proliferation and antibody secretion. The hybridomas are then cloned. This is achieved by suspending low numbers of hybridoma cells in tissue culture medium and placing small volumes into separate wells. At sufficiently low dilutions, most of the wells will be empty although some should contain single cells. Under appropriate growth conditions these single cells will proliferate. This is cloning by **limiting dilution**. Subsequently the antibodies secreted by each clone are screened by testing a little of the culture medium in the desired

affinity purification *n.* the use of antibodies immobilized on a column of beads to capture antigen from a complex mixture.

hybridoma *n.* a **clone** of hybrid cells formed by the fusion of a plasma cell and a myeloma cell (bone marrow-derived cancer cell); used in the production of **monoclonal antibodies**.

Figure 5.8 Production of monoclonal antibodies.

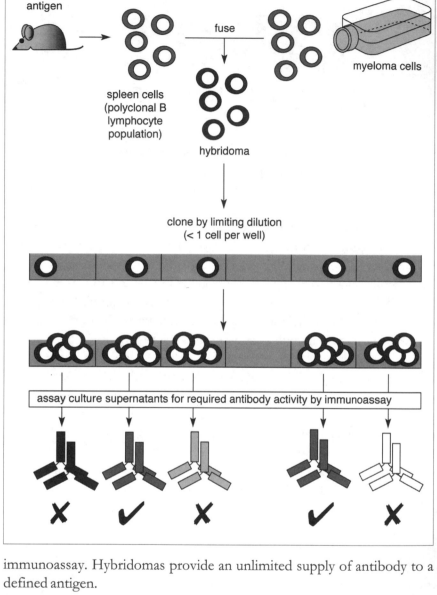

immunoassay. Hybridomas provide an unlimited supply of antibody to a defined antigen.

Summary

- Epitopes, as defined by antibodies, are located on the surface of antigens. The surface of any protein can be regarded as a mosaic of overlapping epitopes, each defined by a B lymphocyte clone.

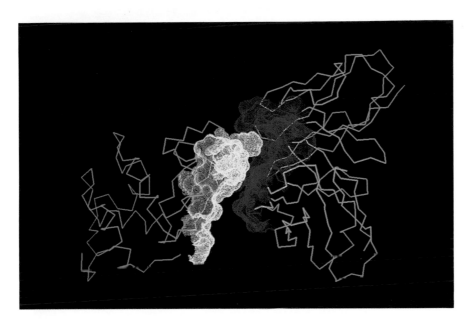

Plate 1 Crystal structure of an antigen–antibody interaction.

This is a complex of lysozyme (left) and a mouse anti-lysozyme monoclonal antibody called Hy HEL-5 (only the variable domains of one antigen-binding site are shown). The blue lines are α-carbon backbones of the polypeptide chains. The surface on the antigen defined by the antigen-binding site is called the epitope (yellow) and the surface of the two variable domains in contact with the antigen (red) are formed by the six complementarity-determining regions. The two molecules have been pulled apart slightly to reveal the complementarity of their surfaces. Data from Sheriff, S. *et al* (1987) *Proc. Natl. Acad. Sci. USA* **84**: 8075–9. This and other plates produced by Paul Hobby from data in the Protein Data Bank at the Brookhaven National Laboratory

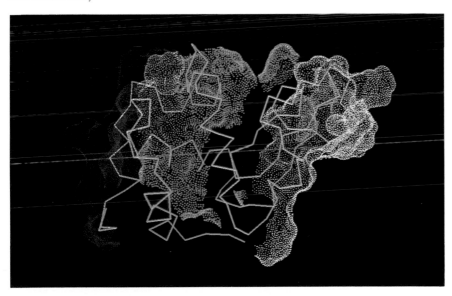

Plate 2 Epitopes on the surface of lysozyme recognized by three different anti-lysozyme monoclonal antibodies.

The epitope recognized by D1.3 is coloured red, Hy HEL-5 coloured in yellow, and Hy HEL-10 coloured in green. The brown area is an overlap between epitopes recognized by Hy HEL-10 and D1.3. Data from Amit, A. G. (1986) *Science* **233**: 747–53; Sheriff, S. *et al* (1987) *Proc. Natl. Acad. Sci. USA* **84**: 8075–9; Padlan, E. A. (1989) *Proc. Natl. Acad. Sci. USA* **86**: 5938–42.

Plate 3 Natural variation in haemagglutinin molecules of influenza virus.

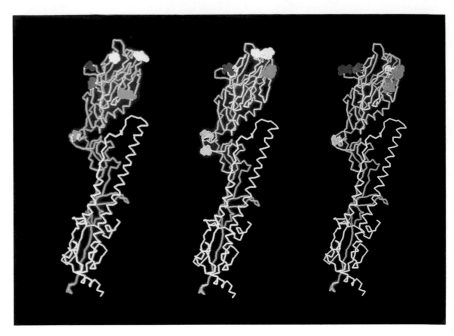

The HA$_1$ chain is shown in blue and the HA$_2$ chain, which anchors the molecule in the viral membrane, in yellow (see BOX 5.2). Marked are the surfaces of amino acids that are different from haemagglutinin of the 'prototype' strain of 1968 and natural variants that caused the epidemics of 1972 (left) and 1974 (centre), and variants caused after selection with neutralizing antibodies *in vitro* (right). These amino acid changes are found to cluster into four discrete epitopes on the globular head formed by the HA$_1$ chain: site A, red; site B, yellow; site C magenta; site D, blue. Data from Wiley, D. C. *et al* (1981) Nature **289**: 373–8.

Plate 4 Three views of the human MHC molecule, HLA-A2.

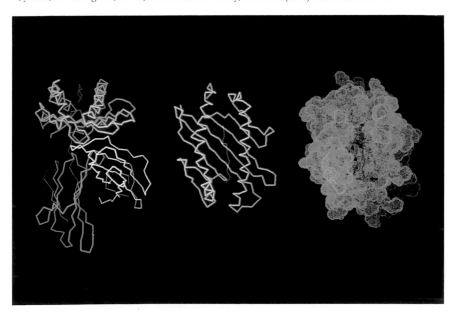

Left is the 'side' view (with the plasma membrane of the antigen presenting cell to the bottom of the plate), and centre is the view of the molecule from above showing the peptide-binding cleft. The α-carbon backbones are depicted: α chain, blue; β2-microglobulin, yellow; antigenic peptide, red. Right shows the surface of the molecule viewed from above. Based on data collected by Madden, D.R. *et al* (1993) Cell **75**: 693–708

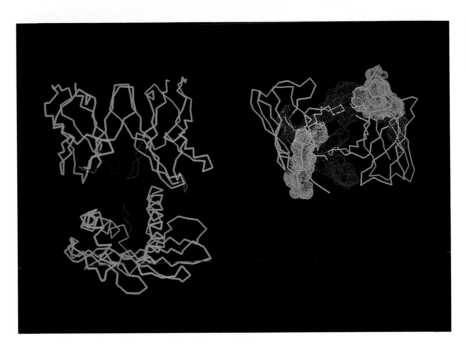

Plate 5 A model for the interaction between the T lymphocyte receptor and a peptide/MHC complex.

A T lymphocyte receptor, left, is represented by an immunoglobulin which is shown positioned over the peptide-binding cleft of an MHC molecule. Only the V domains of the immunoglobulin (V_H and V_L) are shown. The complementarity determining regions (CDRs) of the T lymphocyte receptor are thought to interact with the surface of the peptide/MHC complex as shown; CDR1, green; CDR2, purple; CDR3, red. Right shows the surfaces of the six CDRs. The immunoglobulin V domains have been rotated to show the face of the antigen-binding site.

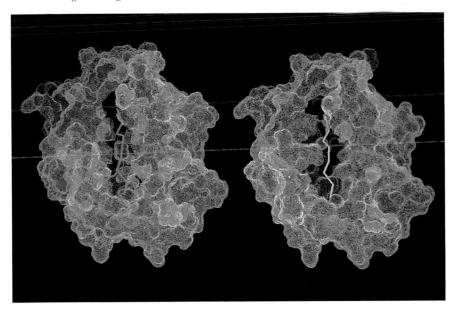

Plate 6 The mouse MHC molecule, K^b, containing two different peptide epitopes.

Left is an epitope from Sendai virus (the same one mapped in Figure 5.14), and right is an epitope from vesicular stomatitis virus. Although the same MHC molecule is involved, the surface topography of the complex is slightly different. See Fremont, D. H. *et al* (1992) *Science* **257**: 919–27.

- The specificity of an immunoglobulin for a particular antigen is dependent on the surface conformation of the epitope. This is lost if amino acids within the epitope are substituted or if the antigen is denatured.

- The ability to produce antibodies that bind to defined antigens has led to the development of many kinds of immunoassays.

5.2 The ligand of the T lymphocyte receptor

We now turn to the antigen recognition by T lymphocytes. Our current understanding of T lymphocyte recognition came almost entirely from studies with mice and humans. Progress has come in part from the development of ways to propagate T lymphocytes and measure or assay their responsiveness to antigen *in vitro*. These assays show that T lymphocytes, like B lymphocytes, display specificity for antigen. In many other respects, however, antigen recognition by T lymphocytes is very different from that by B lymphocytes. Let us first examine the assays that measure antigen recognition by T lymphocytes.

5.2.1 *In vitro* propagation and assay of T lymphocytes

As we saw above, assays that measure antigen recognition by B lymphocytes are based on detecting antigen–antibody interactions. Because T lymphocytes do not secrete their antigen receptors, recognition of antigen is measured instead by the *response* a T lymphocyte makes upon encountering antigen (see below).

In common with antibodies, measuring T lymphocyte responses requires previous immunization to stimulate clonal proliferation. Otherwise the responder frequency is too low to detect any biological activity *in vitro*. For T lymphocytes, this is also accomplished *in vivo* either by deliberate immunization, or by natural infection with a microorganism. Several days after immunization, blood or secondary lymphoid tissue is removed and lymphocytes cultured with the antigen *in vitro* for a few days. This further stimulates the antigen-specific T cells to proliferate. Non-responder cells remain less active and soon die in culture.

Antigen-stimulated T lymphocytes can be maintained *in vitro* for several weeks by periodic restimulation with antigen and essential growth factors, particularly **interleukin 2**. Interleukin 2 is produced by helper T lymphocytes, and it is needed for the antigen-driven proliferation of all types of T lymphocytes. After a few cycles of antigenic stimulation, the population –

Figure 5.9 Principles of proliferation and cytotoxicity assays for mammalian T lymphocytes. (a) T lymphocyte proliferation assay. An animal is immunized with antigen, either by natural infection or deliberately. Cells for proliferation assay are obtained from the secondary lymphoid tissue, such as lymph nodes, draining the site of immunization. Stimulated T lymphocytes are re-exposed to the same antigen *in vitro*, in the presence of accessory cells (explained in Section 5.3.4). As a control, cells are cultured without antigen or with a different antigen. Proliferation is detected by the incorporation of a labelled nucleotide such as [^{3}H]thymidine into the nuclei of daughter cells, which is measured by scintillation counting.

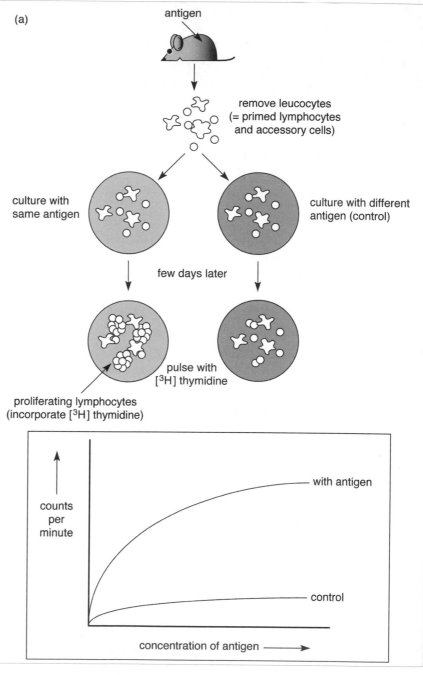

called a polyclonal or bulk T lymphocyte line – becomes progressively more enriched with responding T lymphocytes.

The major types of measurable responses displayed by activated T lymphocytes *in vitro* are **proliferation** and **cytotoxicity**, illustrated in Figure 5.9.

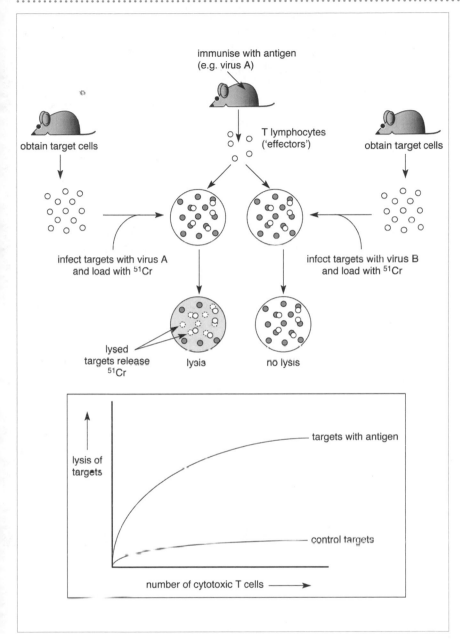

Figure 5.9 continued
(b) Cytotoxicity assay. Virus-specific cytotoxic T lymphocytes are stimulated by immunization *in vivo* with virus. Their presence is detected by culturing spleen cells with target cells. Targets are labelled internally, usually with ^{51}Cr, and infected with the immunizing virus. Control targets are infected with a different virus. The quantity of label released into the culture medium by targets is proportional to the amount of lysis. At the end of the assay the cells are removed by centrifugation and the label in the medium quantified.

- A proliferative response is usually detected by the incorporation of labelled nucleotides into the newly replicated DNA in daughter cells. Tritium ^{3}H-labelled thymidine is often used. After a few hours, the unincorporated label is washed out and the amount of radioactive DNA incorporated into the cells quantitated from the β-emission of the cell cultures. The amount of radioactivity incorporated is proportional to the amount of proliferation.

target cell *n.* the cell lysed by cytotoxic T lymphocytes, NK cells, etc.

- A cytotoxic T lymphocyte response is detected *in vitro* using **target cells** bearing the appropriate antigen. Killing of the target cells by cytotoxic T lymphocytes is measured by labelling the cytoplasm of the target cells, usually with radioactive sodium chromate ($Na_2{}^{51}CrO_4$). This is taken up by cells in a hexavalent form which then binds to low molecular weight intracellular proteins. If a target lyses, a trivalent form of the radioactive chromate is ejected into the surrounding fluid, where the radioactivity can be measured from the β- or γ-emission. The amount of radioactivity released is proportional to the number of targets that have lysed.

A third group of assays is based on the property of all T lymphocytes to secrete soluble hormone-like molecules, called **cytokines**, upon antigen-specific activation. The function of cytokines *in vivo* is to modulate the behaviour of other cells in the immediate vicinity. There are several different cytokines and cytokine assays. For example, the measurement of interleukin 2 released into the culture medium can be as an indicator of helper T lymphocyte activation. As we will see in a later chapter, different subsets of helper T lymphocytes release different mixtures of cytokines, according to their particular functions.

Cytokines define two subsets of helper T lymphocytes: Section 8.2.2

5.2.2 Response *in vitro* correlates with function and phenotype

These two major functionally distinct groups of T lymphocytes also correlate with the two major phenotypes of T lymphocytes, based on the expression of CD4 or CD8 molecules on their surfaces. As we learned in Chapter 3, mature T lymphocytes bear one or other of these molecules, in a mutually exclusive fashion. $CD4^+$ T lymphocytes are predominantly helper cells, whereas $CD8^+$ cells are predominantly cytotoxic T lymphocytes. Thus, proliferation assays generally detect the responses by CD4 lymphocytes and cytotoxicity assays generally measure responses by CD8 cells. This is summarized in Table 5.2.

5.2.3 Antigen specificity

T lymphocytes show specificity for an antigen in proliferation and cytotoxicity assays in much the same way as antibodies in immunoassays. For example, clones of murine helper T lymphocytes stimulated first *in vivo* to an antigen, such as hen egg albumin, will proliferate to the same antigen *in vitro,* but not to a different antigen, such as hen egg lysozyme. Similarly, cytotoxic T lymphocytes that are stimulated *in vivo* by virus A, influenza virus for example, will kill target cells infected with virus A *in vitro*, but not cells infected with virus B, such as rhinovirus. Responses by all the other,

rhinovirus *n.* common cold viruses.

Table 5.2 Properties of T-lymphocyte subsets

	T lymphocyte	
	CD4	CD8
Function	helper cell, mediator of delayed-type hypersensitivity	cytotoxic T cell
In vitro assay	proliferation	cytotoxicity
MHC restriction	class II	class I
Antigen-presenting cell	dendritic cell macrophage B lymphocyte	any nucleated cell

unactivated clones in the bulk cultures remain undetectable. We will revisit the phenomenon of the antigen specificity when we have examined the molecular details of the ligand of the T lymphocyte receptor. Suffice to say here, like the specificity of immunoglobulin, it is based on the conformation of the ligand.

5.2.4 T lymphocytes require antigen-presenting cells to respond

In vitro assays demonstrated a distinctive feature of antigen recognition by T lymphocytes. T lymphocytes have an absolute requirement for the presence of accessory cells. These will be defined in a moment. In the absence of accessory cells, T lymphocytes are unable to respond to antigen.

Manipulating the assays, as depicted in Figure 5.10, showed that T lymphocytes respond normally if the accessory cells are incubated in the antigen first, washed free of antigen and then incubated with the T lymphocytes. In contrast, preincubating the T lymphocytes in antigen, washing it out, and then adding the accessory cells does not elicit a response. Together, these experiments showed that the accessory cells were necessary to present the antigen to the T lymphocytes when the two cells make contact. These accessory cells are therefore called **antigen-presenting cells.** The inability by T lymphocytes to respond to free antigen is a major difference between antigen recognition by T and B lymphocytes.

The requirement for antigen-presenting cells is the same for all T lymphocytes, irrespective of their phenotype. Cells that present antigen to activated CD8$^+$ (cytotoxic) T lymphocytes are lysed in return. As we have seen, these are called 'targets' in cytotoxicity assays. *In vivo*, any nucleated cell in the body can fulfil this role, presumably because they all have the potential to be transformed by viruses or by mutations. The cells able to present antigen to CD4$^+$ lymphocytes are less widespread in the body, and are mainly dendritic cells, macrophages and activated B lymphocytes. These

antigen presentation the display of processed antigenic peptides at the surface of an antigen-presenting cell to the antigen receptors of T lymphocytes.

Figure 5.10 The requirement for antigen presentation to T lymphocytes. Activated T lymphocytes do not respond to the appropriate antigen as shown unless accessory cells are present (compare a and b). The antigen-presenting function of the accessory cells is revealed if they are preincubated with the antigen prior to culture with the T lymphocytes (c). In contrast, preincubation of the T lymphocytes prior to culture with accessory cells has no effect (d). The same principles apply to the cytotoxic T lymphocyte response.

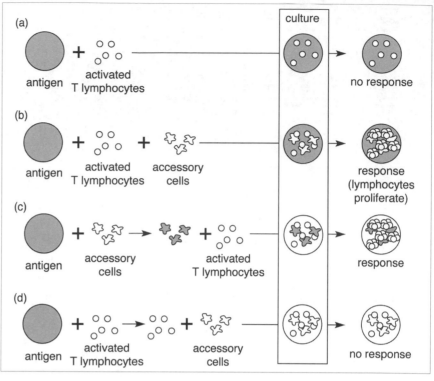

professional antigen-presenting cell antigen-presenting cells that can activate naïve T lymphocytes; constitutively express class I and class II MHC molecules, and T lymphocyte costimulatory molecules; especially cells of the dendritic cell family and activated B lymphocytes.

cells are often described as **professional antigen-presenting cells** to reflect this special status. Presentation of antigen to CD4$^+$ T lymphocytes by professional antigen-presenting cells is *in addition* to their conventional capacity to present antigen to CD8$^+$ lymphocytes.

5.2.5 Antigen is 'processed' by antigen-presenting cells

Why do T lymphocytes require antigen-presenting cells? Several lines of evidence show that antigen-presenting cells do more than simply present the antigen to T lymphocytes. Shown in Figure 5.11 is a proliferation assay in which the antigen-presenting cells are preincubated in antigen, followed by arrest with a metabolic poison after only 5 min. This treatment renders them unable to stimulate proliferation by the activated T lymphocytes. If on the other hand metabolism is arrested 60 min after preincubation with the antigen, they successfully elicit proliferation from the T lymphocytes. This demonstrates that the preparation of antigen suitable for recognition takes time (30–60 min) and requires metabolic energy. A similar situation exists for presentation to cytotoxic T lymphocytes.

We now know that during this lag period, the antigen has to be broken down inside the antigen-presenting cell into peptide fragments. These are then brought back out of the cell and presented at its surface to the antigen receptors of T lymphocytes. There are several lines of evidence for this.

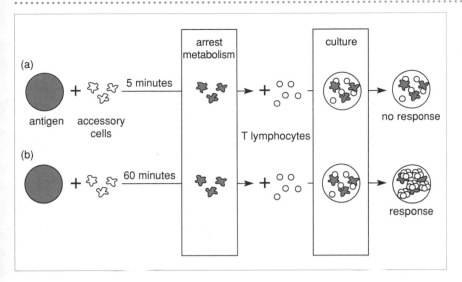

Figure 5.11 Antigen presentation requires time and metabolic energy.

- Unlike B lymphocytes, T lymphocytes are indifferent to the native conformation of the antigen. They respond equally well to native and denatured antigen.

- Proliferation by CD4$^+$ helper T lymphocytes in response to antigen is blocked if the antigen-presenting cells are treated with drugs that inhibit the activity of proteinases. These include the **lysosomotropic agents**, which interfere with proteolysis within endosomes.

- Antigen recognition by CD8$^+$ cytotoxic T lymphocytes also entails digestion of the antigen by proteases. However, this appears to occur within a different intracellular compartment, since lysis of target cells is insensitive to lysosomotropic agents.

- Antigen recognized by a T lymphocyte can be replaced in the assay by peptide fragments of the antigen.

This latter observation was first made in proliferation assays of helper T lymphocytes responding to ovalbumin. The antigen could be replaced in the assay by incubating the antigen-presenting cells in ovalbumin that had been previously cleaved into large fragments with cyanogen bromide. It was subsequently shown that influenza virus-specific cytotoxic T lymphocytes were also to be able to kill target cells incubated with a synthetic peptide that mimicked a fragment of a viral antigen.

5.2.6 The T lymphocyte epitope is a peptide

Shown in Figure 5.12 is an experiment in which synthetic peptides have been tested for recognition by a clone of Sendai virus-specific cytotoxic T lymphocytes. Recall that a clone is a population descended from a single

lysosomotropic agent *n.* substances (chloroquine, etc.) that raise the pH of lysosomes and other compartments of the endocytic pathway; used in experiments to block processing of exogenous antigens.

ovalbumin *n.* main protein component of hen egg white.

antigen processing intracellular events involving the enzymatic degradation of antigens and the formation and transport to the cell surface of complexes between processed peptides and class I or class II MHC molecules.

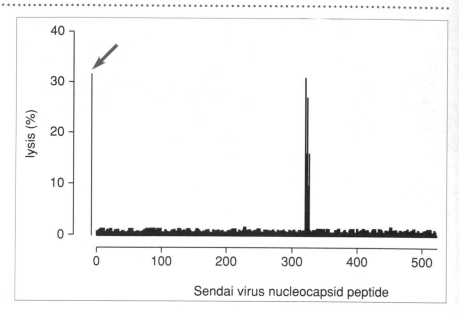

Figure 5.12 Mapping the epitope recognized by a cytotoxic T lymphocyte clone using synthetic peptides. Every possible 12 amino acid-long peptide spanning the protein was synthesized (i.e. amino acids 1–12, 2–13, 3–14 and so on until 513–524) and tested individually by incubating each with uninfected target cells. These targets were then presented to the T lymphocyte clone. Each bar represents a single peptide. A target infected with Sendai virus (arrowed) was used as the positive control. A cluster of peptides around the region 321–336 are recognized by the clone, revealing this region is the epitope recognized by the antigen receptor of this clone. Based on data from Kast W.M., *et al.* (1991) *Proceedings of the National Academy of Sciences USA*, **88**, 2283–7 with permission from the author.

cell, and that each cell has an identical antigen receptor. The clone recognizes an epitope in the nucleoprotein of the virus because the only gene of the virus that, when introduced into targets cells, rendered them susceptible to lysis by the clone, was the gene encoding the viral nucleoprotein.

In the experiment, every possible 12 amino acid-long peptide spanning the nucleoprotein was synthesized (i.e. amino acids 1–12, 2–13, 3–14 and so on, until 513–524) and tested individually by incubating each with uninfected, chromium-labelled target cells and presenting them to the T lymphocyte clone. Notice a cluster of peptides around the region 321–336 were recognized by the clone, thereby identifying this region as the epitope recognized by the clone's receptors.

For helper T lymphocytes too, peptides are used to locate epitopes within antigens that they recognize although the proliferation assay is used.

5.2.7 Antigenic peptides are presented by MHC molecules

It is now necessary to examine the molecules on the surfaces of antigen-presenting cells that present the antigenic peptides to T lymphocytes. These are called **MHC molecules** because their genes are found clustered together in a region called the **major histocompatibility complex.** The discovery of the mammalian MHC and its role in antigen presentation is the subject of the next chapter. Until then, we will jump several decades of research to the late 1980s, when the first MHC molecule was crystallized.

The first to be crystallized was the human MHC molecule called HLA-A2 (Plate 4). Since then, several other different MHC molecules from humans

antigen presentation
the display of processed antigenic peptides at the surface of an antigen-presenting cell to the antigen receptors of T lymphocytes

and mice have been crystallized. On the face of each, pointing away from the plasma membrane of the antigen-presenting cell, is a deep cleft formed between two α-helices surmounting a platform of eight β-strands. Located within this cleft is located a single antigenic peptide. It is the combination of the peptide and the surrounding upper face of the α-helices of the MHC molecule that together form the ligand of the T lymphocyte receptor. The surface of this bimolecular complex has the overall appearance of a flat but irregular topography, with protuberances and depressions, and is somewhat similar in size to an epitope recognized by immunoglobulin.

5.2.8 Modelling the trimolecular complex

The trimolecular complex of T lymphocyte receptor bound to peptide–MHC has yet to be resolved by crystallography, although it is likely this will be achieved soon. Meanwhile it is possible to model the interaction. As we learned in Chapter 4, the homology between T lymphocyte receptors and immunoglobulins suggests that the T lymphocyte antigen receptor and immunoglobulin both fold in a similar way. On this basis, the antigen-combining site of the T lymphocyte receptor should have at least six CDR-like hypervariable loops, three on each chain. Moreover, the face of the peptide–MHC complex is approximately the same size as the face of the antigen-combining site. This has enabled models of how the interaction between T lymphocyte receptor and peptide–MHC complex might look (Plate 5).

Hypervariable regions make contact with antigen: Section 4.2.1

There is evidence that all six CDR-like loops of the T lymphocyte receptor are involved in binding to the peptide–MHC complex, presumably through several non-covalent bonds. However, the binding affinity of the T lymphocyte receptor is much lower than that of antibody (see the table in Box 5.1). This may reflect the different circumstances in which the two receptors operate. While both receptors need to be antigen specific, immunoglobulins also need to be able to capture free antigen, which is often present in the body at low concentration, and to hold onto it tightly. This requires a high affinity receptor. The reasons why the T lymphocyte receptor affinity is so much lower are less clear. Studies in the mouse confirm that developing T lymphocytes with a high affinity for MHC molecules die in the thymus and only those with an intermediate affinity emerge as mature T lymphocytes. This may be a consequence of a way of preventing them from responding to self-antigens.

Tolerance to self-antigens: Chapter 9

The interaction of a T lymphocyte receptor with an MHC molecule is also a transient one, constantly being made and broken as the lymphocyte interacts with different antigen-presenting cells. Only upon encountering the specific antigen does the interaction between the cells become stronger and the separation delayed. This is mediated by the increased expression of a variety of adhesion molecules, which is part of the antigen-specific activation process triggered by the receptor.

5.2.9 Antigen specificity revisited

The molecular principles underlying antigen specificity are very similar for T and B lymphocytes. Like immunoglobulin, antigen specificity by the T lymphocyte receptor is dependent on the surface conformation of its ligand. The only difference between the two kinds of recognition is that antigen is processed first, such that T lymphocyte receptors are insensitive to the *original* conformation of the antigen.

In Plate 6 is shown the structure of the mouse MHC molecule called K^b (pronounced 'K of b') containing the Sendai virus epitope mapped in Figure 5.12. This molecular complex was produced by expressing the MHC gene in bacteria, and allowing the translated polypeptides to assemble in the presence of synthetic peptides of the Sendai virus epitope. In this way, all of the K^b molecules that were crystallized contained an identical peptide. Also shown in the Figure is K^b presenting an entirely different epitope from vesicular stomatitis virus (VSV). Like the Sendai virus epitope, this epitope was mapped by testing synthetic peptides for recognition by VSV-specific cytotoxic T lymphocytes.

The T lymphocyte clone that recognizes the K^b–Sendai peptide complex will not kill targets bearing the K^b–VSV peptide complex and *vice versa*. This antigen specificity occurs because the surface topographies of the two peptide–MHC ligands are different and not complementary to the other T lymphocyte receptor. In part, this is because the peptides themselves have different amino acid sequences and therefore different side chains protrude from the cleft. The crystallography also revealed that the MHC molecule itself, which is the same in our two examples, can be influenced by the particular peptide it holds. Notice for example the bulge that appears on the surface of the K^b molecule when containing the VSV peptide, which is absent when the cleft contains the Sendai virus peptide.

5.2.10 Differences between T and B lymphocyte epitopes

At the beginning of this chapter we used lysozyme as the model antigen to illustrate the nature of epitopes recognized by immunoglobulins. Recall that any protein can be viewed in terms of multiple B lymphocyte epitopes, each defined by a different clone. The same applies to T lymphocyte epitopes. Figure 5.13 shows some of the T lymphocyte epitopes that have been mapped on lysozyme. Notice the differences:

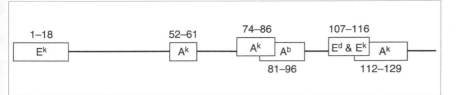

Figure 5.13 Some epitopes in lysozyme recognized by different clones of murine helper T lymphocytes. The antigen is shown as a linear representation with the epitopes marked as boxes. The particular murine MHC molecule that presents each epitope is indicated within the box. Compare with Figure 5.3 in which epitopes recognized by three monoclonal antibodies are depicted. Data from Adorini, L. *et al.* (1988) *Journal of Experimental Medicine*, **168**, 2091–104; Shastri, N. *et al.* (1985) *Journal of Experimental Medicine*, **162**, 332–45; Leighton, J. *et al.* (1991) *Journal of Immunology*, **147**, 198–204.

- While the epitopes for immunoglobulin are surface-located and discontinuous, the amino acids that make up a T lymphocyte epitope are contiguous and not separated in the primary amino acid sequence.

- Because the antigen is processed for recognition by T lymphocytes, their epitopes can be found anywhere within an antigen and are not confined only to its surface.

- The T lymphocyte receptor is specific for the conformation of the complex formed by the processed peptide and the MHC molecule, rather than (as is the case with B lymphocytes) the *original* conformation of the antigen.

Table 5.3 summarizes these differences between epitopes recognized by T and B lymphocytes. This two-tier antigen recognition system allows lymphocytes to gain access to antigen both inside and outside of cells. In the next chapter we will examine the MHC molecules in more detail.

Summary

- T lymphocyte responses can be detected *in vitro* by measuring proliferation or cytotoxicity of cells activated by antigen *in vivo*. Such responses are antigen specific, such that activated T lymphocytes do not respond to non-immunizing antigen.

Table 5.3

	Epitope	
	B lymphocyte	T lymphocyte
	binds whole, free antigen	binds peptide epitope presented by MHC molecule on another cell
Amino acids	discontinuous	linear
Location	surface of antigen	anywhere
Dependent on original conformation	yes	no

- The type of response correlates well with the function and phenotype of the T lymphocyte. Proliferation assays detect responses mostly by $CD4^+$ (predominantly helper) T lymphocytes, whereas cytotoxicity assays measure responses mostly by $CD8^+$ (predominantly cytotoxic) T lymphocytes.

- Both $CD4^+$ and $CD8^+$ T lymphocytes require antigen-presenting cells to process the antigen and display antigenic peptides to them at their surface.

- MHC molecules are molecules on the surface of antigen-presenting cells that present antigenic peptides to T lymphocytes.

The future

- The co-crystallization of a peptide–MHC complex and bound T lymphocyte receptor trimolecular complex will help elucidate the molecular details of this interaction.

Further reading

Amit, A.G., *et al.* (1986) Three-dimensional structure of an antigen–antibody complex at 2.8 Å resolution. *Science*, **233**, 747–53.

Benjamin, D.C., *et al.* (1984) The antigenic structure of proteins: a reappraisal. *Annual Review of Immunology*, **2**, 67–101.

Davis, M.M. and Bjorkman, P.J. (1988) T-cell antigen receptor genes and T-cell recognition. *Nature*, **334**, 395–402.

Wilson, I.A. and Cox, N.J. (1990) Structural basis of immune recognition of influenza virus haemagglutinin. *Annual Review of Immunology*, **8**, 737–71.

Review questions

Fill in the blanks

A technique that allows the three-dimensional structure of proteins to be visualized is _____ _____[1]. The best resolution is achieved when the protein is _____[2]. For immunoglobulins, this problem was circumvented, initially by using _____ _____[3] proteins from _____[4] patients, and later _____[5] antibodies. The latter can be elicited to a defined antigen which permitted the structure of _____[6] and _____[7] complexed with a specific antibody to be studied. The surfaces of an antibody and its antigen have a _____[8] topography at the interface. The strength of the interaction is provided by multiple weak _____[9] bonds between the antigen and six _____ _____ _____[10] on the antibody. The surface on the antigen defined by the immunoglobulin is called the _____[11]. These are always located at the _____[12] of an antigen, and the amino acids involved are usually _____[13] in the primary amino acid sequence. The strength of the interaction between macromolecules, such as an antibody and an antigen, is called the _____[14]. Other factors influence the strength of the interaction, including the _____[15] of the antibody, and the spatial geometry of the _____[16]. Together these factors determine the overall stability, or _____[17] of the interaction. Because antibody molecules have two antigen-binding sites they have the potential to _____[18] antigens. Lattices of antigen and antibody are called _____[19] complexes. Assays of T lymphocyte antigen-specificity differ from B lymphocytes because they require _____ _____[20] cells. Helper T lymphocytes express the CD _____[21] and the activity of these cells is normally measured by _____[22] assay or the release of _____[23]. In contrast, cytotoxic T lymphocytes express CD_____[24] and are measured by their _____[25] of target cells. Antigens recognized by T lymphocytes differ from those recognized by immunoglobulin. T lymphocytes are insensitive to the native _____[26] of the antigen because it is _____[27] into peptides. The antigenic peptides are presented to the T lymphocyte receptor in the context of a molecule encoded in the _____ _____ _____[28]

Short answer questions

1. How do the activities of $CD4^+$ and $CD8^+$ T lymphocytes differ and how can these activities be measured *in vitro*?

2.   What are monoclonal antibodies and how are they produced?

3.   As a marker for previous exposure to a particular virus, you intend to look in human sera samples for antibodies against the virus capsid. You have a supply of virus capsids to use as the detection antigen. Which technique would you expect to best reveal the antibodies – the ELISA or the Western blot? (you will need to refer to Appendix II).

4.   Multiple choice – which of the following are correct: (more than one answer may be correct)?

(a)   Evidence that antigen must be processed for recognition by T lymphocytes came from the observation that in *in vitro* assays:

(1)   denaturation of the antigen by boiling abolishes T lymphocyte responses;

(2)   the antigen can be replaced with antigenic fragments or synthetic peptides;

(3)   proliferation by T lymphocytes is blocked by prior incubation of antigen-presenting cells in lysosomotropic agents;

(4)   mouse T lymphocytes are unable to respond to protein antigens in the absence of antigen-presenting cells;

(5)   all of the above.

(b)   The epitopes recognized by immunoglobulin:

(1)   are often hydrophobic and buried inside the native antigen;

(2)   created from amino acids that are often separated in the primary sequence;

(3)   interact with the hypervariable loops via several non-covalent interactions;

(4)   are usually lost when the antigen is denatured;

(5)   all of the above.

(c)   Antibodies can be used as a tool for:

(1)   calculating the molecular weight of an antigen;

(2)   calculating the concentration of an antigen;

(3)   purifying the antigen from complex mixtures;

(4)   identifying blood group antigens on erythrocytes;

(5)   all of the above.

The major histocompatibility complex

6

In this chapter

- Discovery of the MHC.
- MHC restriction of T lymphocyte responsiveness.
- Mapping the MHC.
- Polymorphism of MHC molecules.
- The cell biology of antigen processing.

Introduction

Before T lymphocytes and their pivotal role in the adaptive immune response were appreciated, the idea that immune 'responsiveness' had a genetic basis was a familiar one. In human populations it is quite common to find individuals who are more susceptible and others more resistant to particular infectious diseases. In this chapter we will examine how the genes that underlie responsiveness were mapped to the **major histocompatibility complex** (MHC) and how molecules encoded in the MHC are involved in the presentation of antigens to the antigen receptors of T lymphocytes.

6.1 Historical overview

Mammals such as guinea pigs, rabbits, rats and mice were the first species in which immune responsiveness to antigen was systematically studied. This

inbreeding *n.* mating between related individuals.

was facilitated by the production of strains of genetically identical individuals by a process of controlled inbreeding (see Box 6.1). Inbred strains allow genes for particular phenotypic traits, such as immune responsiveness to a particular antigen, to be stabilized within the population. Then, matings between low responder and high responder animals can be set up in order to follow the heredity of these phenotypes. It was with inbred strains, in particular of the mouse, that the major breakthroughs in understanding the genetics of responsiveness were made.

Separating an inbred strain into a high or low responder works best, not with large and complex antigens such as microorganisms, but with simple antigens such as synthetic peptides. Among the first experiments of this kind were on responses by inbred guinea pigs to polylysine, which is a peptide consisting of multiple lysines. Proliferation assays of T lymphocytes revealed this particular antigen could split guinea pig populations into high and low responders. By producing inbred strains it was possible to fix the 'responder gene' into an inbred strain (called strain 2).

Similarly in mice, the inbred strain C57 is a high responder to the synthetic peptide (T,G)-A,L – a polylysine backbone with branches of alanines with tyrosines and glycines at the ends. This strain is a low responder to (H,G)-A,L – constructed as before but with histidines and glycines at the ends of the alanine branches. The reverse situation was found for the mouse C3H strain. In all these systems, crossing high responders with low responders yielded high responder progeny. This demonstrated that immune responsiveness showed Mendelian inheritance and had therefore a genetic basis.

Mendelian inheritance inheritance of traits, now known to be mediated by nuclear genes, according to the laws defined by Gregor Mendel.

6.1.1 Tumour research leads to the discovery of histocompatibility

Progress in understanding this genetic basis was made more rapidly in the mouse than any other species, mainly because several different inbred strains had already been produced by researchers for another purpose. Before tissue culture technology was developed, the only way to obtain tumour cells for cancer research was from a spontaneous or chemically induced tumour growing *in vivo*. Attempts made to propagate the tumour cells after the animal died by transplanting the cells into the bodies of other (genetically unrelated) animals failed. This was because the tumour cells were consistently eradicated by the recipient's immune system. Moreover, normal tissue and cells were also rejected when transplanted from one individual into another.

Cells are rejected by recipients because of the genetic disparity between donor and recipient. Genetically unrelated members of the same species are

Alleles

BOX 6.1 Inbreeding

Somatic cells have two copies of each chromosome, derived from the fertilization of the parental gametes. Therefore there are two copies of a given gene present in the chromosomes of a somatic cell. On one chromosome is the paternally derived gene and in the same position (**locus**) on the homologous chromosome is the other, maternal gene. A given gene may exist as several different variants, called **alleles**, in the gene pool of the population. Consider a locus we will call A (Figure 1). The gene is **homozygous** if the alleles at the locus are identical, which can be represented as A*1/A*1 or A*2/A*2. The gene is **heterozygous** if they are different, such as A*1/A*2. In a heterozygous gene, one allele may be dominant over the other, recessive allele, in which case only the dominant allele is expressed. This is described as complete dominance, although conditions of incomplete dominance and codominance can occur. Recessive genes are only seen expressed in the phenotype when they are homozygous.

Population genetics

The frequency of alleles within a population is influenced by two main factors:

- *Natural selection.* This operates on the different phenotypic trait each allele encodes. Certain alleles of a given gene may influence the 'fitness' of an individual, i.e. influence its chances of passing its genes to offspring. Alleles that increase fitness in a given environment increase in frequency over subsequent generations, while those that decrease its fitness tend to decrease.

- *Genetic drift.* Some alleles have no effect on the fitness of an individual and are not subject to the forces of natural selection. At each generation, the frequency of these 'neutral' alleles has an equal chance of increasing, decreasing or remaining constant. Thus, over the generations the frequency of neutral alleles can change up and down at random, a process called genetic drift.

There is a tendency, particularly in small populations, for the frequency of an allele to reach a frequency of 1.0 (i.e. become present in a homozygous state in every member of the population). Once an allele reaches a frequency of 1.0 it is 'fixed' in the population and it remains homozygous (until a new mutant emerges or if an alternative allele is introduced into the population by immigration). Progression to homozygosity can be accelerated by repeated inbreeding of close relatives.

Homozygosity is achieved when two copies of the same allele, one present in each parent, are combined in the offspring at fertilization. For both parents to have the same allele means they must share a common ancestor. In small populations, where opportunities for random mating are few, matings between related individuals –

homozygous

A*1/A*1

A*2/A*2

heterozygous

A*1/A*2
complete dominance
(A*1 over A*2)

A*1/A*2
incomplete dominance

A*1/A*2
co-dominance

Figure 1 Phenotypes of homozygous and heterozygous genes, represented in the diagram as the colours of flower petals. The locus for petal colour is called 'A' and the different alleles by the numbers.

inbreeding – may occur and the rate of genetic drift toward homozygosity occurs more quickly than in large populations. In large populations, random mating can occur, which, together with instinctive avoidance mechanisms that prevent mating with close relatives, maximizes the heterozygosity of genes.

The major effect of homozygosity is the expression of recessive genes. Mutated genes are often recessive and are usually masked by their dominant wild-type counterparts in outbred populations. Many mutated genes are deleterious to the normal functioning of the protein it encodes. Thus, inbred populations often feature rare recessive phenotypes and sometimes debilitating diseases. Well-known examples are the thalassaemias of humans. These are diseases caused by mutations in the α and β chains of haemoglobin that are often fatal in homozygotes.

There are many cases in nature where individuals that are heterozygous at certain loci are more likely to survive than homozygotes (**heterozygote advantage**). The heterozygous condition may mask or dilute the effect of a deleterious recessive gene, as just described, or it may have provide a selective advantage against an environmental pathogen, predator or toxic substance (see Box 6.3).

Inbred strains of animals

Man has used inbreeding programs to fix phenotypic traits in many different strains of animals and plants. Breeds of the domestic dog are a good example, in which the several 'desirable' phenotypic traits have been fixed by selective inbreeding. The genes for these traits are homozygous and identical in every member of a breed, and provided mating is confined to within the same breed, and no mutation occurs, the traits are stable. Several of these genes are recessive and would otherwise be silent in heterozygotes in outbred populations. Unfortunately, less desirable recessives also emerge during selective breeding.

Fully inbred strains of animals are produced by several generations of inbreeding, usually by brother/sister matings. Apart from the sex chromosomes, all the individuals of an inbred line are virtually identical and homozygous at most loci, and their lineages can be traced back to a single breeding pair. Fully inbred strains have been established in several animal species. The strains of the house mouse used by immunologists are fully inbred strains. It takes 10–20 generations of brother/sister matings to produce a fully inbred strain of a mammal such as the mouse. The process is not without problems since lethal recessive genes often emerge and terminate the line. Each strain has a unique and stable combination of traits, the most conspicuous of which is coat colour. The recessive albino phenotype, for example, has been fixed in the BALB/c strain. Albinism is very rare in outbred animals, occurring in human populations in 1/20 000 individuals. Inbred strains have been produced in laboratories to stabilize other less obvious traits, in particular susceptibility to certain diseases such as diabetes, arthritis and leukaemia. These provide useful models of human diseases. Inbred mice have also been used in cancer research and immunological studies for over 60 years.

allogeneic *a.* two genetically dissimilar individuals of the same species, such as would be found in an outbred population; allogeneic mammals reject exchanged tissue grafts (see **syngeneic**)

mixed lymphocyte reaction (MLR) the activation of T lymphocytes by **allogeneic** MHC molecules *in vitro*.

described as being **allogeneic** to each other, and reject exchanged grafts (Figure 6.1a). An *in vitro* correlate of graft rejection is the **mixed lymphocyte reaction** (Figure 6.1b). This occurs when leucocytes from two allogeneic animals are cultured together *in vitro*. This causes a mutual stimulation of T lymphocytes, and cytotoxic and proliferative responses are induced.

The problem of propagating tumour cells was resolved by producing inbred strains of animals. Since members of the same inbred strain are genetically homogeneous, grafts of tumour cells (or, indeed, normal cells) could be exchanged without being rejected. Genetically related individuals

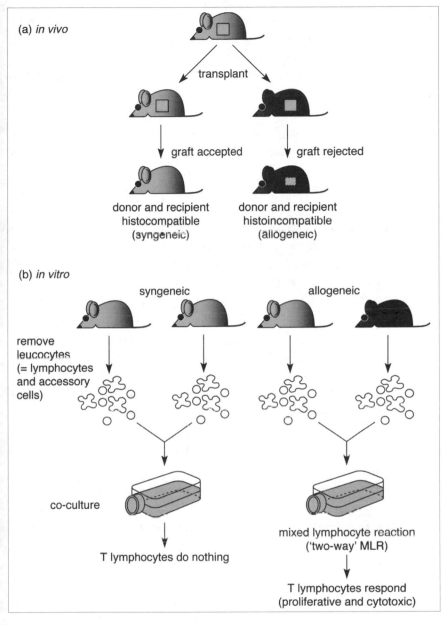

Figure 6.1 Allorecognition phenomena. Animals that have different MHC alleles are allogeneic or histoincompatible. This leads to graft rejection *in vivo* (a) and T cell stimulation *in vitro* (b). Animals that have the same MHC molecules are syngeneic or histocompatible and do not reject exchanged grafts or respond in mixed lymphocyte cultures.

that accept exchanged grafts of cells or tissues are described as **syngeneic** or **histocompatible**.

6.1.2 Histocompatibility genes underlie graft rejection

It was suspected that grafts between allogeneic individuals (**allografts**) triggered immune responses by virtue of cell-surface antigens recognized as foreign by the recipient. One approach to study these **histocompatibility antigens** or **alloantigens** was to produce antibodies to the protein

syngeneic *a.* two genetically identical individuals of the same species, such as would be found in an inbred strain of mice; syngeneic mammals accept exchanged tissue grafts because of identity at the MHC (see **allogeneic**).

allograft *n.* a graft of cells, tissue or organs between two allogeneic individuals.

alloantigen *n.* molecules encoded in the major histocompatibility complex recognized by alloantisera.

products of the histocompatibility genes. These antibodies could then be used as tools to identify the antigens in different offspring and follow the patterns of inheritance of the histocompatibility genes.

In the first of these studies, rabbits were immunized with mouse cells to produce several different antisera. One particular antiserum, serum II, was particularly interesting because some inbred strains of mouse had antigens recognized by serum II, while others did not. Antigen II turned out to be a the product of a histocompatibility gene because animals with antigen II would accept grafts from other strains that were antigen II-positive (i.e. histocompatible) but rejected grafts from mice that were antigen II-negative.

A complementary approach came from genetic studies. Several ingenious inbreeding systems were developed with the aim of isolating the histocompatibility genes. Consider two allogeneic inbred strains of mice, say X and Y. Let us also let H represent the locus where the histocompatibility genes of mice are located. The alternative versions of a gene that can be found at a given genetic locus are called **alleles**. Clearly, the histocompatibility alleles of strain Y (we can call H^Y, pronounced 'H of Y') are different from the equivalent genes from strain X (H^X) because these strains reject exchanged tissues. Replacing Y's histocompatibility genes with those of X creates a new strain with the same **background** as Y but with different alleles at the H locus. Two strains that only differ by one or a few neighbouring loci are called **congenic**. This is achieved, as illustrated in Figure 6.2, by a controlled system of inbreeding.

With approaches like this, several loci controlling histocompatibility in mice were revealed, called H-1, H-2, H-3 and so on. Of all of these loci, grafts exchanged between mice that had different H-2 alleles were rejected the most rapidly. Grafts exchanged between mice that differ elsewhere are rejected much more slowly. For this reason H-2 is the *major* histocompatibility locus of mouse, whereas the remainder are termed *minor* histocompatibility loci. Antigen II, detected by the serological approach, was later found to correspond to the product of the H-2 genes and was thus renamed H-2.

Subsequent genetic mapping (discussed later) revealed that H-2 consists of a cluster of several closely linked genes, what geneticists call a *complex*. For this reason, H-2 of mouse, and its equivalent in other species of vertebrates, is termed the **major histocompatibility complex** or **MHC**. Graft rejection and mixed lymphocyte reactions, called **allorecognition phenomena**, are widely used to demonstrate the existence of MHCs in other vertebrate species (Chapter 7).

allele *n.* one of a number of alternative forms of a gene that can be found at a particular locus on a chromosome.

congenic *a.* inbred strains of animals that are genetically identical except at a particular locus; produced by superimposing an allele or cluster of alleles from one inbred strain onto the genetic background of another strain by controlled inbreeding.

H-2 *n.* the major histocompatibility complex of the mouse.

allorecognition *n.* recognition by T lymphocytes of the MHC molecules on an allogeneic individual's antigen-presenting cells; results in cell-mediated immune responses as manifested by allograft rejection *in vivo* or the mixed lymphocyte reaction *in vitro*.

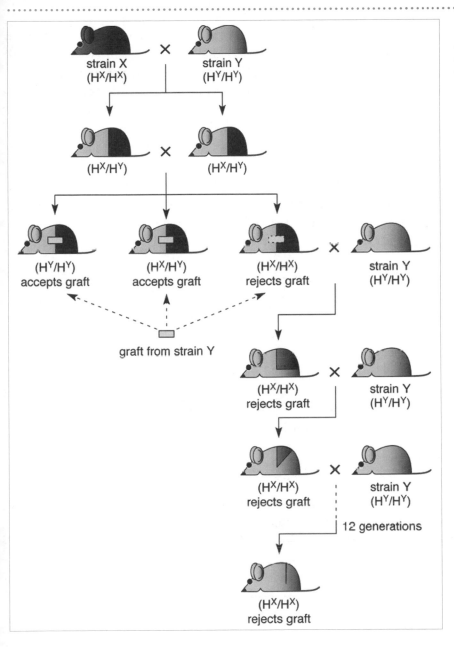

Figure 6.2 Schematic representation of the production of congenic mice strains. Two congenic inbred strains are genetically identical except for a short segment of a chromosome. One method to produce two strains that are congenic at a given locus (*such as a histocompatibility locus, H*) is to cross two inbred mice that have different alleles at the locus. One parent, inbred strain Y, contains the genetic background but the unwanted alleles (*in this case, H^Y/H^Y*). The other parent, X, has the particular alleles (H^X/H^X) we wish to swap into strain Y. The F1 progeny of this cross are all heterozygous at the locus (i.e. H^X/H^Y). These are then intercrossed; around a quarter of the progeny will be homozygous for the required allele, and the rest of their genomes will be a mixture of X- and Y-derived genes. A homozygote is then selected. (*In our example, the H^X/H^X homozygotes would be detected by their ability to reject grafts expressing H^Y/H^Y. Heterozygotes or H^Y/H^Y homozygotes will not reject the grafts.*) The homozygote is then backcrossed with another Y strain mouse and the progeny again screened. This is repeated for around 12 generations; at each generation the offspring with the desired allele (H^X/H^X) are backcrossed with the Y strain. Thus at each generation the selected alleles are retained, while owing to random crossovers at meiosis the proportion of X-derived genes in the 'background' becomes progressively more diluted with Y genes. The resultant strain is virtually devoid of any strain X genome, other than the locus that was selected at each generation (H^X/H^X). The name given to a new inbred strain is derived from both parental names, or abbreviations of them, background first followed by source of new allele(s) (*in our example, Y.X*). Several congenic strains of *M. musculus* have been produced that have enabled the properties of genes from different inbred strains to be compared without interference from the different backgrounds.

6.1.3 Alloantisera reveal multiple MHC alleles

Inbred strains greatly facilitated the identification of histocompatibility antigens in animals, especially mice. As discussed above, cross-immunizing one inbred mouse with cells of an allogeneic strain elicits in the recipient antibodies against foreign antigens, including the histocompatibility antigens. Sera produced in this way that specifically recognize MHC antigens are called **alloantisera**.

Serological studies of H-2 revealed that there are in fact many different H-2 alleles in the mouse population. In the mouse, these are assigned different superscripts (H-2^b, H-2^k, H-2^d and so on), although this terminology has not been adopted for other mammals (see later). Thus, mice with H-2^b vigorously reject grafts from mice with H-2^k, H-2^d and so on, whereas a more gradual rejection occurs if the donor and recipient share the same H-2 alleles but have a different background (i.e. have different minor histocompatibility antigens).

In humans, deliberate immunization with cells from another person is not possible, although the sera of multiparous women (i.e. who have had several pregnancies) are a good source of antisera that are specific for allogeneic MHC molecules present on the foetus inherited from the father.

multiparous *a.* birth of multiple offspring.

6.1.4 The 'real' function of histocompatibility genes

Graft rejection studies led to the discovery of MHC genes and allorecognition phenomena have been widely used to demonstrate their existence in other species. To this day, precisely how an allogeneic MHC molecule stimulates T lymphocytes remains unclear, although understanding this phenomenon is important to reduce the incidence of graft rejection. However, it is important to understand that T lymphocytes are not normally exposed to allogeneic MHC molecules. Thus, allograft recognition is not the normal function of the MHC. What then is its more usual function?

The first clues to this came in the 1960s when the genes controlling histocompatibility and immune responsiveness were found to be one and the same. Several inbred and H-2 congenic mice strains that had been produced for the cancer studies were tested for responsiveness to short peptide antigens. All the strains that responded to (T,G)-A,L, all shared H-2^b. The allele for responsiveness to (H,G)-A,L turned out to be H-2^k.

This association between responsiveness and a particular histocompatibility allele was also demonstrated using alloantisera. In crosses between high and low responders, *all* the high responder offspring expressed the H-2^k allele, whereas none of the low responders did. Similarly in the guinea pigs, crosses between strain 2 (a high responder to polylysine) with a low responder strain yielded a mixture of responder phenotypes in the offspring. Only those that were high responders had strain 2 MHC molecules. In rats and rabbits a similar finding was associated with the response to the antigen lactic acid dehydrogenase. These were the first indications that MHC genes controlled responsiveness.

6.1.5 Important *in vitro* studies

Clues to how the MHC controls responsiveness came from manipulating *in vitro* assays. Initial studies were conducted with strain 2 guinea pigs. It had been established by proliferation assays *in vitro* that this strain responded well to polylysine. If, however, the antigen-presenting cells in the assay are replaced by antigen-presenting cells from an allogeneic strain, the T lymphocytes fail to respond to the antigen.

However, it was the availability of H-2 congenic mouse strains that quickly pinpointed the genetic differences between high and low responders to the MHC. Mouse T lymphocytes failed to proliferate in response to antigen *in vitro* if they and the antigen-presenting cells were obtained from H-2 congenic mice (i.e. had different MHC alleles) whereas a response did occur if the two strains shared the same H-2, irrespective of the background.

A similar requirement applies to cytotoxic T lymphocytes (Figure 6.3). Here is shown an experiment in which cytotoxic T lymphocytes are elicited by infection with virus. These cells kill syngeneic targets infected with virus, but not H-2 congenic or allogeneic targets infected with virus. In general terms therefore, both $CD4^+$ and $CD8^+$ T lymphocytes respond to antigen only if they and the antigen-presenting cells are syngeneic.

This simple rule may not seem so remarkable, since under normal circumstances *in vivo*, T lymphocytes and their antigen-presenting cells *are* from the same individual, and as a consequence there is no problem with mismatched histocompatibility genes. However, these early experiments with inbred mice did reveal that MHC molecules influenced responsiveness through an effect on T lymphocytes.

6.1.6 MHC restriction

The reason for this dependence on MHC molecules is in part due to the way in which thymocytes develop into mature T lymphocytes in the thymus. Inside the thymus, thymocytes rearrange their antigen receptor genes and express the receptor molecules at the cell surface. Studies in the mouse show that before leaving the thymus, the thymocytes are then exposed to MHC molecules like those to be found elsewhere in the body. Only those thymocytes bearing a receptor with an inherent affinity for the MHC molecules are allowed to emerge from the thymus as mature T lymphocytes.

As a consequence, the entire repertoire of mature T lymphocytes in an animal is skewed towards recognizing antigen presented by cells bearing self-MHC alleles. This dependency by lymphocytes on self-MHC molecules for recognizing antigen is called **MHC restriction**.

Notice the difference between allorecognition and 'normal' antigen

Thymocyte ontogeny:
Section 9.2

MHC restriction the dependence by T lymphocytes on MHC molecules for the recognition of antigen.

Figure 6.3 Demonstration that the MHC influences the recognition of antigen by T lymphocytes. Cytotoxic T lymphocytes are produced in an inbred strain of mouse by infecting it with a virus. In a cytotoxicity assay these T lymphocytes only kill virus-infected target cells if they are obtained from a syngeneic mouse (i.e. genetically identical), but not targets from an allogeneic or H-2 congenic strain (i.e. have different H-2 alleles). This pinpoints the genes at the MHC in influencing T lymphocyte recognition of antigen. Based on the work of Zinkernagel, R.M. and Doherty, P.C. (1974) *Nature*, **248**, 701–2.

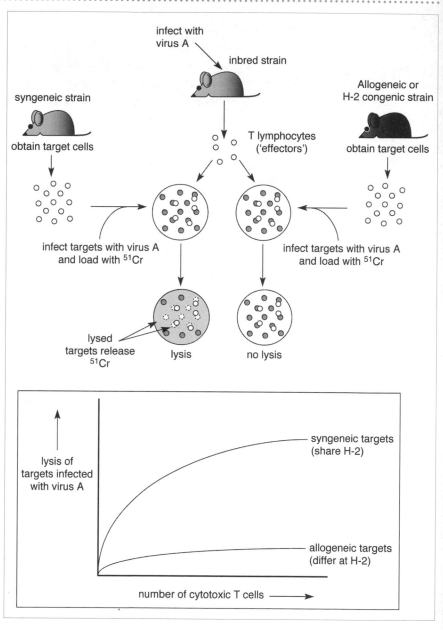

recognition. At first sight there appears to be a paradox. Why in experiments, such as the allogeneic targets shown in Figure 6.3, is a response to alloantigen not seen? In fact a mixed lymphocyte reaction *is* stimulated under these conditions, but the outcome is not seen in readout of the assay because the duration of a T lymphocyte assay, such as the cytotoxicity assay shown in Figure 6.3, is much shorter than the time necessary to mount a response to allogeneic cells *in vitro*.

6.1.7 The duality of the ligand

The dependence of T lymphocytes on MHC molecules for responding to antigen raised much speculation over the molecular basis of recognition. Were there two separate receptors on the T lymphocyte, one for MHC molecule and one for antigen, or did the T lymphocyte antigen receptor recognize both? Important clues came from the creation of a T lymphocyte clone with two different antigen receptors. Normally of course, a clone of T lymphocytes expresses a unique receptor. A clone with two receptors can be produced by fusing two different clones, or by introducing the α and β genes of the T lymphocyte receptor from one clone into a different clone. This was achieved very soon after the first α and β genes had been cloned.

Figure 6.4 illustrates the principle. A clone of T lymphocytes that recognizes antigen 'A' is obtained from a mouse with the H-2^d MHC. The genes for the αβ chains of the antigen receptor are obtained and then

Figure 6.4 Experiment that demonstrates the dual specificity of the T lymphocyte receptor for antigen *and* MHC. See text for details. Based on the experiments of Dembic, Z., *et al.* (1986) *Nature* **320**, 232–8.

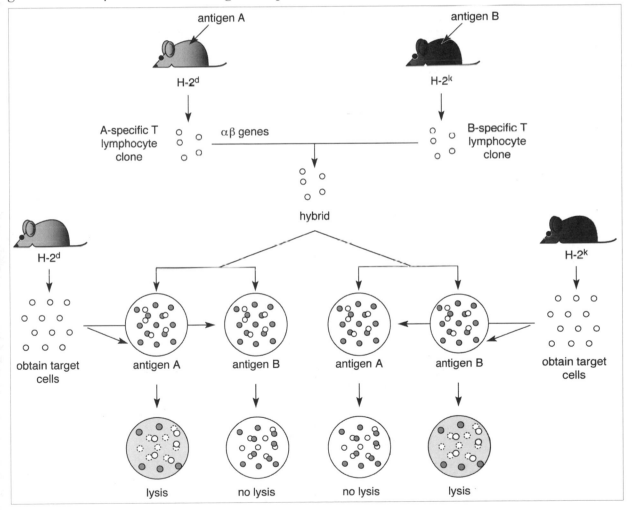

introduced into a different clone. This second clone is obtained from a different strain mouse (H-2^k), which recognizes antigen 'B'. The resultant hybrid cell has both antigen specificities. It retains the original H-2 requirements of each receptor (A with H-2^d and B with H-2^k) but does not respond to new combinations (B with H-2^d and A with H-2^k). This experiment indicates the T lymphocyte receptor recognizes antigen and MHC molecules *simultaneously*.

Recall from the previous chapter that antigens are processed into peptides by antigen-presenting cells before presentation to T lymphocytes (Sections 5.2.5–5.2.7). Combining this with the observation that antigen and MHC molecule are recognized simultaneously suggests that a T lymphocyte receptor binds to a molecular complex of MHC molecule and antigenic peptide. There was considerable speculation as to how this complex might look. A dramatic resolution to this issue came in 1987 when the first MHC molecule was crystallized (Plate 4). In the centre of the distal face of the molecule could be seen a deep cleft containing an antigenic peptide.

Summary

- Genes that control histocompatibility and responsiveness to antigen map to the MHC.

- The 'normal' function of MHC molecules is presentation of antigen to T lymphocytes. The receptor binds to a complex of antigenic peptide and self-MHC molecule.

6.2 The genes and molecules of the MHC

Classically, the existence of a MHC in a vertebrate species is usually indicated first by allorecognition phenomena. If inbred strains are available, its existence is substantiated by MHC-restricted T lymphocyte responses *in vitro*. Ultimately the genes that underlie these cellular phenomena are cloned.

6.2.1 Mapping the MHC
Pioneering work in vertebrate MHCs has come largely from studies of the house mouse, *Mus musculus,* and of humans. Today, the MHCs of these two mammals – H-2 in the mouse and **HLA** (for **human leucocyte antigen**)

in humans – are now almost completely mapped and sequenced. As the details of the MHCs of more species become available it seems likely that all vertebrates have a MHC that functions to control T lymphocyte responses in essentially the same way as it does in the two model vertebrates, mouse and man.

Before the DNA technology revolution of the 1980s, maps of the linear organization of MHC genes were drawn using the established method of **crossover** (or **recombination**) **frequencies** (see Box 6.2). Crossing-over is the process of genetic recombination that occurs between homologous chromosomes during meiosis.

Genes on the same chromosome, such as those in an MHC, are described as being **linked** since they are sufficiently close to each other that they are usually inherited *en bloc*. Occasionally, however, crossovers can occur within the complex. Then the alleles that were linked become segregated into different chromosomes and ultimately into separate gametes. The chance of a crossover occurring between two linked genes increases proportionally with the distance between them. The frequency with which linked genes become unlinked in the progeny, caused by crossovers in the parents, enables maps of their relative positions to be constructed.

The H-2 complex of mouse was the first MHC in which mapping by crossover frequencies was used. Alloantisera were used to detect the frequency of new combinations of alloantigens among the progeny of matings between two allogeneic inbred mice strains. In one of the first experiments of this kind, two different antibodies that detected histocompatibility antigens specific to the A inbred strain were used to follow their segregation in the progeny of an A × C57 cross. In practice, one of the genes must be heterozygous (see Box 6.2), thus the F_1 of this mating were used by back crossing with a C57 parental strain mouse. Of 194 offspring, two expressed one antigen or the other, instead of both, indicating the markers had become segregated by crossovers. The loci encoding the two antigens were closely linked (in fact, $2/194 \times 100 = 1$ centiMorgan) and were designated K and D. These loci are often regarded as the opposite ends of the H-2 complex, and several other loci have since been mapped in between.

The first linkage maps of HLA were established by examining the inheritance of alleles in human families. Crossovers that occur in children are revealed when they display their parental histocompatibility antigens in new combinations. However, breeding in human populations cannot be controlled in the same way as in laboratory animals, so this approach was supplemented by the use of **somatic cell hybrids**. Somatic cells are fused *in vitro* to mimic the fertilization of gametes, and the 'inheritance' of MHC

HLA (human leucocyte antigen) *n.* the human major histocompatibility complex.

crossing-over the process of exchange (or recombination) of genetic material during meiosis; caused by the breakage of homologous chromatids in the four chromatid stage and exchange of homologous regions of DNA.

linkage *n.* the tendency of two or more parental alleles to be inherited together because of their localization on the same chromosome.

Recombination and gene conversion: Box 2.1

homozygous *a.* a gene in which the paternal and maternal alleles are the same.

heterozygous *a.* a gene in which the maternally and paternally derived alleles are different.

F_1 hybrid first filial (son or daughter) hybrids arising from a first cross, with subsequent generations denoted by F_2, F_3, etc.

BOX 6.2   Mapping genes

The linear order of gene positions (**loci**) is fixed on a chromosome. Gene mapping is concerned with identifying the position of a locus in relation to other loci on the same chromosome. One widely used method is based on the frequency of **crossing-over** that occurs between genes on the same chromosome. Crossing-over is an event that occurs during meiosis, the process by which chromosomes are duplicated prior to separation into the gametes. Breaks may occur in the chromosomes after duplication, in the four chromatid stage, causing exchange of homologous regions of DNA (see Box 1.2).

The probability of a crossover occurring between two loci on a chromosome is proportional to the distance between them. The closer this distance, the less likely they are to become separated or **segregated** into different gametes at meiosis. Genes that segregate together are described as **linked**.

The frequency of crossover between two linked loci allows their relative positions to be calculated. This is assuming the phenotypic effects of the genes, often called genetic 'markers', can be traced in the progeny. Crossovers occurring in a parent's gametes can be seen in the offspring. If a crossover occurred between two marker genes, the phenotypic traits they specify will appear in a new combination in an offspring. In a large number of progeny, the proportion of the offspring showing signs of the crossover between parental markers is proportional to the distance between them. Another requirement for mapping is that at least one of the genes is heterozygous, as the effect of a crossover between two homozygous genes would be invisible in the phenotype.

Figure 1 shows an example in which we want to know the distance between two

Figure 1.

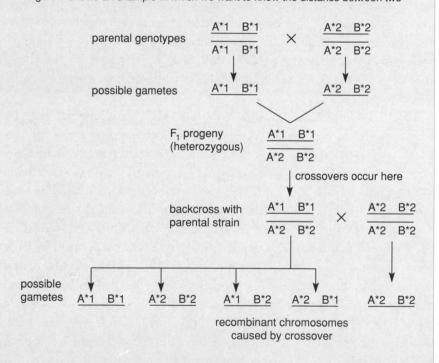

linked loci, A and B. Two inbred strains are chosen with different alleles at the two loci. In both strains the loci are homozygous, so in order to generate heterozygotes for mapping purposes we will cross them. This renders both loci heterozygous in the F_1 offspring. We now wait until the gamete-producing cells in the offspring undergo meiosis and for crossovers to occur. To determine the frequency of crossovers that occur between the A and B loci, an F_1 animal is now 'back-crossed' with a homozygous parental phenotype. Most of the progeny will have parental phenotypes, although some, owing to crossovers between the two loci, will show new or **recombinant** combinations.

Over short chromosomal distances, the frequency of new recombinants is proportional to the distance between the loci. The greater the distance, the greater the likelihood of a crossover event occurring between them, and consequently a larger proportion of the progeny will shown recombinant phenotypes. One **map unit** (or 1 **centiMorgan**) is defined as the distance between two loci that yields recombinants in 1% of the progeny chromosomes. Thus in the example, if 2% of the progeny expressed the recombinant phenotypes, the A and B loci would be 2 map units apart. If the distance between B and another linked locus, C, is 3 map units, the distance between A and C would be around 5 map units. The relationship becomes less accurate with increased distance between the loci. When loci are 10–20 map units apart the incidence of double crossovers, where two loci are brought back onto the same chromosome, thus giving the impression that no crossover has occurred, becomes significant.

In recent times, the application of recombinant DNA technology has been used to obtain maps of HLA and H-2. Overlapping cloned fragments spanning the entire length of each complex have been placed in order by **chromosome walking** and both complexes are now almost fully sequenced. See Box 4.1

alleles determined by using allele-specific antibodies to detect cell-surface MHC molecules. Linkage maps are constructed by counting the hybrid cells in which rearrangements of DNA (equivalent to crossover in meiosis) have occurred.

6.2.2 Cloning MHC genes

In addition to helping construct linkage maps, antibodies to MHC molecules also provided the first handle on the MHC genes themselves. Two general strategies were used (represented schematically in Figure 6.5).

One approach (Figure 6.5a) involves preparing mRNA from cells and using it to synthesize proteins by translation *in vitro*. Those mRNAs coding for MHC molecules are identified by testing whether their protein products are recognized by an antibody to MHC molecules. In practice, mRNA molecules from human cells were first reverse-transcribed back into cDNA using reverse transcriptase, and the different cDNA molecules produced were individually ligated into plasmids to produce a cDNA library. Recombinant plasmids were then introduced into bacteria, which were then cloned and expanded into large cultures.

To identify the bacterial clone containing the cDNA coding for an HLA molecule, plasmids from each were extracted and immobilized on nitrocellulose filters. These were then used to capture complementary mRNAs from human cells by hybridization. Each captured mRNA species

linkage map *n.* a map of the relative positions of genetic loci on the same chromosome; produced by analysing recombination frequencies (the frequency with which linked alleles are segregated into different gametes owing to crossing-over).

somatic cell hybrid a hybrid cell formed by the fusion of cells, often from different species of animals.

Cloning new genes: Box 4.1

Figure 6.5 Schematic representation of how antibodies to MHC molecules were used in the first cloning of MHC genes. See text for details. Based on the experiments of (a) Ploegh, H. *et al.* (1980) *Proceedings of the National Academy of Sciences USA,* **77**, 6081 and (b) Reyes, A.A., *et al.* (1981) *Immunogenetics,* **14**, 383–92.

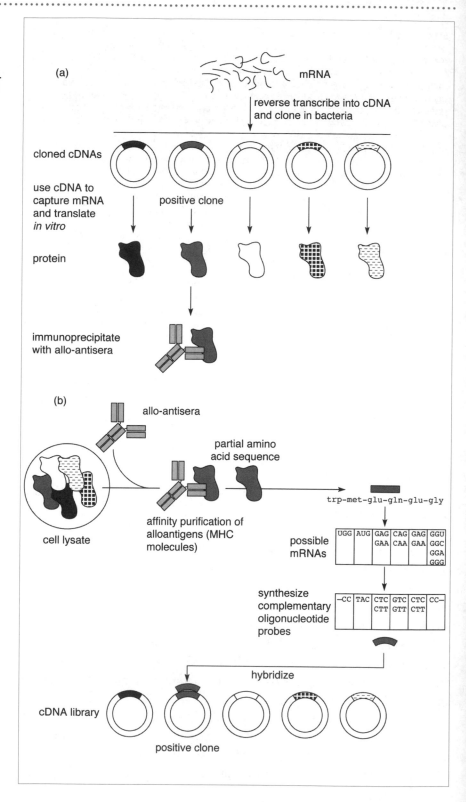

was then eluted from the filter and separately translated into protein *in vitro*. Radioactive amino acids were incorporated into the translation reaction to label any proteins synthesized. Finally, the presence of HLA molecules present in any of the reactions were precipitated with the specific anti-HLA antibody (immunoprecipitation: see Appendix B *Commonly used immunoassays*).

In this way, a cDNA coding for an HLA molecule was first cloned. Sequencing the cDNA revealed that it corresponded exactly with a short stretch of known amino acid sequence from the HLA B7 molecule that been determined previously by biochemical means from HLA B7 molecules purified with the help of anti-HLA B7 antibodies.

A slightly different strategy (represented in Figure 6.5b) was used in the cloning of the first mouse MHC gene (which was H-2 K^b). Here, antibodies to MHC molecules were used to capture the MHC molecules from cell lysates and the sequence of a few amino acids obtained by biochemical means. Next, short DNA molecules (**oligonucleotides**) were designed having the appropriate nucleotide sequence to code the known amino acid sequence. In practice, several oligonucleotides need to be synthesized because several amino acids are encoded by more than one codon. These oligonucleotides were used as probes to screen a mouse cDNA library (see Box 4.1).

oligonucleotide *n.* a short, synthetic DNA molecule used as a probe or a primer of DNA synthesis in the laboratory.

Once a cDNA fragment of MHC has been obtained, by whichever route, it may itself then be used as a probe to identify other clones containing homologous or partially overlapping sequences in cDNA or genomic DNA libraries. Probes for one allele rapidly led to the cloning of other alleles of the same locus and other related loci. Moreover, because mouse and human MHC genes are homologous, probes derived from human DNA were used to probe mouse libraries and *vice versa*. This 'cross-hybridization' approach expanded the search for MHC molecules in other mammals.

Thus while antibodies provided the original handle on the genes, the rapid expansion in understanding MHC was propagated mainly by molecular means. More recently, neighbouring genes within HLA and H-2 have been mapped by a systematic process called **chromosome walking** (also outlined in Box 4.1). Using these strategies, the entire H-2 and HLA complexes have been cloned in several partially overlapping genomic DNA clones.

chromosome walking a technique for mapping genes from a collection of partially overlapping cloned restriction fragments.

DNA sequencing of HLA has revealed that there are many more genes present in the MHC than the 'classical' MHC genes revealed by mapping using crossover frequencies. This is because some are too closely linked to be revealed by this method, and many more of these new genes do not encode alloantigens. As we will learn in the following chapter, cross-

classical MHC molecule *n.* polymorphic class I or class II MHC molecules involved in antigen presentaion to T lymphocytes (see **non-classical MHC gene**).

class I MHC molecule
vertebrate cell-surface heterodimers of polymorphic heavy (or α) chains encoded in the MHC and monomorphic β₂-microglobulin encoded elsewhere; function to present peptide antigens to CD8⁺ T lymphocytes; expressed in varying levels on all nucleated cells.

class II MHC molecule
vertebrate cell surface heterodimers of polymorphic α and β chains encoded in the MHC; function to present peptide antigens to CD4⁺ T lymphocytes; constitutively expressed only on specialized antigen presenting cells (dendritic cells, macrophages and B lymphocytes).

hybridization has worked less successfully as probes for the MHC genes in non-mammalian species. Instead, the synthesis of 'retrospective' oligonucleotide probes on the basis of amino acid sequence data (as depicted in Figure 6.5b) has provided the points of entry into the MHCs of several of these animals.

6.2.3 Two classes of MHC molecules

The following is an overview of the MHC molecules and genes of mouse and humans. Unfortunately, the names for homologous genes and their loci are different in H-2 and HLA. The HLA nomenclature is the most logical of the two, and the human nomenclature is being adopted for the MHCs of other mammals. The terminology of H-2 is somewhat archaic and is retained for historical reasons. Despite the differences in nomenclature of HLA and H-2, the products of the genes are homologous, and the organization of the loci within is very similar.

Classical MHC molecules are classified according to their structure and function as **class I** or **class II MHC molecules**. In most mammalian MHCs, the genes that encode these two classes are not jumbled up, but organized into large and well separated class I and class II 'regions'.

Figure 6.6 Relative positions of class I, II and III genes in the MHCs of different animals. For simplicity, only the classical class I and II genes are shown, and class II and III are shown as regions representing several loci. Not to scale. Adapted with permission from Figure 3 in Trowsdale, J. (1995) *Immunogenetics*, **41**, 1–17, copyright Springer-Verlag.

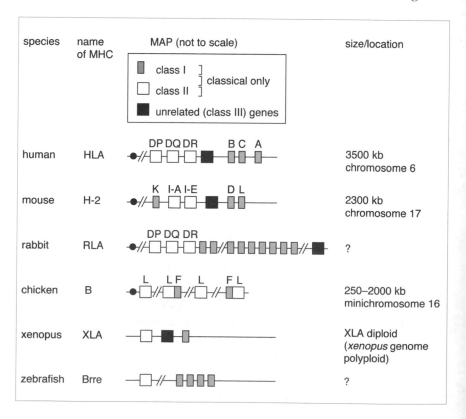

Diagrammatic representations of some vertebrate MHCs are shown in Figure 6.6.

Class I molecules of man are called **A**, **B** and **C**, and their equivalents in mouse are called **K**, **D** and **L**.

Class II molecules of man are called **DP**, **DQ** and **DR**, and **I-A** and **I-E** in mice. DQ and I-A are homologous, as are DR and I-E. The mouse has no equivalent of the human DP molecule.

The overall arrangement of class I and class II genes is similar in different animals. In well-characterized mammals, the class II region is situated between the centromere and the class I region. The H-2 complex of mouse is slightly unusual, since some of the class I genes have moved to the other, centromeric side of the class II region.

Figure 6.7 shows how the products of the class I and II genes are assembled into the class I and class II molecules.

- Class I molecules are heterodimers. One chain is called the α (or **heavy**) **chain** and is encoded by one of the class I genes of the MHC. The other chain is a smaller component called β_2-**microglobulin**, which is always encoded outside the MHC on a different chromosome. The number of different heavy chain loci varies in different vertebrates: *Xenopus*, rabbits and dogs have one, humans and mice have three, and domestic pigs have six.

- Class II molecules are also $\alpha\beta$ heterodimers, although the terminology surrounding the genes is a little more complex. This is because *both* chains of a class II molecule are encoded in the class II region of the

> **classical MHC molecule**
> *n.* polymorphic class I or class II MHC molecules involved in antigen presentation to T lymphocytes (see **non-classical MHC gene**).

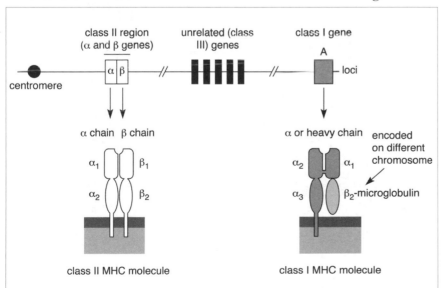

Figure 6.7 A 'generic' mammalian MHC complex and the products it encodes.

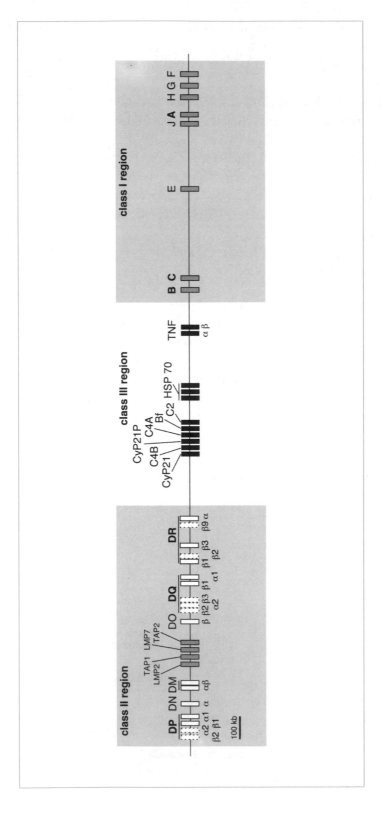

Figure 6.8 Class I, II and III genes of human MHC (HLA). Non-functional pseudogenes are shown as boxes with dotted borders and genes that encode classical MHC molecules are designated by the bold letters. The cluster of genes in the class II region (hatched) encode proteasome and transporter genes (see Section 6.5.1). CyP, cytochrome P; C2 and C4, components 2 and 4 of the complement cascade; Bf, factor B of complement; HSP 70, heat shock protein 70; TNF, tumour necrosis factor. Adapted from Campbell, R.D. and Trowsdale, J. (1993) *Immunology Today* **14**(7), 349–52.

MHC. Like the class I molecules, there are usually several classical class II molecules in a given species. In each case, however, each molecule is encoded by a *pair* of MHC genes, one encoding the α chain and one encoding the β chain.

A more detailed map of HLA is shown in Figure 6.8. In the MHCs of different animals there are usually more than one class II loci, whose products can be used interchangeably in the assembly of class II molecules. In HLA for example, there two functional loci encoding β chains of DR. Each can assemble with the corresponding DR α chain. However there is no significant mixing of α and β chains between isotypes (i.e. between DP, DQ and DR in man, or between I-A and I-E molecules in mouse).

The three dimensional structure of class I and class II molecules is remarkably similar (Figure 6.9) reflecting a common function of presenting peptide antigens to T lymphocyte receptors. Both classes of molecule possess four discrete extracellular domains. The α and β chains of class II molecules are roughly the same length, each having two domains. The class I α chain has three domains, with β_2-microglobulin being the fourth. In both classes of molecule, the two membrane-proximal domains (α_3 and β_2-microglobulin of class I, and α_2 and β_2 of class II) are typical immunoglobulin folds. This places both class I and class II MHC molecules as members of the immunoglobulin gene superfamily.

Immunoglobulin gene superfamily: Section 3.8.1

The two membrane-distal domains of each class (α_1 and α_2 of class I, and α_1 and β_1 of class II) are not immunoglobulin folds. These domains each consists of an α-helix supported on a platform of four anti-parallel β-strands. When the two distal domains associate in the heterodimer, a row of hydrogen bonds unite the two β-sheets into a single platform. This platform is surmounted by the two α-helices which lie roughly parallel with each other. Between the α-helices is formed the peptide-accommodating cleft.

6.2.4 CD4 T lymphocytes are restricted by class II molecules and CD8 are restricted by class I

Perhaps the most significant distinction between class I and class II molecules lies in the phenotype of T lymphocyte to which they present antigen (Figure 6.10). This can be demonstrated by *in vitro* assays of T lymphocytes. CD8[+] cytotoxic T lymphocytes only kill antigen-expressing targets with syngeneic class I MHC molecules, irrespective of the class II molecules the targets have. Similarly, CD4[+] helper cells will only proliferate in response to antigen on antigen-presenting cells with syngeneic class II molecules, irrespective of the class I molecules they have. In other words, CD8[+] T

Figure 6.9 Ribbon diagrams of class I and class II MHC molecules. Left is the human class I molecule, HLA-Aw68, and right is the human class II molecule, HLA DR1. The plasma membrane would be beneath each molecule. The antigenic peptide is shown bound within the cleft (top) formed between the two α-helices above the β-sheet.

Redrawn with permission from Stern, L.S. and Wiley, D.C. (1994) *Structure*, **2**, 245–51.

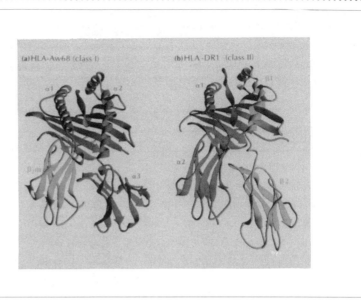

lymphocytes are restricted by class I MHC molecules and CD4$^+$ T lymphocytes are restricted by class II MHC molecules.

The CD4 and CD8 molecules are important in antigen recognition by T lymphocytes. When a T lymphocyte receptor binds an appropriate peptide-MHC complex, the CD4 or CD8 molecules (depending on the type of T lymphocyte) move in the plane of the plasma membrane to join the antigen receptors. CD8 molecules bind to a site on the α3 domain of class I MHC

Figure 6.10 CD4$^+$ T lymphocytes are restricted by class II molecules and CD8$^+$ are restricted by class I.

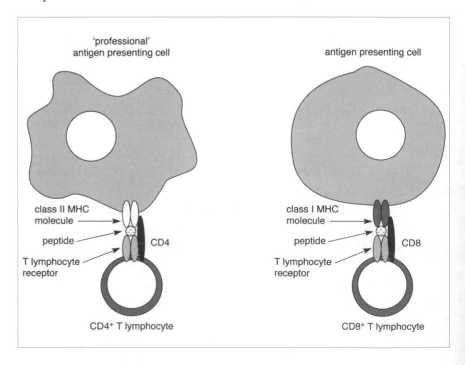

molecules, whereas CD4 binds to a site between the α_1 and α_2 domains of a class II molecule. This serves to strengthen the interaction between the receptor and the MHC molecules. As we will learn in Chapter 8, the CD4 and CD8 molecules also have enzymes associated with their cytoplasmic domains that are an important component in the activation process. Because CD4 and CD8 are important accessory molecules of the antigen receptor of T lymphocytes, these molecules are often referred to as **coreceptors**.

Signal transduction by antigen receptors: Section 8.1.2

6.2.5 Tissue distributions of class I and class II molecules

All cells are potential targets for transformation, by viruses or by mutations, and must be susceptible to lysis by CD8$^+$ cytotoxic cells. Consequently class I molecules are found expressed on the surface of virtually every nucleated cell in the body. The actual levels of expression of class I vary enormously, being highest on professional antigen-presenting cells and lymphoid cells, and lowest on cells such as neurons of the central nervous system and skin keratinocytes. Also, expression of class I molecules is elevated in response to cytokines, such as interferon-γ produced during inflammatory reactions.

cytokines: Box 8.2

Class II molecules, in contrast, are normally only expressed on professional antigen-presenting cells – dendritic cells, macrophages and activated B lymphocytes. On these cells, class II molecules are expressed *in addition* to the class I molecules. In mammals, several cells of epithelial origin are known to express class II molecules at the sites of local inflammation. This may allow them to participate in localized inflammatory responses.

6.2.6 Both alleles are expressed

Classical MHC genes are unusual genes because both the maternal and paternal alleles are transcribed and translated into polypeptide. Neither allele is dominant or recessive to the other. Thus, a mouse that is homozygous at each locus has three different class I molecules (K, D and L) on the surface of every nucleated cell. This is the case in a fully inbred strain. However, if the loci were heterozygous, as might be the case in outbred animals, *all six* molecules would be found on each cell.

A mouse also produces two different class II molecules, I-A and I-E. The situation is a little more complex with the class II molecules. Firstly, as described earlier, there are multiple loci whose products are interchangeable in the assembled heterodimer. Secondly, there is simultaneous expression of maternal and paternal alleles. This allows hybrid molecules of maternal α and paternal β chains, and *vice versa*, to be created. Thus, heterozygous individuals will have a greater variety of class II MHC molecules than homozygotes. These hybrid molecules may possess new peptide-binding

codominance *n.* a state in which both alleles of a heterozygous gene are expressed to give a phenotype that is different from that produced by either allele in a homozygous state.

properties and expand the range of antigens that can be exposed to the T lymphocytes. A situation when two alleles present in a heterozygous state produce a different phenotype from that caused by either allele in the homozygous state is termed **codominance** (see Figure 1 in Box 6.1 *Inbreeding*).

6.2.7 Other genes in the MHC

In addition to the classical MHC molecules, the genes for many other proteins are found in the MHCs of animals (Figure 6.7). In HLA, for example, there are at least 100 genes, of which only 10 or so encode classical MHC molecules. Around half of the genes in HLA have no known function. The remaining genes can be classified into three categories.

The first are the **non-classical MHC molecules**. These have sequences that are related to the classical molecules, although their products are not detected by allorecognition phenomena or alloantisera. Non-classical molecules include the abundant Qa and Tla class I molecules of H-2, or the DM, DN and DO class II molecules in HLA. These molecules are probably also involved in antigen presentation to T lymphocytes in some way, although as yet, little is known about their function.

The second category are genes that are completely unrelated to the classical and non-classical MHC genes. Some interesting genes, which in HLA are located in the class II region between the DP and DQ genes, are thought to be important in the intracellular processing and transport of antigen: we will return to these in Section 6.5.1.

class III MHC molecules heterogeneous proteins encoded in the MHC that are distinct from class I and II proteins, and include tumour necrosis factor and some complement components.

The third category, also unrelated to classical MHC genes, are found in several mammals as a large cluster of genes between the class I and class II regions (see Figure 6.6). These are called **class III genes**, although unlike the classical class I and class II genes, the class III genes are mostly unrelated to each other and specify a heterogeneous group of proteins with diverse functions. In humans, these proteins include cytochrome P450, tumour necrosis factors α and β (which are cytokines), components C2, C4 and factor B of the complement cascade, and heat shock protein 70 (whose precise function is unclear). In many animals the class III genes are clustered between the class I and II regions.

Summary

- There are two classes of MHC molecule, I and II. The model animals for the study of MHC genes and molecules are mouse and man.

- Class I molecules are heterodimers consisting of a heavy or α chain, encoded at the A, B and C loci in the human MHC, or K, D and L loci in the mouse MHC, and a light chain called β_2-microglobulin that is encoded outside the MHC.

- Class II molecules are heterodimers of α and β chains, each encoded at loci within the DP, DQ and DR regions in the human MHC, and in the I-A and I-E regions of the mouse.

- Class I molecules are found on all nucleated cells and present peptide epitopes to $CD8^+$ T lymphocytes. Class II molecules are found only on professional antigen-presenting cells and present peptides to $CD4^+$ T lymphocytes.

- Maternal and paternal MHC alleles are expressed codominantly.

6.3 Polymorphism of MHC proteins

A polymorphic protein is encoded by a gene that exists as two or more different alleles that are present in a population at frequencies too high to be due to spontaneous mutation alone. Usually natural selection, operating on the phenotypic differences that the different alleles encode, accounts for the higher than expected frequencies.

Classical MHC molecules are the most polymorphic proteins known. While most polymorphic genes may have up to five different alleles, many classical MHC genes have ten-fold this number. In man for example, there are at least 40 *different* alleles that can be found at the A locus and at least 60 at the B locus.

6.3.1 Nomenclature of MHC alleles

Human

The original method for identifying the products of the different human MHC alleles was the use of antibodies as probes. This is called **serological typing** and is used for both class I and class II alleles. Because alloantiserum contains the antibodies from many different clones, which recognize a variety of different epitopes on the antigen, alloantisera often define a number of distinct, but antigenically related, HLA molecules. Sera for HLA typing are therefore carefully chosen to have the narrowest range of antigen specificities, although even the best will recognize several different alleles.

serological typing the identification of histocompatibility antigens (class I and II MHC molecules) on cells by using antibodies.

alloantiserum *n.* antiserum raised in one animal against the cells of an unrelated member of the same species; such sera contain antibodies to cell-surface antigens that differ between the two animals, especially alleles of genes in the major histocompatibility complex.

BOX 6.3 Polymorphism

Origins of polymorphism

Individual members of a population vary for virtually any morphological characteristic we wish to measure. Some characteristics fall into discrete visible categories, such as the well-known white and black dimorphism of the peppered moth, *Biston betularia* (Figure 1a), or the multiple colours and banding patterns of the terrestrial snail *Cepaea nemoralis* (Figure 1b). Any characteristic that is present in more than one recognizable form in a species is **polymorphic**.

Polymorphism also exists at the protein level, which may or may not be manifested as a visible trait. Less conspicuous protein polymorphisms may require analytical techniques that measure biochemical properties, such as molecular weight or isoelectric point (the characteristic pH at which a protein carries no net charge). Alternatively, the protein can be cleaved into peptide fragments using proteinases. A polymorphic protein obtained from different individuals produces variable patterns of peptide sizes, depending on whether the differences between the variants are found at enzyme cleavage sites. This is called **peptide mapping.** Peptide mapping is likely to underestimate the true number of different variants since variable amino acids located outside the enzyme recognition site would go unnoticed. Antibodies can also be used to discriminate the products of different alleles of a given protein.

Ultimately, polymorphisms originate from genetic alterations, such as mutations, gene rearrangements, etc. Only mutations in the germline are inherited and can become established in populations through the forces of natural selection (see below). As a general rule, the variation between individuals increases as we descend from the morphological, through the protein level to the DNA level. At the DNA level, individuals are genetically unique.

Figure 1 Polymorphism of the peppered moth *Biston betularia* (a) and the snail *Capaea nemoralis* (b). From Kettlewell, H.B.D. (1959) *Scientific American*, March, and Riley, G., Kerney, M.P. and Cameron, R.A.D. (1979) *A Field Guide to Land Snails of Britain and Northwest Europe*. HarperCollins, London.

(a)

(b)

Although genetic polymorphisms can be identified by direct sequencing of DNA from different individuals, a quicker method is to compare the patterns of fragment sizes generated when DNA is cleaved at precise sites using restriction endonucleases (**restriction fragment length polymorphisms**). In this technique (Figure 2), a restriction enzyme is used to cleave the DNA from different individuals at short specific sequences. Any polymorphism will alter the number of these sites, and consequently the pattern of fragments generated.

Maintenance of polymorphism

A mutation that has a beneficial effect on the fitness of the organism is favoured by natural selection and will increase in frequency in subsequent generations. In the absence of natural selection a mutation usually remains at low frequency, although this frequency may drift in very small populations. A mutant gene 'qualifies' as an allele when it reaches a frequency in the population of 0.05. This is substantially higher than would occur without selection. The genetic definition of polymorphism therefore is a locus for which there at least two alleles, the rarest of which is present in the population at a frequency too high to be accounted for by mutation alone. A locus with two alleles, with the rarer of the two below this frequency, is considered monomorphic.

It has been estimated that 60–75% of all loci show genetic variation and on average 20–30% of all loci in an individual may be heterozygous. This reservoir of genetic variability allows populations to adapt to a change of environmental conditions. Dark or melanic forms of moths that have emerged in industrialized countries are an example we used in Figure 1. Prior to the widespread burning of coal in industrialized areas of England, the white form of *B. betularia* was predominant. This colouration provides excellent camouflage from insectivorous birds during daylight as the moths rest on lichen-covered branches and rocks. The darker form was more conspicuous. Over several generations, consistently higher levels of predation of the darker forms gradually decreased the frequency of the melanic allele. In areas of industrialization, the situation is reversed; atmospheric pollution killed the lichens and blackened the trees with soot. This environment favours the darker form and the white coloured moths were disadvantaged.

Although more recent studies have suggested that the situation is more complex than first assumed (three alleles are involved, and the effects of other ill-defined selective forces, and of moth migration, have been raised), melanism of *B. betularia* is a classic example of a single locus model of natural selection. In the case of the banded snails it has been

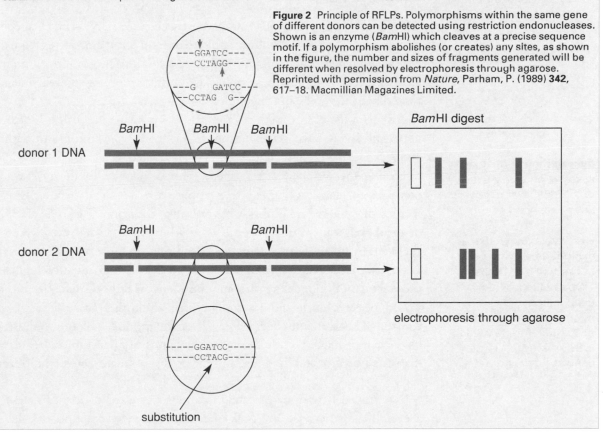

Figure 2 Principle of RFLPs. Polymorphisms within the same gene of different donors can be detected using restriction endonucleases. Shown is an enzyme (*Bam*HI) which cleaves at a precise sequence motif. If a polymorphism abolishes (or creates) any sites, as shown in the figure, the number and sizes of fragments generated will be different when resolved by electrophoresis through agarose. Reprinted with permission from *Nature*, Parham, P. (1989) **342**, 617–18. Macmillian Magazines Limited.

hypothesized that this is also influenced by predation by birds, the rarer phenotypes being evidently more conspicuous than the more abundant and better camouflaged phenotypes. Other data suggest the selective force comes from the influence of different patterns on the ability of the snail to regulate its body temperature.

Human adult haemoglobin is an example of a less conspicuous polymorphic protein. This protein of is a tetramer of two α chains and two β chains ($\alpha\alpha\beta\beta$) plus a haem group. Over 150 different forms of these polypeptides can be resolved if the haemoglobins from different people are compared by peptide mapping. Probably the best known mutant is that which causes sickle cell anaemia. This mutation causes a substitution at position 6 in the β chain from the normal valine to a glutamic acid. The result is a conformational change in the molecule and its properties, with the consequence that the morphology of the erythrocyte changes to a sickle shape. Sickle cells are very fragile and easily destroyed. Individuals that are homozygous for the sickle cell allele develop severe haemolytic anaemia and in 80% of cases die before reproductive age. In contrast, heterozygotes with both wild-type and sickle cell alleles have a mild form of anaemia called sickle cell trait caused by their 'codominant' expression (see Box 6.1).

Ordinarily therefore, the sickle cell mutation would be eliminated from the population by the forces of natural selection. However, this particular mutation is found at high frequency in certain populations of Africa. The reason is that people that are heterozygous are less susceptible to the effects of falciparum malaria. There is thus a balance in which the heterozygotes have an advantage compared to homozygotes (an example of **heterozygote advantage**). As a consequence, in areas of indigenous malaria the frequency of the sickle allele can exceed 10%, whereas it is normally below 2% elsewhere in the world.

microlymphocytotoxicity assay *n.* an assay used in the typing of HLA antigens.

vital dye *n.* a dye that is excluded by living cells but not by dead cells. Trypan blue is widely used in light microscopy, whereas propidium iodide (which is fluorescent) is widely used in flow cytometry.

In practice, a large panel (50–100) of these well-characterized HLA-specific antibodies are tested for binding to blood mononuclear leucocytes from the donor. One of the best sources of HLA-specific antisera comprises multiparous women; naturally these only exist in finite supplies and reagents are shared between collaborating typing laboratories.

An established procedure for HLA typing is the **microlymphocytotoxicity assay.** Pure lymphocytes from the donor and typing antibodies are mixed in small wells and incubated to allow the antibodies to bind to the surface HLA antigens. This is followed by the addition of complement, usually obtained from rabbit serum. This leads to complement fixation and cell lysis if antibody has bound to the cell surface. The quantity of dead cells can be measured by the addition of a vital dye, which is excluded by live cells and stains only dead cells. By observing the overlapping patterns of reactivity of the antibodies it is possible to identify the alleles present on the donor cells with some confidence. The hope that HLA typing would be revolutionized by monoclonal antibodies has not yet been realized. This is mainly because many of the monoclonal antibodies that have been produced bind to conserved regions of HLA molecules, rather than the polymorphic regions, and fail to discriminate between different alleles.

The mixed lymphocyte reaction can also be used to identify class II alleles. This is because the bulk of the lymphocytes respond to class II antigens rather than class I. Here, reference cells with known class II alleles are used to stimulate the cells from the donor to be typed. The presence of a proliferative response in the mixed lymphocyte reaction indicates a disparity of one or more class II alleles, whereas the absence of a response indicates the class II alleles of both parties are identical. Using a sufficiently large panel of reference cells it is possible to identify the class II alleles a person possesses in this way.

Currently, techniques for identifying the HLA directly by their genes are being introduced that may obviate the need for protein-based HLA typing. Restriction fragment length polymorphisms (RFLPs) were used initially but this added little over conventional serological typing.

Several new methods, based on the specific amplification of allelic DNA by the polymerase chain reaction (PCR) (see Box 6.4), are changing the way HLA typing is performed. One approach is to design oligonucleotide primers that will hybridize only to specific DNA sequences shared by HLA alleles found at a particular locus, such as HLA DQA – which codes for the α chains of DQ molecules. The PCR product is then spotted on nitrocellulose membranes. The spots are then probed with other, radiolabelled oligonucleotide probes that have been designed to bind specifically to different DQA alleles. Sequencing of PCR products have revealed several new alleles (or 'sub-types') that were not known from conventional serological typing. This is because serological typing cannot discriminate between antigens that differ by amino acids that are inaccessible to the antibody, such as those located inside the folded MHC molecule.

On the basis of emerging DNA sequence data, the current nomenclature used for naming different human HLA alleles is in a period of transition. The modern nomenclature places the locus before a unique allele number, the two being separated by an asterisk: HLA-A*0201, HLA-A*0202 and so on. The older terminology based on serological typing would call these alleles HLA-A2.1 and HLA-A2.2. A recent list of different alleles of classical class I and II genes in the HLA, given in Table 6.1, emphasizes the considerable polymorphism.

RFLPs: see Polymorphism: Box 6.3

RFLP (restriction fragment length polymorphism) genetic polymorphism as revealed by the sizes of fragments generated with a particular restriction endonuclease.

PCR (polymerase chain reaction) a technique that replicates a specific sequence of DNA.

polymorphism *n.* a gene that exists as two or more different alleles that are present in a population at frequencies too high to be due to spontaneous mutation alone.

Mouse

As we learned above, the production of specific anti-H-2 antibodies has been greatly facilitated by having inbred strains. Superscripts were originally assigned to the inbred strains. Thus, an inbred strain such as BALB/c has an MHC designated H-2^k, to distinguish it from an entirely different collection of alleles that is found in another strain, such as DBA/2 (which is H-2^d). The individual alleles of the inbred strain are given the same superscript. Thus H-2^k is shorthand for K^k, D^k, L^k, Eα^k, Eβ^k, Aα^k and

BOX 6.4 The polymerase chain reaction

The polymerase chain reaction (PCR) is a laboratory technique for amplifying a segment of DNA lying between two regions of known sequence. Two oligonucleotides that hybridize to these known flanking regions are used as primers for DNA synthesis by a DNA polymerase. The oligonucleotides are chosen such that they hybridize to opposite strands of the template DNA. In the presence of the template and the four dNTPs, the DNA polymerase uses the intervening region between the primers as the template for synthesis. The reaction proceeds in cycles, starting with the separation of the double-stranded template DNA into single-stranded DNA by heating (denaturation phase). The reaction is then cooled to allow the oligonucleotide primers to hybridize to their target sequences (annealing phase), which are then extended by the DNA polymerase (primer extension phase).

The cycle of denaturation, annealing and extension is then repeated, usually 25 to 35 times. The first cycle produces long double-stranded DNA molecules whose length exceeds the target sites of the primers. With the second cycle, products of the desired size (whose length is defined by both primers) are produced. Because these reaction products can also act as templates for DNA synthesis in subsequent cycles, there is doubling of the yield of new DNA with each cycle (see Figure 1) which then increases at an exponential rate with subsequent cycles. Although the longer forms are also produced they accumulate at a linear rate and represent a negligible fraction of the end product.

Various enhancements have been made to the original protocol. Originally the Klenow fragment of *Escherichia coli* DNA polymerase I was used for the DNA synthesis phase. This enzyme is inactivated at the high temperature required to denature DNA, and a fresh aliquot of enzyme was needed for each cycle. The problem was solved by the use of a thermostable DNA polymerase from the bacterium *Thermus aquaticus*. This is a thermophilic bacterium that has adapted to life in hot springs. The enzyme could be added once at the beginning of the reaction. Automation has also been introduced. PCR reactions are conducted in automatic heating blocks, or 'thermal cyclers', that automatically raise and lower the temperature of the reaction.

PCR has proved to be an extraordinarily useful technique. By careful design of the primers so that they bind specifically to target DNA sequences, PCR can be used as a diagnostic tool for detecting genetic disorders, the presence of pathogens, or polymorphisms or matching individuals with forensic samples. For example, primers that anneal with the genomic DNA of certain pathogens, such the human immunodeficiency virus (HIV), can be used to determine whether infection is present in a blood sample or biopsy. In these cases, the presence of a PCR product indicates the presence of template DNA in the sample. HLA typing is another example of this kind of application. The reaction is able to amplify as little as a single molecule of DNA. When used as a diagnostic tool PCR can be very sensitive.

Another use for PCR is for obtaining useful quantities of a particular piece of genomic DNA for further manipulation such as sequencing. This is particularly useful if the desired DNA is present in minuscule amounts in the source, or part of a complex mixture of genomic sequences. PCR allows the desired DNA to be 'extracted' in a pure form.

The reaction is not without problems; no two applications are the same and time must be spent optimizing the technique. The ability of the technique to amplify vanishingly small amounts of DNA requires that precautions are taken to prevent the template DNA from contamination which may cause 'false positives' or non-specific background bands. Moreover, mutations are sometimes introduced into amplified DNA. It is important therefore that if the DNA is being amplified for sequencing, the sequences of amplified DNA from at least two PCR reactions are compared.

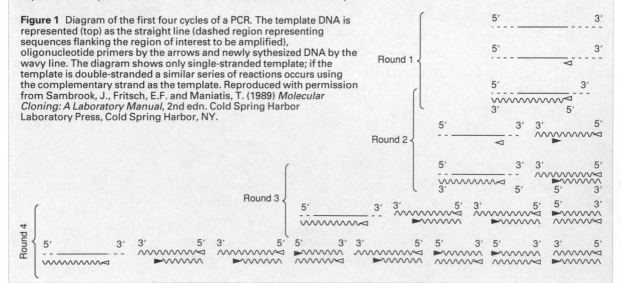

Figure 1 Diagram of the first four cycles of a PCR. The template DNA is represented (top) as the straight line (dashed region representing sequences flanking the region of interest to be amplified), oligonucleotide primers by the arrows and newly sythesized DNA by the wavy line. The diagram shows only single-stranded template; if the template is double-stranded a similar series of reactions occurs using the complementary strand as the template. Reproduced with permission from Sambrook, J., Fritsch, E.F. and Maniatis, T. (1989) *Molecular Cloning: A Laboratory Manual*, 2nd edn. Cold Spring Harbor Laboratory Press, Cold Spring Harbor, NY.

Further Reading

Innis, M.A., Gelfand, D.H., Sninsky, J.J. and White, T.J. (eds) (1989) *PCR Protocols. A Guide to Methods and Applications.* Academic Press, San Diego, CA.

McPherson, M.J., Quirke, P. and Taylor, G.R. (1991) *PCR: A Practical Approach.* IRL Press, Oxford.

White, T.J., Arnheim, N. and Erlich, H.A. (1989) The polymerase chain reaction. *Trends in Genetics,* **5**(6), 185–8.

$A\beta^k$. Remember that inbred strains are homozygous at every locus, so in reality H-2^k stands for K^k/K^k, D^k/D^k and so on. Normally of course, animals are not inbred, but belong to outbred populations and are usually heterozygous at each locus.

6.3.2 Inheritance of MHC alleles

Figure 6.11 shows the patterns of inheritance of MHC genes in two humans. (For simplicity the older nomenclature that applied to serological typing has been used in Figure 6.11.) Human cells have 23 pairs of homologous chromosomes, with the HLA complex located on the sixth pair. Shown are the main class I loci (A, B and C) and class II regions (DP, DQ and DR) of HLA.

The particular set of alleles found in a cluster, such as that in the HLA complex, borne *on one chromosome* of a homologous pair, is called a **haplotype**. Thus the particular combination of HLA alleles on each chromosome is the HLA haplotype. The combination of both haplotypes in an individual is usually called the 'HLA type'; for example, the male in Figure 6.11 has the following HLA type: A2, 24/B8, 62/Cw3, 5/DP3, 4/DQ 8/DR4, 11. Notice that some loci are homozygous, such as DQ, and others are heterozygous.

> **haplotype** *n.* the particular combination of alleles in a linked group, such as the major histocompatibility complex, present on one chromosome of a homologous pair.

The offspring usually have one of four possible pairs of haplotypes, as shown in Figure 6.11. Obviously, two siblings may have identical HLA haplotypes as each other and could donate grafts of tissue or cells to each other should the need arise. The loci in the MHC are clustered close together and are normally inherited *en bloc*. However, as discussed earlier, crossovers do occasionally occur within the HLA complex. This enabled geneticists to produce the first maps of HLA and the MHCs of other animals. Crossovers also create novel HLA haplotypes.

6.3.3 Responsiveness depends on the MHC alleles inherited

An important consequence of the remarkable polymorphism of animal MHCs is that the particular set of MHC haplotypes an individual inherits is likely to be unique among unrelated individuals in the population.

To a large extent, the MHC haplotype inherited by an individual deter-

Table 6.1   Different alleles of nine polymorphic human classical MHC genes.

					Locus			
DPB	DPA	DQB	DQA	DRB1	DRA	B	Cw	A
*0101	*0101	*0501	*0101	*0101	*0201	*2702	*0101	*0201
*02011	*0103	*0502	*0102	*0102	*0202	*2703	*0102	*0202
*0202	*0102	*05031	*0103	*0103		*27051	*1401	*0203
*0301	*0201	*05032	*0104	*0104		*27052	*0301	*0204
*0401	*02011	*0504	*0201	*1501		*2707	*0302	*0205
*0402	*02022	*06011	*03011	*1502		*4001	*0303	*0206
*0501	*0301	*06012	*03012	*1503		*4002	*0304	*0210
*0601	*0401	*0602	*0302	*1601		*4003	*0401	*0211
*0801		*0603	*0401	*1602		*4004	*0402	*0212
*0901		*0604	*05011	*03011		*4005	*0201	*68011
*1001		*0605	*05012	*03012		*4006	*02021	*68012
*1101		*0606	*05013	*0302		*4101	*02022	*6802
*1301		*0607	*0601	*0303		*4701	*0501	*6901
*1401		*0608		*0401		*1301	*0801	*2501
*1501		*0201		*0402		*1302	*0802	*2601
*1601		*0301		*0403		*4401	*0803	*3401
*1701		*0302		*0404		*4402	*0601	*3402
*1801		*03031		*0405		*4403	*0602	*4301
*1901		*03032		*0406		*1401	*1201	*6601
*20011		*0304		*0407		*1402	*1202	*6602
*2101		*0305		*0408		*3801	*1301	*2901
*2201		*0401		*0409		*39011	*1501	*2902
*2301		*0402		*0410		*39013	*1502	*31011
*2401				*0411		*3902	*1503	*31012
*2501				*0412		*3903	*0701	*3201
*2601				*0413		*0701	*0702	*3301
*2701				*0414		*0702	*0703	*7401
*2801				*0415		*0703		*2301
*2901				*0416		*0801		*2401
*3001				*11011		*4201		*2402
*3101				*11012		*4801		*2403
*3201				*1102		*4802		*0101
*3301				*1103		*4501		*0301
*3401				*11042		*4901		*0302
*3501				*1105		*5001		*1101
*3601				*1106		*5401		*1102
*3701				*1201		*5501		*3001
*3801				*1202		*5502		*3002
*3901				*1301		*5602		*3003
*4001				*1302		*5601		*3601
*4101				*1303		*5602		
*4201				*1304		*1801		
*4301				*1305		*3701		
*4401				*1306		*3501		
*4501				*1307		*3502		
*4601				*1308		*3503		
*4701				*1401		*3504		
*4801				*1402		*3505		
*4901				*1403		*3506		
*5001				*1404		*3507		
*5101				*1405		*3508		
				*1406		*5301		
				*1407		*5101		
				*1408		*5102		
				*1409		*5103		
				*1410		*5104		
				*1411		*52011		
				*0701		*52012		
				*0702		*7801		
				*0801		*1501		
				*08021		*1502		
				*08022		*1503		
				*08031		*1504		
				*08032		*1505		
				*0804		*1506		

DPB	DPA	DQB	DQA	DRB1	DRA	B	Cw	A
				*0805		*1507		
				*0806		*4601		
				*09011		*7901		
				*09012		*5701		
				*1001		*5702		
				*5801				
				DRB3				
				*0101				
				*0201				
				*0202				
				*0301				
				DRB4				
				*0101				
				*0102				
				DRB5				
				*0101				
				*0102				
				*0201				
				*0202				
				*0203				
				DRB2				
				*0101				
				DRB6				
				*0101				
				*0201				
				*0202				
				*0101				
				DRB8				
				*0101				
				DRB9				
				*0101				

These different alleles are defined by their nucleotide sequences. Each locus is occupied by one maternally derived and one paternally derived allele. Refer to Figure 6.6 for the relative positions of these loci.

mines responsiveness. Synthetic peptides and inbred animals provide a useful insight. Recall from Section 6.1 that a strain 13 guinea pig fails to respond to polylysine. This is because it lacks an appropriate MHC allele to present it to T lymphocytes. C57 mice are poor responders to (H,G)-A,L because the H-2^b alleles it has cannot present this peptide. High responders in contrast have the required MHC alleles. Thus strain 2 guinea pigs and DBA-2 mice are able to present polylysine and (H,G)-A,L, respectively, to their T lymphocyte repertoire to elicit a response.

It is more difficult to find evidence of this above phenomenon in outbred populations, particularly for more relevant antigens such as microorganisms. Microorganisms contain a myriad of antigens that offer a much wider variety of potential epitopes for presentation to T lymphocyte receptors than synthetic peptides. Yet examples do exist and they provide strong evidence that the polymorphism of the MHC is the result of selective pressure from microorganisms.

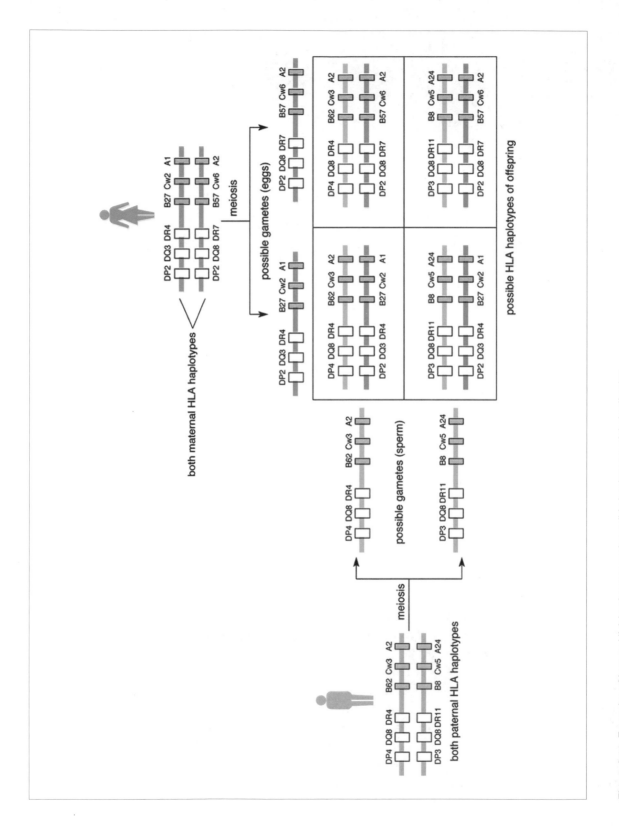

Figure 6.11 Examples of human HLA haplotypes and possible haplotypes of their offspring.

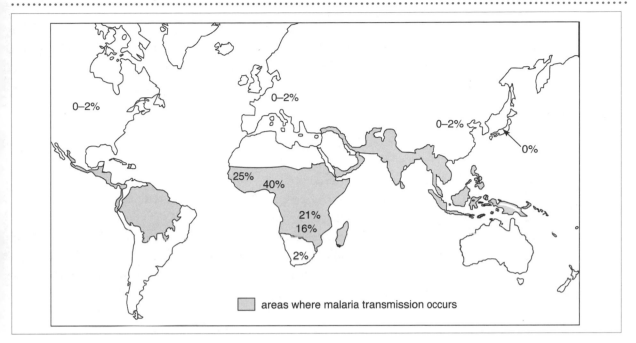

areas where malaria transmission occurs

6.3.4 MHC polymorphism is maintained by natural selection

Alleles are, by definition, present within the population at frequencies too high to be accounted for by spontaneous mutation alone. A key question in immunobiology concerns the selective forces that maintain MHC polymorphism at levels above the rate of mutation, and the reasons why there are so many MHC alleles. It has been suggested that each allele has been maintained by its usefulness in protecting the animal against a particular pathogen.

Evidence for this has come from a studying infectious diseases, such as malaria in West Africa. Several class I and class II HLA alleles seem to be important for protecting humans against infection by the malaria-causing parasite, *Plasmodium falciparum*. In the Gambia, for example, the frequency of the HLA B*5301 allele is noticeably lower among those who succumb to the disease (16.9% compared with 25.4% in controls). Similarly, the frequency of a class II allele, HLA DRB1*1302, is also lower among cases of malarial anaemia (8.7% versus 16.75% in healthy sub-Saharan Africans). These figures are much higher than elsewhere in the world. In non-malarial regions, the B*5301 allele is normally present in less than 2% of the population, whereas its frequency can be as high as 40% in some malarial regions of Africa (Figure 6.12).

Clearly these particular alleles are protective in some areas where malaria occurs. The influence of malaria on their frequency is an example of

Figure 6.12 Correlation between high frequency of HLA-B*5301 and malaria. The map shows areas where malaria transmission occurs (shaded), and the frequency of the population possessing the HLA-B*5301 allele where known. In Africa, HLA-B*5301 is present at the highest frequency in sub-Saharan zones, but it is less frequent in black South Africans, rare in Caucasian and Oriental populations, and absent in Pacific Islanders and Amerindians. Data from *WHO World Malaria Situation in 1989* Parts I and II (1991); *Weekly Epidemiological Record*, **66**(22/23), 157–163/167–170; Hill, A.V.S., *et al.* (1991) *Nature*, **352**, 595–600 and **360**, 434–9.

Plasmodium n. a genus of parasitic protozoa, including *P. falciparum*, the causative agent of malaria in humans.

Epstein–Barr virus (EBV) a herpes virus that causes infectious mononucleosis (glandular fever) in humans, and is implicated in the development of several malignancies of B lymphocytes; can immortalize B lymphocytes *in vitro*.

Darwinian natural selection. Those with the allele have an increased fitness (better prospects of producing offspring) and its frequency in the population rises. This is the same principle behind the elevated frequencies of other alleles in areas of endemic malaria, such as the sickle cell allele, discussed in Box 6.3. But why are these alleles protective? Studies have shown that the HLA B*5301 allele is particularly suited for presenting a particular peptide derived from the *Plasmodium* parasite while it is living inside liver cells. It is thought this display allows infected liver cells to be recognized and killed by cytotoxic T lymphocytes.

A similar situation appears to exist for the class I allele HLA-A*1101 and resistance to strains of Epstein-Barr Virus (EBV). This virus causes many diseases and is implicated in nasopharyngeal carcinoma. In some human populations, including the Chinese and islanders of Papua New Guinea, the A*1101 allele is present in over half the population. As such, strains of EBV to which this allele confers protection are virtually absent in these localities. As might be expected, however, other strains of EBV have emerged in these areas, each with amino acid substitutions within epitopes presented by the A*1101-presented epitopes.

The extraordinary polymorphism of the MHC is thought to be the combined result of similar situations occurring for other alleles, and combinations of alleles working together. The relative abundance of each allele in the population reflects their usefulness for protecting individuals against pathogens found in their locality. *Plasmodium* and other pathogenic organisms are likely to have been important in shaping the polymorphism of MHC molecules of man during the millions of years that they have been co-evolving. This is discussed in more detail in the following chapter.

Some mammals such as the Syrian hamster, the cotton-top tamarin, whales and the Florida panther appear to display very limited MHC polymorphism. Some have argued that this is because these animals have been exposed to very few pathogens, resulting in little selective pressure to maintain extensive polymorphism. In some species, such as the panther, the reduced polymorphism has been caused by dwindling population sizes and the extinction of particular alleles. In small populations, the lack of polymorphism is exacerbated by inbreeding, and in many species threatened with extinction, demise is likely to be due to the emergence of deleterious recessive genes as well as an inability to fight infection. Many examples of limited polymorphism may be due to such 'evolutionary bottlenecks'.

Summary

- Classical MHC molecules are the most polymorphic proteins known.

- An individual inherits a particular combination (haplotype) of classical MHC alleles from each parent.

- The maternal and paternal MHC haplotypes in an offspring are expressed codominantly. Their products, the MHC molecules, together play a large part in dictating the efficiency with which peptides can be presented and, ultimately, whether an individual can respond to a particular antigen.

- Like other polymorphisms (see Box 6.3) the variation originates in the germline as mutations, which are then subjected to natural selection. In the MHC, alleles that confer resistance to infection by a pathogenic organism increase in frequency in the population by natural selection. Malaria in humans is a naturally occurring example of this.

6.4 How the structure of MHC molecules relates to function

By definition, each allele of any given gene must have provided the protein with a property that is, or was at some stage, beneficial to the organism. To understand how different MHC alleles are useful to an animal we need to first examine allelic differences at the molecular level.

6.4.1 Differences between alleles are concentrated in the peptide-binding cleft

The A*0201 molecule of man is the most abundant MHC allele in the Caucasian population, present in around 40% of individuals. This was the first MHC molecule to be crystallized, followed soon after by HLA A*6801. These are two different alleles of the HLA-A locus. Consequently they are homologous proteins, differing by only 13 amino acids out of the total of 273 that make up the three extracellular domains. As shown in Figure 6.13, 10 of these differences are at positions that line the peptide-binding cleft.

Several other class I and class II molecules have since been crystallized, and each folds in a similar fashion. In each case, the differences between

Figure 6.13 Positions of amino acids that differ between HLA-A*0201 and HLA-Aw68.

Shown is the 'above' view of the consensus structure showing the peptide-binding cleft formed between the α_1 and α_2 domains. Reprinted with permission from Parham, P. (1989) *Nature*, **342**, 617–18, copyright 1989 Macmillan Magazines Limited.

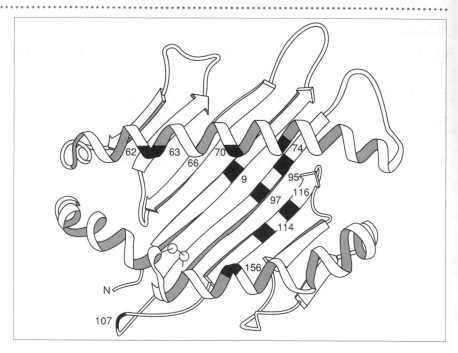

MHC alleles are not randomly scattered throughout the molecule, but focussed in and around the peptide-binding cleft, leaving the remainder of the molecule relatively well conserved. Since amino acids differ by the size and charge of their side chains, it follows that surface topography and charge of this region of molecule are different between alleles. This means that while the overall shape of different MHC molecules is very similar, the surface topography of the cleft of each allele is different.

Since the interior contours of the cleft are different for each allele, it follows that each is able to accommodate different peptides. In reality, the binding is extraordinarily promiscuous. Figure 6.14 shows the heterogeneous mixture of peptides extracted from the DR1 molecules on a B lymphocyte clone. Such an experiment requires MHC molecules from around 10^9–10^{10} cells; each cell has in the order of 10^5 class II molecules. Peptides eluted from MHC molecules in this way are called 'naturally processed peptides', to distinguish them from synthetic peptides used by researchers to mimic them. It is likely that each allelic form of classical MHC molecule can accommodate any one of several hundred or even thousand different peptides.

The sizes of the peptides eluted from MHC molecules differ depending on the class of MHC molecules. Peptides found associated with class I MHC molecules are 8–10 amino acids long, whereas those from class II are typically longer and more heterogeneous in size, ranging from 12 to 18 amino acids. While the peptides bound to class I fit snugly inside the cleft,

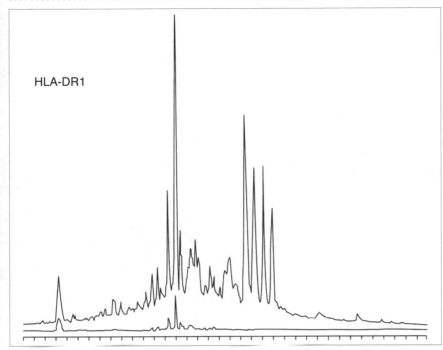

HLA-DR1

Figure 6.14 Naturally processed peptides eluted from purified HLA-DR1 molecules.
The peptides are shown separated by high performance liquid chromatography (HPLC). From Chicz, R.M. *et al.* (1992) *Nature,* **358**, 764–8.

the longer class II-bound peptides tend to drape out of the ends. The peptide is held in the cleft by hydrogen bonds that form between the MHC molecule and the backbone of the peptide. In class I molecules, the bonds form predominantly at the ends of the peptide, whereas in class II the bonds form along the length of the peptide.

6.5 The cell biology of antigen processing

Despite the genetic complexities of the MHC, the *raison d'être* of classical MHC molecules is simply to present peptides at the surface of an antigen-presenting cell to the antigen receptors of T lymphocytes. Cells in the body are constantly degrading antigens that arise from within, or are taken in from, the extracellular fluid, and displaying the resultant peptides to T lymphocytes. It is likely that all proteins produced by a cell are represented by their processed peptides, several thousand of which are on display to the immune system at any one time. It is through these two pathways, summarized in Figure 6.15, that the T lymphocyte repertoire is constantly appraised of the presence of antigen, both surrounding cells and within them.

In this final section, we will examine the origin of these peptides, and the intracellular routes that are taken by antigen during processing. In these respects, class I and class II molecules are very different.

Figure 6.15 Schematic view of antigen processing. Compare these intracellular compartments with Figure 2 in Box 4.2.

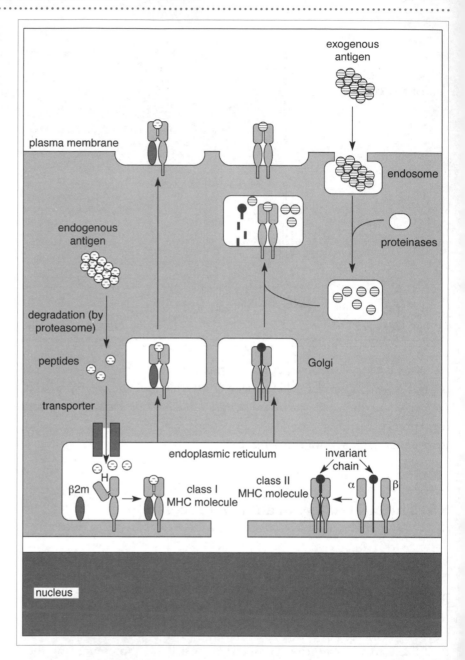

6.5.1 The class I pathway of antigen processing

In all mammals studied, class I molecules are present in varying amounts on nucleated cells. Their function is to present peptides to CD8^{+} cytotoxic T lymphocytes. The export of MHC class I molecules from the site of their synthesis in the rough endoplasmic reticulum to the cell surface follows the same general route as many other exported or plasma membrane associated proteins.

Export of proteins: Box 4.2

The current model for the class I pathway is as follows (Figure 6.15). Proteins synthesized within the cytoplasm of a cell (**endogenous antigens**) are degraded into peptides. This includes self proteins, as well as foreign proteins such as those that are produced by intracellular parasites such as viruses. The involvement of several enzymes in degradation has been postulated, including a huge multisubunit complex of proteinases called the **proteasome** and enzymes of the **ubiquitin degradation pathway**. Very little is known about how many peptides are produced from a protein, or where in a protein the processing enzymes cleave.

The peptides are then transported into the endoplasmic reticulum. The major route into the endoplasmic reticulum appears to be through a transmembrane tube consisting of two polypeptide chains called **TAP-1** and **TAP-2**, which stands for *t*ransporter *a*ssociated with *a*ntigen *p*rocessing. In mammals, the genes encoding TAP-1, TAP-2 and two proteasome polypeptides are all located within the class II region of the MHC (see Figure 6.8). However, the significance of this remains to be determined.

In addition to the heavy chain and β_2-microglobulin, a peptide is a third major component of the class I molecule. Without it, the empty molecule rarely leaves the endoplasmic reticulum. Binding causes a conformational change as the two α-helices envelop the peptide. This change promotes its assembly with β_2-microglobulin, and subsequently its movement out of the endoplasmic reticulum and on its way to the surface.

This particular pathway of antigen processing is vital for exposing intracellular pathogens to cytotoxic T lymphocytes. Indeed, many viruses have evolved ways to prevent the export of class I MHC molecules to the cell surface to remain invisible to the immune system (Box 6.5).

6.5.2 The class II pathway of antigen processing

The expression of class II MHC molecules is confined to professional antigen-presenting cells. The major difference between the class I and class II pathways is the source of the antigen. Class II molecules are loaded with peptides that are processed from proteins of extracellular origin (**exogenous antigens**). These include proteins in the extracellular fluid that are ingested by endocytosis (phagocytosis and pinocytosis) and also plasma membrane proteins that are endocytosed.

Antigen that is bound by antibody or complement components are efficiently ingested by professional antigen-presenting cells that bear receptors for complement, or the Fc region of immunoglobulins. Similarly, B lymphocytes can ingest antigen via their membrane-bound immunoglobulin and process it prior to presentation by class II molecules to CD4$^+$ helper T lymphocytes.

---

endogenous antigen *n.* an antigen that originates from within a eukaryotic cell by translation of self or foreign (e.g. viral) RNA in the cytosol; peptides derived from the processing of endogenous antigens are normally presented at the cell surface by class I MHC molecules.

proteasome *n.* a large, multisubunit complex of proteinases found in the cytoplasm of eukaryotic cells.

ubiquitin *n.* a protein found in prokaryotic and eukaryotic cells that is attached to abnormal proteins and which targets them for degradation by proteinases.

Effector function is determined by the Fc region: Section 4.6.1

BOX 6.5 Some viruses interfere with antigen processing

In view of the importance of class I MHC molecules in the recognition of virus-infected cells by cytotoxic T lymphocytes, it is not surprising that some viruses have evolved ways of blocking the export of class I MHC molecules to the cell surface to escape recognition (see Table 1). Many viruses can persist in their hosts without being eliminated for long periods. The down-regulation of class I MHC molecules is likely to play a major role in this process. Whenever such viruses become 'visible' again, such as when they replicate and emerge from the infected cell, they are quickly cleared away. In this way, persistent viruses are prevented from causing disease symptoms by the immune system, but not entirely eliminated. A balance called **latency** is achieved, typified by a long-term but asymptomatic infection. In immunosuppressed individuals, however, latent viral infections may re-emerge to cause disease symptoms.

Table 1   Some human viruses that interfere with the export of class I MHC molecules

Virus	Name of viral protein	Cellular target protein	Mechanism
Herpes simplex	ICP47	TAP-1/TAP-2	prevents transport of peptides into endoplasmic reticulum
Cytomegalovirus	UL18 (homologue of class I heavy chain)	β_2-microglobulin	thought to sequester β_2-microglobulin in the endoplasmic reticulum and prevent assembly with host heavy chains
Adenovirus type 2	E3/19K	class I heavy chain	binds to class I heavy chains; has an endoplasmic reticulum retention signal and blocks export

Processing of exogenous antigen occurs within the endocytic vesicles that are pinched off from the plasma membrane during endocytosis (Figure 6.15). As the endosomes are transported deep into the cytoplasm, vesicles containing various proteinases fuse with the endosomes. In this way the contents of the endosome are gradually digested into smaller and smaller peptide fragments. The endpoint of the endocytic pathway is the lysosome, in which the peptides are degraded into individual amino acids.

Meanwhile, newly synthesized class II MHC molecules assemble in the endoplasmic reticulum. A third chain, called the **invariant chain** (so-called because it is not polymorphic), also assembles with the $\alpha\beta$ heterodimer of the class II molecule at this stage. The functions of the invariant chains are threefold.

exogenous antigen *n.* an antigen that is ingested into endosomes; originate extracellularly, or may be derived from cellular membrane-bound proteins; peptides derived from the processing of exogenous antigens are normally presented at the cell surface by class II MHC molecules.

- First, it acts as a template for the correct folding of the newly synthesized α and β chains.

- Second, it blocks the peptide-binding site from binding peptides transported into the endoplasmic reticulum by the TAP molecules.

- Third, it directs the export of vesicles containing newly synthesized class II molecules to intersect incoming vesicles of the endocytic pathway.

Vesicles containing the class II–invariant chain complexes fuse with endocytic vesicles containing enzymes and digested antigens. The precise location of this compartment is not clear, although it is likely to be before the lysosome. When the contents mix, the proteinases of the endosomes digest the invariant chain away from the class II molecules. The class II molecules are then free to bind peptides found within the endosome. The export vesicles then continue to the cell surface.

invariant chain *n.* a third chain of class II MHC molecules that promotes correct assembly of the α and β chains in the endoplasmic reticulum (ER), prevents binding of peptides while in the ER, and serves to direct exported class II molecules to the endocytic pathway.

6.5.3 Immune surveillance

As can be appreciated from Figure 6.15, the sources of antigen that gain entry into either of these pathways are somewhat different. Antigens that find their way into the class I processing pathway for recognition by CD8$^+$ cytotoxic T lymphocytes are endogenously derived. These may be self proteins such as ribosomal proteins, nuclear proteins or enzymes, or foreign proteins such as the products of viral gene expression. Antigens that find their way into the class II pathway for recognition by CD4$^+$ T lymphocytes are exogenously derived. These may be self membrane-bound proteins that are recycled into endosomes, such as MHC molecules and transferrin receptors, or self proteins in the extracellular fluid. Also foreign proteins captured by surface immunoglobulin, Fc receptors or complement receptors enter this pathway, as do foreign proteins in the extracellular fluid. It is through these two pathways, summarized in Figure 6.16, that the T lymphocyte repertoire is constantly exposed to self and foreign antigens, both surrounding cells and within them. This is called **immune surveillance**. In Chapter 8 we will discuss how the repertoire is 'educated' to ignore the self component of the peptides on display.

Summary

- While the overall shape of different MHC alleles is very similar, the surface topography of the cleft of each allele is different.

- An MHC molecule is able to accommodate any one of a large number of different peptides.

- Antigens produced within the cytosol of a cell are processed therein

Figure 6.16 Immune surveillance. Endogenously produced proteins enter the class I processing pathway, whereas exogenous proteins enter the class II processing pathway.

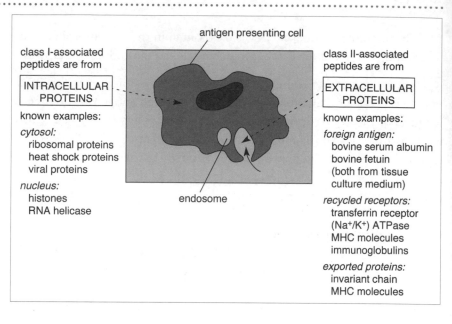

antigen presenting cell

class I-associated peptides are from

INTRACELLULAR PROTEINS

known examples:

cytosol:
    ribosomal proteins
    heat shock proteins
    viral proteins

nucleus:
    histones
    RNA helicase

endosome

class II-associated peptides are from

EXTRACELLULAR PROTEINS

known examples:

foreign antigen:
    bovine serum albumin
    bovine fetuin
    (both from tissue
    culture medium)

recycled receptors:
    transferrin receptor
    (Na+/K+) ATPase
    MHC molecules
    immunoglobulins

exported proteins:
    invariant chain
    MHC molecules

and the peptides enter the class I pathway. Antigens that gain entry from outside enter the endocytic pathway and the peptides generated therein are carried back out by exported class II MHC molecules.

The future

• Many questions remain about the genomic organization of the MHC. Why in different animals are the class I and class II MHC genes always clustered together? Is there any significance attached to the presence in the MHC of other genes needed for antigen processing (encoding proteasome polypeptides and the TAP-1/TAP-2 transporter genes)?

Further reading

Books

Klein, J. (1986) *Natural History of the Major Histocompatibility Complex.* J. Wiley, London.

Reviews

Barber, L.D. and Parham, P. (1993) Peptide-binding to major histocompatibility molecules. *Annual Review of Cell Biology,* **9**, 163–206.

Gooding, L.R. (1992) Virus proteins that counteract host immune defenses. *Cell,* **71**, 5–7.

Jackson, M.R. and Peterson, P.A. (1993) Assembly and intracellular transport of MHC class I molecules. *Annual Review of Cell Biology,* **9**, 207–36.

Klein, J. and OhUigin, C. (1993) Composite origin of major histocompatibility complex genes. *Current Opinion in Genetics and Development,* **3**(6), 923–30.

Koup, R.A. (1994) Virus escape from CTL recognition. *Journal of Experimental Medicine,* **180**, 779–782.

Trowsdale, J. (1993) Genomic structure and function in the MHC. *Trends in Genetics,* **9**, 117–22.

Review questions

Fill in the blanks

Discovery of the MHC was facilitated by the use of _____ [1] strains of mice. These were produced originally for _____ [2] research. It was found that tissue grafts from some strains would be rejected by other strains, in which case the donor and recipient strains are described as _____ [3]. On the other hand, strains that accept exchanged cells or tissue grafts are called _____ [4]. Special _____ [5] strains were also produced that differed only in the genes that controlled graft rejection. In the mouse this locus is called _____ [6]. By immunizing a strain against cells from an _____ [7] strain, antibodies against the donor's cell-surface _____ [8] antigens were obtained from the recipient's serum. These sera are called _____ [9]. Different strains were also found to differ in their immune responsiveness to simple antigens, such as _____ _____ [10]. Further studies showed that responsiveness to a particular antigen and the presence of certain _____ [11] antigens were inherited together, demonstrating that the genes that control responsiveness and graft rejection were the same. These genes are located in the MHC so it is more appropriate to refer to _____ [12] antigens as MHC molecules.

The 'real' function of MHC molecules was revealed by using _____ [13] of T lymphocytes. T lymphocytes respond to antigen when they and the antigen-presenting cells are _____ [14]. This phenomenon is called MHC _____ [15] and it is a consequence of the way in which immature _____ [16] develop in the _____ [17] (explained in Chapter 9). Subsequent experiments showed that the T lymphocyte antigen receptor recognizes both _____ [18] and _____ [19] simultaneously. X-ray crystallography of purified MHC molecules confirmed that the ligand of the receptor is a complex consisting of an antigenic _____ [20] buried within a _____ [21].

Short answer questions

1. What are the main differences between classical and non-classical MHC molecules?

2. What are the similarities and differences between the *structures* of class I and class II MHC molecules?

3. What are the similarities and differences between the *functions* of class I and class II MHC molecules?

4. Define *polymorphism* and explain what is meant by *MHC polymorphism*.

5. Suggest how polymorphism at the MHC influences the responsiveness of an animal to a particular antigen.

6. Below are the H-2 haplotypes of three inbred strains of mouse.

strain	H-2 haplotype
BALB/c	$H\text{-}2^d$
DBA/2	$H\text{-}2^d$
CBA	$H\text{-}2^k$

    Will the following mixed lymphocyte reactions result in a strong T lymphocyte response *in vitro*?
    (i)   DBA/2 and CBA
    (ii) DBA/2 and BALB/c
    (iii) (DBA/2 × CBA)F_1 and CBA

7. A cytotoxic T lymphocyte clone that recognizes an epitope of influenza virus has been obtained from a CBA mouse. Will the following target cells infected with influenza virus be killed by the clone in a ^{51}Cr-release assay?
    (i)   Cells from a DBA/2 mouse.
    (ii) Cells from a CBA mouse.
    (iii) Cells from a (DBA/2 × CBA)F_1 mouse.
    (iv) Cells from a (DBA/2 × BALB/c)F_1 mouse.

8. Most virus infections are cleared by both cytotoxic T lymphocytes and antibodies in their hosts. Several vaccines against viruses are produced by heat-inactivating the virus, thereby preventing it from infecting cells and causing disease. The inactivated virus can then be

used safely to immunize the host to provide immunity. However, while an inactivated virus vaccine elicits antibodies, it often fails to elicit a good cytotoxic T lymphocyte response. Offer an explanation, with reference to antigen processing pathways, why this might happen.

7 Evolution of the adaptive immune system

In this chapter

- A comparison of the immune systems of different vertebrate classes.
- Evolution of the immunoglobulin superfamily leading to MHC molecules and antigen receptors.
- Sources and maintenance of MHC polymorphism.

Introduction

We saw in Chapter 3 how the structure and physiology of vertebrate haemopoietic and lymphoid tissues have changed during the course of evolution. Here we turn to the evolution of the system. Unlike palaeontology, which can draw from the fossil record, the evolution of the adaptive immune system can only be studied from contemporary vertebrates. Mapping immunologically important characters on the evolutionary tree of vertebrates (see Box 3.1) allows us to suggest which are more recently derived and which are ancestral. For example, molecules that are present both in mammals and in the vertebrates that evolved before, such as the fishes and amphibians, are probably ancestral. In contrast, molecules found only in mammals evolved more recently.

7.1 Comparative immunology

Comparative immunology is the study of the immune systems of all vertebrates. By mapping this knowledge onto an evolutionary tree, we can

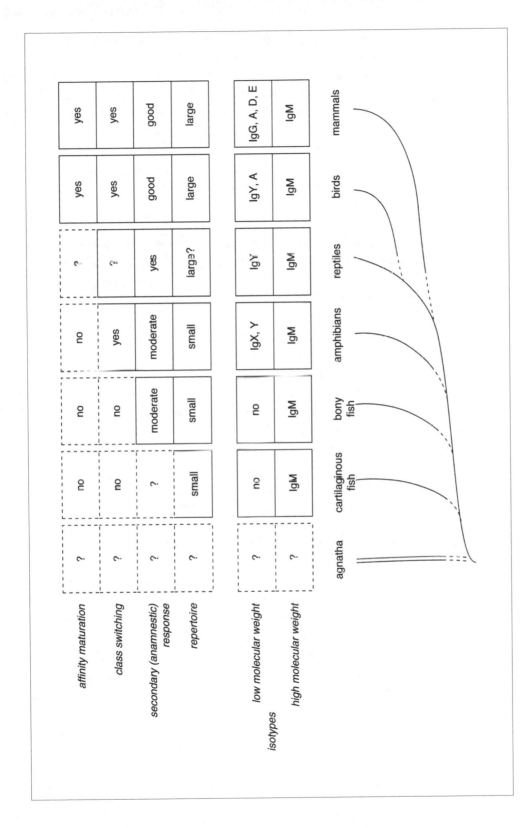

Figure 7.1 Phylogeny of immunoglobulins.

reconstruct events in the evolution of the immune system. Figures 7.1 and 7.2 show how the molecules associated with antigen recognition by B and T lymphocytes are represented across the different classes. Compare these with the phylogenetic progression of lymphoid tissues, shown in Figure 3.9.

It is evident that there has been an increase in the complexity during the course of evolution, built by the addition, piece by piece, onto a pre-existing functional immune system. It is important to stress at this point that there is no evidence that the simpler systems of modern fishes or amphibians are any way inferior or any less able to eliminate pathogens than the more complex systems of mammals.

Of all vertebrate immune systems, that of the eutherian (placental) mammal has been extensively investigated, and it provides the basis for the comparative studies in other species. We will now briefly survey the immune systems of the main groups of vertebrates, highlighting particularly the features that differ from the mammalian system.

Synopsis of vertebrate evolution: Box 3.1

7.1.1 Fishes

Fishes are the first vertebrates to appear on the fossil record. Of all modern vertebrates, therefore, the immune system of fishes is likely to most closely resemble that of the earliest vertebrates. In all respects, the immune system of the Agnatha is the most primitive among the vertebrates. These animals have only diffuse lymphoid tissues, and there is some debate over whether T and B lymphocytes exist as separate phenotypes in these animals. Antibody-like molecules are produced in response to immunization, although they are produced in small amounts and it is not known whether they are genetically related to the immunoglobulins of higher vertebrates.

The antibody-dependent (classical) pathway of complement is also absent from lampreys, hagfish and some cartilaginous fishes, suggesting this pathway arose later in the vertebrate lineage. The lytic phase of the complement pathway probably also arose around this time, as it is absent from the lamprey and from any of the invertebrates. In contrast, the C3 component has been found in representatives of all major vertebrate groups, including the hagfish and lampreys. Studies of the Agnatha, therefore, suggest that the pathway that is initiated by C3 (the alternative pathway) predates the classical pathway. The alternative pathway may have acted in the earliest vertebrates mainly as a source of opsonins and other inflammatory molecules.

In contrast, all jawed (gnathostome) vertebrates possess a true thymus and true secondary lymphoid tissues for collecting and presenting antigen to lymphocytes. In fishes, this is performed by the spleen, head kidney and trunk kidney. Jawed fishes also have diffuse lymphoid tissues associated

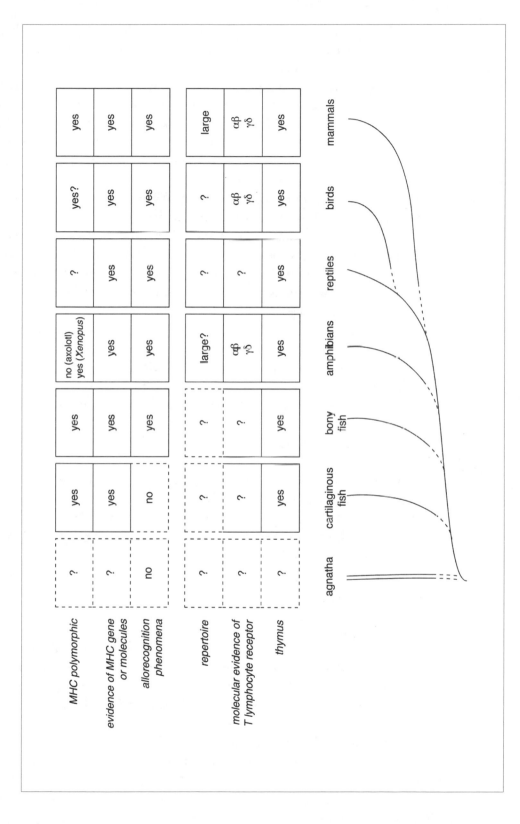

Figure 7.2 Phylogeny of T lymphocyte receptor and the major histocompatibility complex.

with the digestive tract and gills, although these animals do not have solitary lymphoid follicles or multifollicular structures such as the lymph nodes and Peyer's patches of mammals.

Fishes and all other vertebrate groups produce immunoglobulins based on the two heavy chains/two light chains structure, indicating this molecular format is ancient. The only antibody isotype seen in fishes is called IgM. In fishes, as in the vertebrates that evolved later, IgM is a high molecular weight immunoglobulin isotype, which exists as cell-surface monomers, or as polymeric forms in the serum. In sharks and rays, pentameric IgM is found, whereas in bony fish it exists in tetrameric form. In general, fish IgM shows low diversity and low affinity, and there appears to be no affinity maturation, perhaps reflected by the lack of lymphoid follicles and germinal centres. However, teleosts, the bony fishes, do show moderate secondary antibody response upon re-exposure to a particular antigen. Gnathostome fishes also have both the alternative and classical complement pathways.

As yet, nothing is known of T lymphocyte receptors in fishes, although the presence of a thymus (in all but the Agnatha) suggests T lymphocytes should be present. Moreover, class I and II MHC genes have been found in elasmobranchs (sharks and rays) and teleosts (bony fishes such as the carp, salmon and zebrafish), although strong rejection of grafts is only seen in the latter. The class II genes of cichlid fishes are known to be particularly polymorphic. So far, no one has managed to clone MHC genes from the Agnatha.

As in most other vertebrates, immune responsiveness of fishes is influenced by seasonal variations in temperature and length of daylight (**photoperiod**). In general, lower levels of peripheral blood lymphocytes occur in the winter and higher levels in the summer. There is also a trend towards regression of the thymus and other lymphoid tissues in winter. Even at constant temperatures these seasonal changes occur, suggesting the effects are hormonal.

7.1.2 Amphibians

Amphibians are a particularly important group of animals in the study of immune system evolution, largely because these animals were the first to emerge onto land. A significant proportion of amphibian development occurs during a free-living aquatic larval stage (which subsequently metamorphoses into an adult). A characteristic of amphibian immune systems, therefore, is its relatively early development in the embryo. This is thought to be necessary to provide the larva with protection from pathogens. The major lymphoid organs of larval *Xenopus* tadpoles are the thymus and

> **elasmobranchs** *n. plu.* cartilaginous fish including sharks, skates and rays.
>
> **photoperiod** *n.* duration of exposure to daylight.

spleen, with diffuse lymphoid tissues associated with the kidney, liver and in the pharynx next to the gills.

In all amphibians, the metamorphosis from the larva to adult is associated with a profound change in body plan from an aquatic to a terrestrial mode of life. In *Xenopus,* the spleen and thymus are retained, as well as the lymphoid tissues in the kidney and liver, whereas gills and associated pharyngeal lymphoid tissues are lost. Haemopoietic bone marrow and diffuse gut and skin associated lymphoid tissues appear for the first time after metamorphosis.

In the adult form, the lymphoid tissues of the simplest amphibians (apoda and urodeles) show little advance over that of the fish, with the major lymphoid tissues being the thymus and the spleen. As in other vertebrates, the thymus of urodeles (salamanders) is the site of T lymphocyte maturation, whereas the spleen is the major haemopoietic tissue and the predominant site of trapping antigens and other debris from the blood.

The lymphoid architecture of anurans (frogs and toads) shows several advances from the apoda and urodeles. Solitary lymphoid nodules, believed to be a forerunner of the multifollicular lymph node of mammals, appear for the first time in anurans. These supplement the spleen by trapping antigens in lymphatic fluid. The role of bone marrow as a major site of haemopoiesis also emerges for the first time in anurans. As with fish, frog immune responsiveness varies with season. The thymus is most active in summer but involutes in winter.

Amphibian IgM exists on the B lymphocyte membrane as monomers. When secreted it assembles into polymers. In *Xenopus,* for example, secreted IgM is an assembly of six IgM molecules (hexameric IgM). Amphibians, and all other tetrapods, also have one or more additional low molecular weight antibodies. This is in contrast to fishes, which only produce IgM. The major low molecular weight isoform in amphibians, reptiles and birds is IgY. *Xenopus* also has a another distinct isoform, IgX. IgX and IgY are slightly different from mammalian low molecular weight isoforms because the former have four C domains. Amphibians, like all the animals that evolved after them, also have both the classical and alternative complement pathways.

As with other non-mammalian vertebrates, little is known about the amphibian T lymphocyte repertoire, although a gene for a β-chain-like protein has been cloned from the axolotl. Antibodies are available to anuran αβ- and γδ-like structures, and the cells that express these molecules are reduced in number if the tadpoles are thymectomized. Allorecognition phenomena, such as graft rejection, are more vigorous in anurans than urodeles, although both have class I and class II MHC molecules. The XLA

Apoda *n.* an order of limbless, burrowing amphibians (see **Anura** and **Urodela**).

Urodela *n.* one of three orders of extant amphibians containing the newts and salamanders (see **Anura** and **Apoda**).

Anura *n.* one of three orders of extant amphibians containing the frogs and toads (see **Apoda** and **Urodela**).

involution *n.* a reduction in size or functional activity, such as with age.

IgY immunoglobulin Y; the predominant antibody found in birds, reptiles and amphibians.

IgX immunoglobulin X; an isotype found in some amphibians.

isoforms *n. plu.* different forms of a protein encoded in the same gene produced by alternative RNA splicing.

axolotl *n.* an aquatic, newt-like salamander.

complex of *Xenopus* is the most thoroughly studied non-mammalian MHC. Like the MHC of most mammals, XLA is very polymorphic. The axolotl MHC, in contrast, shows limited polymorphism.

An interesting feature of *Xenopus* MHC expression is the absence of class I molecules until after metamorphosis. *Xenopus* also seems to express class I heavy (α) chains on the surface of erythrocytes and leucocytes in the absence of β2-microglobulin. The function of the *Xenopus* class I molecule is not known. *Xenopus* also has invariant chain associated with its class II MHC molecules.

7.1.3 Reptiles

Reptiles are the closest living relatives of the ancestors of birds and mammals. Despite their pivotal position in the ancestry of higher vertebrates, very little is known about reptilian immune systems. Reptiles, and the birds and mammals that evolved from them, are amniotes. Unlike the amphibian embryo, the amniote embryo is surrounded by amniotic fluid contained within an amniotic membrane. Consequently, amniote embryos are not dependent on an aquatic habitat and do not undergo metamorphosis. Also, the amniote embryo is relatively well protected and less exposed to potential infections. In comparison with the rapid development of the immune system by the free-living amphibian embryo, therefore, the immune system of the amniotic embryo develops more gradually.

Reptiles have provided much of the information we have regarding seasonal changes in immune systems. Reptilian antibody responses and allorecognition phenomena are generally more vigorous in summer than in winter. Involution of the reptilian thymus occurs in winter, although there is considerable variation between different reptile species when the thymus is active during spring and summer.

Reptiles, like the amphibians, also have the high and low molecular weight antibody isoforms IgM and IgY. So far, however, very little is known about their T lymphocyte receptors or about the reptilian MHC. Nevertheless, reptiles have well-developed thymuses and they show good allorecognition phenomena.

7.1.4 Birds

Birds and mammals evolved from different reptile ancestors, and modern birds retain many reptilian features. Despite obvious differences between birds and mammals, both are endotherms ('warm-blooded') and can generate their own body heat. This condition evolved independently in these two classes after their divergence from the reptiles. Endothermy enables animals to maintain a constant body temperature and reduces their depen-

amniotes *n. plu.* vertebrates whose embryos are surrounded by an amniotic membrane; reptiles, birds and mammals.

endotherm *n.* an animal that derives all or most of its heat content from endogenous metabolic heat rather than from the external environment (see **exotherm**).

dency on the ambient temperature of the environment for physiological activity. This stability has allowed birds and mammals to exploit more diverse environmental conditions, but a constantly warm bodily environment has also provided new opportunities for infection by microorganisms. This is likely to have contributed to the evolution of several characteristics that are less apparent in the immune systems of ectotherms (reptiles, amphibians and fish). These are:

- A rapid immune response.
- Large antigen receptor repertoires.
- A diversification of immunoglobulin isotypes.
- Complex secondary lymphoid tissues.

Although birds and mammals evolved from reptilian stock, several unique and interesting features have evolved independently in birds. One is the site of maturation of B lymphocytes. Avian B lymphocytes mature in a specialized organ of primary lymphoid tissue, the bursa of Fabricius. A bursa is absent in reptiles and mammals and must have arisen after birds diverged from reptiles. The importance of the bursa was discovered in the mid-1950s, when surgical removal of these organs from chicks (**bursectomy**) abolished antibody responses, without having an effect on cell-mediated responses such as graft rejection. Indeed it was this observation that first led to the discoveries of the division of lymphocytes into T and B cells.

A second unique feature in birds is the mechanism by which diversity of their immunoglobulins is generated. Immunoglobulin heavy and light chain loci in the chicken have a single functional V gene segment and several ΨV pseudogenes at each locus (around 30 ΨV_L and around 90 ΨV_H). After a single rearrangement, diversity is then generated by gene conversion as the lymphocytes proliferate in the bursa. In the chicken, the population expands from 10^5 to $1-2 \times 10^9$ mature B lymphocytes, which can accumulate as many as 10 gene conversion events in the rearranged V gene. Gene conversion is an unusual method of generating antigen receptor diversity, though it is also found in the rabbit.

As in other amniotes, the immune system of birds develops slowly, and a full repertoire develops only after hatching. Birds (and, as we will see, mammals) confer protection to their developing embryos during this time by the **passive transfer** of ready-made maternal antibodies. Birds (chickens) achieve this by the passive transfer of maternal IgY into the egg yolk. This passes through the follicular epithelium of the oviduct while the egg

Recombination and gene conversion: Box 2.1

is developing. IgM is also transferred into the egg albumin as the egg moves down the oviduct. The developing chick embryo then absorbs these antibodies while inside the egg. IgY is absorbed as the yolk is consumed, whereas IgM is ingested into the alimentary canal from the amniotic fluid.

As in reptiles, the predominant antibody isotype produced during secondary responses in birds is the low molecular weight species, IgY. Interestingly, Anseriform birds (waterfowl) are able to gradually replace the IgY antibodies during affinity maturation with a form of IgY which lacks the Fc region, called IgY(ΔFc). As we saw in Chapter 4, this is achieved by alternative splicing of the heavy chain pre-mRNA (see Figure 4.15). Some reptiles may also do this. Although mammalian immunoglobulins switch forms from membrane-bound to soluble by a similar process, immunoglobulins lacking Fc regions have not been found in mammals, and the role of these truncated antibodies in birds remains a mystery.

B complex *n.* the major histocompatibility complex of the chicken.

The MHC of the chicken, called the **B complex**, is another departure from the mammals. The functions of chicken class I (called B-F) and class II (B-L) MHC molecules are similar to those described in mammals. However, the genes of B complex are very tightly linked, and the class I and II regions are not as well delineated as they are in mammals. There are also 'class IV' genes of unknown function, and the 'class III' genes seen in mammalian MHC seem entirely absent from bird MHCs. Birds show rapid allograft reactions *in vivo,* and mixed lymphocyte reactions *in vitro,* and considerable progress is being made in characterizing their $\alpha\beta$ and $\gamma\delta$ T lymphocyte receptors.

7.1.5 Mammals

Much of what we know about mammals comes from placental (or eutherian) mammals. In comparison, much less is known about the immune systems of monotremes and marsupials. The marsupial gestation period is characteristically short and the young, born at a very immature stage, are carried in a pouch. Eutherian mammals, in contrast, have a comparatively longer gestation period and are born at a more mature stage of development. Nevertheless, as in other amniotes, the immune systems of both marsupial and eutherian mammals develop relatively slowly. Newborns are largely dependent on passively transferred antibodies for protection, until their antigen receptor repertoires are acquired in early postnatal life.

colostrum *n.* a clear fluid secreted by mammary glands before milk is produced; contains a higher concentration of antibodies than milk, which are passively transferred to newborns immediately after birth.

Passive transfer of antibodies occurs either across the placenta and/or after birth in **colostrum** or **milk**, depending on species. During pregnancy, transfer of IgG occurs either across the placenta as in humans, or from the yolk sac as in rodents and rabbits. Other mammals such as ungulates (hoofed herbivores) and marsupials do not provide pre-natal antibodies.

The mammary gland is a uniquely mammalian innovation which produces secretions that are rich in antibodies. These are passively transferred to newborns during suckling.

In eutherians, the first meal consists of colostrum, which is very rich in immunoglobulins. In human colostrum, for example, 97% of the protein content is IgA. Thereafter milk is produced, of which 10–25% of the protein is IgA. The class of immunoglobulin secreted differs between species. For example, in humans, rabbits and rodents, IgA is found in both milk and colostrum, whereas in sheep and cows it is IgG. In addition to IgA, human milk also contains additional microbicidal components including complement proteins, interferons, lysozyme and lactoferrin – which sequesters iron that is needed by aerobic bacteria for growth.

Eutherian mammals are also unique among the vertebrates for the variety of immunoglobulin isotypes produced (Figure 7.3). As we learned in Chapter 4, each isotype has a different physiological function. In addition to the phylogenetically ancient IgM, eutherians have IgG, IgA, IgE and IgD (see Figure 4.12 for structural differences). IgG is the predominant low molecular isotype produced in secondary responses to antigen. IgG is absent from non-mammalian vertebrates and therefore evolved after the divergence of mammals from the reptiles. IgA and IgE have also been found in marsupials, whereas only IgM and IgG are present in monotremes. IgD and IgG are closely related to the predominant low molecular weight isoform of birds, reptiles and amphibians, IgY.

Giving birth to live young (**viviparity**) has evolved independently in several vertebrate classes, but the relationship between foetus and mother is at its most intimate in eutherian mammals. The relatively long gestation periods, of what is essentially allogeneic tissue expressing paternal MHC

viviparous *a.* the condition of giving birth to live young; all mammals except monotremes (lay eggs) and a number of exceptions in other classes of vertebrates.

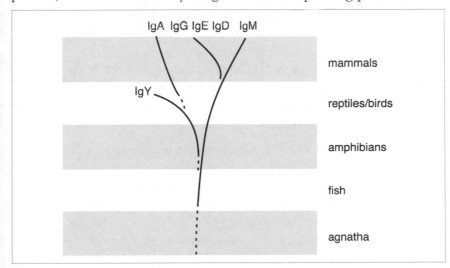

Figure 7.3 Evolution of immunoglobulin isotypes.

molecules, has led to the development of several uniquely mammalian mechanisms to prevent 'rejection' of the foetus and possible abortion.

trophoblast *n.* the outer layer of cells surrounding the embryo.

- The foetus of eutherians is surrounded by a membrane called the **trophoblast**. In man and rodents the trophoblast expresses very low or undetectable levels of class I and class II MHC molecules and is immunologically silent.

- Pregnancy in some mammals is also associated with an overall suppression of cell-mediated immunity. Humans, for example, become temporarily less able to control latent viral infections during pregnancy. It is thought immunosuppression is mediated by T lymphocytes with suppressive properties. Immunosuppressive cytokines, steroid hormones and prostaglandins secreted by the placenta are also involved. Moreover, the surface expression of T lymphocyte receptors that recognize foetal alloantigens is decreased. Together these mechanisms help reduce the incidence of allorecognition of the foetus.

Summary

- IgM probably evolved before other isotypes of immunoglobulin because it is found in all vertebrates except the Agnatha. Additional low molecular weight forms emerge first in amphibians and reach their most diverse in eutherian mammals.

- A polymorphic major histocompatibility complex has been found in bony fishes and in representatives of all vertebrate classes that evolved afterwards.

- Little is known about the T lymphocytes of non-mammalian vertebrates, although the presence of a thymus in all gnathostome vertebrates, and allorecognition phenomena in all but the elasmobranchs, suggests they exist.

- In fish, amphibians and reptiles, immune responsiveness is regulated by cyclical seasonal changes in photoperiod and temperature. This is thought to be under hormonal control. In general, decreased humoral and cell-mediated responses, decreases in numbers of circulatory lymphocytes, and involution of the thymus occur in winter months, although there are exceptions, particularly in the reptiles.

- The mammalian and avian immune systems have several characteristics that are less apparent in reptiles, amphibians and fishes. These include rapid immune response, increase in size of antigen receptor repertoires, radiation of immunoglobulin isotypes and complexity of secondary lymphoid tissues.

- These characteristics are likely due in part to endothermy. This has enabled mammals and birds to maintain a stable body temperature and an independence from seasonal changes. However, endothermy has offered new opportunities for exploitation by pathogens.

- Amphibians have a free-living larval stage and immunocompetence is acquired early. In contrast amniotes (reptiles, mammals and birds) are protected from microorganisms until birth or hatching. Consequently, their immune systems develop more gradually and do not reach competence until after hatching or birth.

- Mammals and birds have evolved mechanisms of passive transfer of maternal antibodies to protect their offspring. Hens transfer antibodies across the follicular epithelium of the oviduct into eggs as they descend the oviduct. Mammals transfer antibodies during gestation, either across the placenta or in the yolk sac, depending on species. All mammals transfer antibodies to newborns in mammary gland secretions (colostrum and milk).

- In contrast to egg laying, viviparity raises the potential for allorecognition by the mother of paternal alloantigens on the embryo. In eutherian mammals, which have long gestation times, this is prevented by immunological silence of the trophoblast and the release by the placenta of immunosuppressive cytokines, steroid hormones and prostaglandins.

7.2 Evolution of the immunoglobulin superfamily

It is clear from the preceding section that all modern vertebrates possess the hallmarks of the adaptive system. Antibodies, T lymphocyte receptors, classical class I and class II MHC molecules, and a host of other receptors involved in the immune systems, are members of the immunoglobulin superfamily. As we will see below, members of the family are present in modern invertebrates, suggesting that the common ancestral molecules were in existence at the time of the divergence of vertebrates and

The immunoglobulin superfamily: Section 3.7.1

invertebrates. However, the rearrangement of antigen receptor genes is an innovation which appears to have arisen only in the vertebrates.

Each member of the superfamily has one or more common structural units – the immunoglobulin fold. The amino acid sequences of different members now vary quite widely, although the three-dimensional shape of a domain is well conserved. It has been speculated that the shape of the immunoglobulin fold has been conserved during the course of evolution because it is particularly resistant to extracellular proteinases. Tracing the evolution of the immunoglobulin superfamily is fundamental to understanding the origins of the adaptive immune system.

The majority of contemporary members of the superfamily are found on cell surfaces participating in cell-to-cell interactions. In vertebrates these molecules are involved in cellular interactions in the nervous system as well as the immune system. The first members of the family to be cloned and sequenced were the heavy and light chains of mammalian immunoglobulins. The first non-immunoglobulin member discovered was β_2-microglobulin, obtained originally from the urine of patients with kidney disease. Since then over 100 members of the family have been described. Several have been found in invertebrates, including molluscs and arthropods where they are involved in the nervous system and in the control of muscle structure and elasticity. These invertebrate examples indicate that the basic unit of the immunoglobulin superfamily was in existence at the divergence of the protostomes (which gave rise to many of the major invertebrate phyla) and deuterostomes (giving rise to other invertebrates and the chordates) around 500 million years ago.

7.2.1 Gene duplication and the origins of homophilic and heterophilic interactions

homophilic *a.* the binding between two similar molecules (e.g. cell-surface adhesion molecules). See **heterophilic**.

gene family set of homologous genes created by duplication and divergence of a single ancestral gene; some families with large numbers are superfamilies (see **immunoglobulin gene superfamily**).

The gene for the immunoglobulin fold may have first evolved from a 'half domain' gene encoding three or four β-strands, which duplicated to form a full domain. This idea is supported by the existence today of a few immunoglobulin superfamily genes that have a centrally located intron. Initially the ligand for this ancestral molecule may have been another copy of itself, expressed on the other cell. This is a **homophilic** molecular interaction (Figure 7.4). Several homophilic interactions exist today, such as between N-CAM and N-CAM (neural cell adhesion molecule).

The possible evolution into a gene family is also represented in Figure 7.4. The first step is the duplication of the ancestral gene and genetic divergence of the two copies by the gradual accumulation of mutations. This allows two different genes to evolve along separate lines and into different members of the same family. This leads to the clustering together

BOX 7.1 Molecular evolution

Genetic variation between individuals provides the raw materials upon which natural selection can operate, allowing organisms to evolve. New genetic variation is created by a variety of mechanisms that either involve recombination of pre-existing genetic information, or mutations – alterations in the sequence of DNA caused by errors in replication or other sources (chemical, radiation, etc.). Natural selection does not operate directly on the gene, but on the phenotype it expresses in the organism. Therefore, for genetic changes to influence the course of evolution, they must:

- Change the protein encoded in some way. For example, mutations within the coding sequence of the gene may alter the amino acid sequence it encodes, whereas mutations outside may influence the expression of the protein by some other indirect means.

- Be inheritable. In practice this means it must arise in the germline because somatic mutations are not inherited.

Recombinations and mutations that lead to changes in the amino acid sequence of a protein are particularly important in evolutionary processes. The majority of such genetic changes are detrimental to the function of the protein because they change its three-dimensional structure and consequently its biological function. Such changes may also be detrimental to the welfare of the organism, and, through the effects of natural selection, will tend to remain at low frequency in the population.

On the other hand some mutations can, in certain environmental conditions, increase the fitness of an individual and increase in frequency in subsequent generations (see the examples in Box 6.3). A similar effect is achieved by genetic drift, particularly in small populations. Because a given gene may acquire different mutations independently in different members of the population, a given gene can exist in several different variants within a population. This reservoir of genetic variation between individuals is the means by which populations can adapt to changing environmental conditions, and for certain mutations to become established as different alleles in the population.

Mutations

Mutations that arise in genes spontaneously during DNA replication take the form of nucleotide substitutions, deletions, duplications and insertions. Mutations occur at uniform rates within most eukaryotic genes, which are particularly useful for constructing evolutionary genealogies of related species of animals. This is achieved by comparing the sequences of a homologous gene (i.e. a gene arising from a common ancestry) in different species. The longer the elapsed time of divergence of two species, the greater the number of independent mutations that can accumulate.

These data can be displayed as **phylogenetic trees** (Figure 1). A given gene and its nearest neighbour on the tree have fewer differences, and presumably diverged more recently, than the given gene and another gene located further away. Such analyses are useful for reconstructing the possible lineages of related species and estimating the rates of evolution. Phylogenetic reconstructions based on sequence data often correlate very well with more traditional means of classifying species, such as on morphological criteria.

Recombination

If left to mutation alone, the appearance of entirely new genes encoding novel proteins would be a very gradual process. However, recombinational mechanisms are particularly important in accelerating the creation of new genetic information from pre-existing genetic material. These include recombination by crossing over at meiosis and gene conversion (see Box 2.2).

An additional source of variation are **transposable elements** (or **transposons**), which are small DNA sequences found in all prokaryotic and eukaryotic cells. Transposons move or 'transpose' to new locations in the genome by either a recombinational mechanism, in which the DNA is excised and recombined elsewhere, or by a replicative process in which the DNA is copied into another DNA molecule which then recombines elsewhere in the genome. The simplest transposons carry only the

BOX 7.1 *continued*

Figure 1 A phylogenetic tree of primates. This tree was constructed from the sequences of the η globin pseudogene. The numbers represent the number of substitutions per 100 nucleotides. Redrawn from Goodman, M., *et al.* (1990) *Journal of Molecular Evolution*, **30**, 260–266.

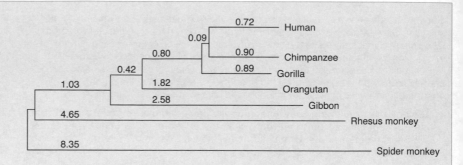

genetic information to transpose, while others carry longer sequences.

An additional process that is important in the creation of new genes and proteins is called **exon shuffling.** We learned in Box 6.1 that eukaryotic genes consist of coding regions (**exons**) interspersed with non-coding regions (**introns**). While exons are all roughly 150 nucleotides long, the length of introns is much more variable. The origin and functions of introns still attracts much debate. These are ancient structures, thought to have been inserted into coding sequences prior to the divergence of eukaryotes and bacteria. In eukaryotic genes, the site of intron insertion appears to be non-random. Figure 4.11, for example, shows the leader sequence, variable, constant, transmembrane and cytoplasmic domains of immunoglobulin heavy chains are all encoded in separate exons. This modular nature of protein genes has enabled exons from different genes to be 'shuffled' about, by the mechanisms described below, to create entirely new genes and novel proteins. The **selectins**, for example, are cell-surface lectins composed of carbohydrate recognition domains, epidermal growth factor-like domains and short consensus repeats that are homologous to the repeats of complement receptors (see Section 3.8.3).

A vital contributor to molecular evolution is **gene duplication** and **exon duplication** caused by unequal crossovers during meiosis (see Box 2.2). Exon duplication can result in repeated domain structures within a given protein, such as the immunoglobulin domains in an antibody molecule. The duplication of whole genes is a major route to the evolution of gene families and superfamilies. The importance of this latter process is that one of the copies of the duplicated gene can undergo mutation in the absence of natural selection, while the other copy of the gene can continue to uphold its function in the cell. Released from the constraints of having to perform a vital function the gene is left to diverge, by the mechanisms described above, from the functional copy. Occasionally, a useful new gene may arise. However, most changes will be deleterious, leading to a gene that fails to produce a functional protein. Such genes are called **pseudogenes** and are designated by the prefix Ψ (psi).

Gene duplication is the first step in the evolution of gene families such as the globin genes, the immunoglobulin superfamily genes and MHC genes. Once a gene has duplicated, exchange of material between the copies, by gene conversion and unequal crossovers, occurs more readily. This can accelerate the divergence of members of gene families. An example of this is in the MHC, where gene conversion is a major source of MHC polymorphism. Multigene families, such as the globin gene family or the MHC, are continually expanding and contracting through the actions of unequal crossovers.

Pseudogenes can have their uses in other ways. Genetic diversity in functional genes is often acquired from pseudogenes by a recombinational mechanism such as gene conversion. This is the means by which birds and rabbits generate diversity in their immunoglobulin repertoires during B lymphocyte maturation. Since pseudogenes do not produce a protein product, natural selection is blind to them. Consequently pseudogenes tend to accumulate mutations more rapidly than their functional counterparts. These are therefore particularly useful genes for determining the phylogenetic relatedness of very closely related species (such as humans and chimpanzees) in which functional genes are generally too similar.

Further reading

Maeda, N. and Smithies, O. (1986) The evolution of multigene families: human haptoglobin genes. *Annual Review of Genetics*, **20**, 81–108.

of genes of superfamily members. Modern relics of this include the V and C genes of antigen receptors, MHC genes, and in mice a gene cluster encoding N-CAM, Thy-1 and chains of the CD3 complex (which are associated with the antigen receptor of T lymphocytes).

Divergence can give rise to interactions between two different molecules. These are called **heterophilic** interactions and are the type in which the majority of modern immunoglobulin superfamily members engage. The expression of a particular gene in one kind of cell but not another (**differential expression**) allows the evolution of one way interactions between specific cell types. The variety of immunoglobulin superfamily members we see today, each with a different function, can be accounted for by the generalized process of gene duplication and subsequent divergence. Exon shuffling (see Box 7.1) has played only a minor role in shaping the immunoglobulin superfamily.

heterophilic *a.* the binding between two dissimilar molecules (e.g. cell surface adhesion molecules).

exon shuffling the evolutionary process of creating novel genes by the recombination of exons from pre-existing genes.

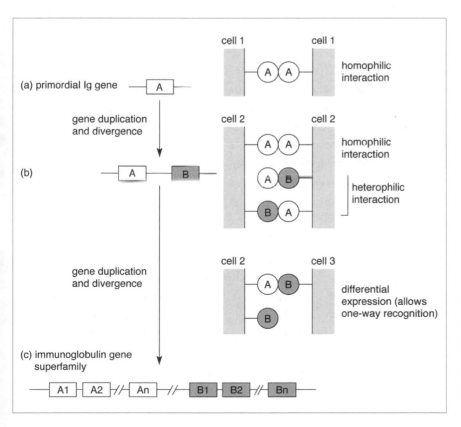

Figure 7.4 Schematic representation of the evolution of a typical receptor superfamily. The primordial immunoglobulin (A) gene probably encoded a cell-surface homotypic receptor. Gene duplication by unequal crossover (Figure 3 in Box 2.2) would allow the two copies to gradually diverge by the independent accumulation of mutations. These will also evolve different physiological functions. This enables heterophilic interactions to evolve (B). If the different receptors are expressed in some cells and not others, one-way recognition of cell types can evolve (C). Cell 2 can interact with both cell 2 and cell 3, whereas cell 3 can only interact with cell 2. Although single domain molecules are shown in the diagram, similar gene duplication events give rise to multiple domain molecules. Adapted, with permission, from the Annual Review of Immunobiology, Volume 6, ©1988, by Annual Reviews Inc.

7.2.2 Relics?

The non-polymorphic MHC and MHC-like genes of the modern mammalian immune system may be relics of more primordial aspects of the system. The Tla and Qa-2 molecules of the MHC of mice, for example, are known to be involved in the presentation of a limited set of antigens to T lymphocytes. Another family of non-polymorphic MHC class I-like molecules, the **CD1** family, are found on the surface of antigen-presenting cells and are thought to play a role in presentation of microbial antigens. The CD1 family members are also assembled with β_2-microglobulin, although the heavy chain is encoded outside of the MHC and it is only weakly homologous with the heavy chains of classical MHC molecules. Recently, the function of the CD1b molecule in humans has been shown to be the presentation of uniquely microbial lipids to T lymphocytes. CD1 molecules are particularly abundant in mucosal and cutaneous epithelia where they present peptide or lipid antigens derived from microorganisms to intraepithelial T lymphocytes.

7.3 Evolution of the antigen receptors

The raw material for antigen receptors, in the form of the immunoglobulin fold or domain, was already in existence by the time the first vertebrates emerged. The uniquely vertebrate feature of these molecules, and the property that underlies the unique features of the adaptive immune system, is the way in which antigen receptor genes are constructed from gene segments by a process of rearrangement during lymphocyte development.

T and B lymphocyte receptors have related structures and sequences, and almost certainly evolved from a common ancestor. The use by both receptors of the same recombinase enzymes to rearrange their gene segments supports this notion. The affinity of T lymphocyte receptors for another immunoglobulin superfamily molecule (i.e. MHC) is reminiscent of other heterophilic interactions within the superfamily, suggesting the T lymphocyte is more like the ancestor of the modern antigen receptors. Immunoglobulin appears to have lost its dependency on another immunoglobulin superfamily member as a ligand. In addition, B lymphocyte receptors have diverged in a number of other ways from T lymphocytes, as demonstrated by their ability to switch from one class to another and to be secreted from the cell. T lymphocytes show neither of these phenomena.

7.3.1 The origins of antigen receptor diversity

The transition of an antigen receptor from a conventional molecule to one of immense diversity is likely to have occurred in concert with the evolution of polymorphism in its ligand, the MHC molecule. Antigen receptor diversity and MHC polymorphism are two different ways that together enhance the ability of the immune system to recognize pathogens. These may have evolved in response to the variety of pathogens in the environment (see below).

In the case of antigen receptors, diversity is achieved in contemporary vertebrates by having large numbers of gene segments that are brought together by a process of DNA rearrangement. The large numbers of different – but closely related – gene segments can be explained by a process of gene duplication and divergence. However, a turning point in the evolution of antigen receptors would have been the capture of a **transposable element** (or **transposon**) by the primordial antigen receptor gene (Figure 7.5). The insertion of a transposon into the gene would create two

> **transposable elements** (or **transposons**) *n.* a sequence of DNA that moves by translocational or replicative means to new locations within eukaryotic or prokaryotic genomes.

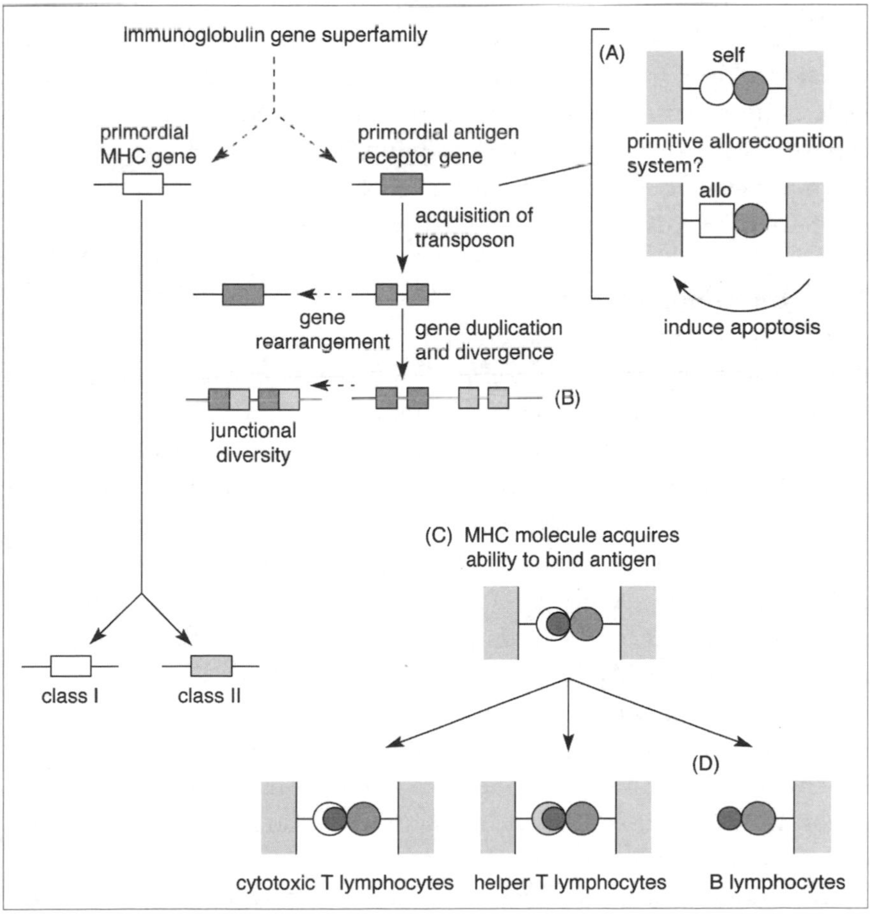

Figure 7.5 A speculative view of how modern MHC molecules and antigen receptors may have co-evolved from a primitive allorecognition phenomenon. Allorecognition might have been mediated by two immunoglobulin superfamily members (A), one declaring 'self' (the primordial MHC molecule), the other its receptor on the surface of an effector immunocyte. Diversity among the receptors may have increased the spectrum of allogeneic self molecules they could recognize. Capture of a transposon by the receptor would have been the first step in the change to a rearranging gene and, ultimately, of junctional diversity (B). Another key event would be the self marker acquiring the ability to become modified with proteins or peptide fragments of intracellular protein (C). This would allow the receptor to change its function to an antigen receptor. This ability to alert T effector cells to foreign antigen promoted the complexity of the antigen receptor repertoire and, in parallel, the polymorphism of the self marker (MHC molecule). B lymphocytes may have split from the T lymphoid line through having lost the dependence for the MHC molecule (D).

segments (the primordial V and J gene segments) which could be recombined by the excision on the transposon. This would be the first step in the transition to a rearranging gene and the capacity to produce an antigen receptor repertoire.

7.3.2 T lymphocyte receptors

All jawed vertebrates possess differentiated thymuses, and all but the agnathous and elasmobranch fish show T lymphocyte-mediated responses or allorecognition phenomena. The ligands of T lymphocyte receptors – the MHC molecules – seem to be present in all vertebrates. As yet, T lymphocyte receptor genes have only been cloned from a handful of species (human, mouse, chicken and Mexican axolotl). It is currently impossible from such sparse data to construct phylogenetic trees or to map the evolution of T lymphocyte receptors on established phylogenies.

The evolutionary relationship between $\alpha\beta$ and $\gamma\delta$ T lymphocyte receptors is controversial. It has been speculated that $\gamma\delta$ cells represent a more primitive form of T lymphocyte. If this is the case, the $\alpha\beta$ genes probably emerged by a process of gene duplication and divergence of a common ancestor of the $\gamma\delta$ receptor. Whatever the origins, both $\alpha\beta$ and $\gamma\delta$ cells have undergone extensive gene duplication and divergence to arrive at their current complexity. Examine the diagrammatic representations of murine T lymphocyte receptor genes in Figure 4.7, which has been drawn to emphasize these duplications.

7.3.3 Immunoglobulins

More can be said about the evolution of immunoglobulins than of T lymphocyte receptors. Immunoglobulins first appear in elasmobranchs. In sharks, IgM is the major immunoglobulin isotype produced, reflecting its early evolution in the vertebrates. In these, and the other vertebrates that evolved subsequently, IgM forms high molecular weight polymers. In amphibians and reptiles we first see the emergence of an additional non-μ, low molecular weight isoform, called IgY, which is often the predominant isoform found as soluble antibody. In general the number of different non-μ isotypes has increased during the course of evolution, being most numerous in mammals.

The basic organization of introns and exons in immunoglobulin V gene segments has been remarkably well conserved during the radiation of vertebrates. This is important because it tells us that this basic plan was established early, probably in the elasmobranchs and before the other major classes of vertebrates evolved.

All modern vertebrates have several V gene segments in their im-

phylogenetic tree a branching diagram showing the evolutionary history of a group of related species.

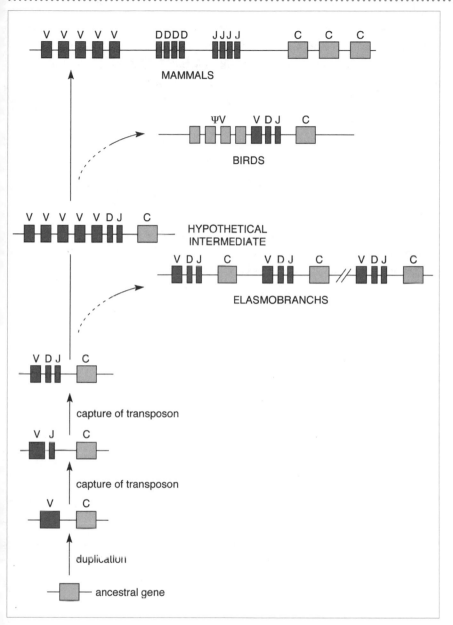

Figure 7.6 The three main patterns of immunoglobulin gene segments found in extant elasmobranchs (sharks and rays), birds and mammals, and their possible evolutionary relationships.

munoglobulin heavy and light chain loci, which almost certainly arose by gene duplication. Heavy chain genes also feature D and J gene segments whereas light chain genes feature J gene segments. The overall organization of these segments has evolved along at least three different lines resulting in distinctive, almost diagnostic, patterns. This is shown in Figure 7.6. In modern elasmobranchs (sharks and rays), the gene segments are organized into 'cassettes' separated by long spaces (VDJC-VDJC-VDJC). The shark *Heterodontus franscisci,* for example, has hundreds of these cassettes which occupy a significant proportion of its genome. Rearrangements are confined to within each cassette, which means that there is probably very little combinatorial diversity in shark immunoglobulin. The shark genes also lack the typical promoter, and instead employ a promoter that is more reminiscent of that of a T lymphocyte receptor.

In the remaining vertebrates (except birds) a different pattern has emerged. Here, like gene segments are clustered with like (VVV-DDD-JJJ-CCC). This correlates well with the idea, based on morphological criteria, that the elasmobranchs diverged from the common lineage before teleosts and the amphibians, reptiles and mammals. Birds have evolved a third distinct pattern in which there is a *single* functional V_H gene but several V pseudogenes. After the segment rearranges, diversity is generated by multiple gene conversions between the functional V_H gene and the reservoir of pseudogenes. It is possible that this pattern evolved from the teleost/tetrapod pattern and all but one of the V_H segments became non-functional.

Summary

- Members of the immunoglobulin superfamily are found in vertebrates and invertebrates, showing that the basic unit, the immunoglobulin fold, pre-dates the divergence of vertebrates from the other phyla.

- The immunoglobulin superfamily has evolved by extensive gene duplication and divergence.

- Vertebrate antigen receptors are immunoglobulin superfamily members that are encoded in rearranging genes. This may have evolved by the capture of a transposon into the primordial antigen receptor gene after vertebrates diverged from invertebrates.

- The T lymphocyte receptor probably pre-dates the emergence of antibody.

7.4 Evolution of the MHC

Like the antigen receptors, MHC molecules belong to the immunoglobulin superfamily. Although two classes of MHC molecule have evolved, their shapes are remarkably similar. This reflects a common function in peptide presentation and presumably an evolutionary descent from a common ancestor. The domains that form the peptide-binding cleft are unlike immunoglobulin folds and it is possible that they have been acquired by exon shuffling from a different gene.

There is much debate as to the origins of a MHC–T lymphocyte receptor interaction. Sequence studies suggest a common ancestor would have been more like the modern class II molecule, with two chains each having two domains. The purpose of the ancestral interaction is not known and it may have been very different to its modern role of antigenic peptide presentation. One theory is that the primordial MHC molecule was a marker of self identity. Some colonial invertebrates, such as sponges and tunicates, are able to distinguish cells to which they are genetically related and thereby prevent entry to the colony of unrelated cells. The rejection of unrelated cells is a process that is superficially like allograft rejection in vertebrates. In the colonial tunicate, *Botryllus schlosseri,* the locus that controls this acceptance or rejection response is very polymorphic. This has obvious similarities with the vertebrate MHC, although the genes that underlie allorecognition in *Botryllus* have yet to be cloned to determine whether they are related to vertebrate MHC molecules (see Box 7.2).

Metazoans that are non-colonial are not threatened by allogeneic cells. In these organisms, invading microorganisms, or aberrant cells that could lead to cancers, are a more vital threat. It is possible the primordial MHC molecule, perhaps hitherto involved in a primitive allorecognition system, evolved the ability to modify itself with cellular material and display it at the cell surface to the effector cells. This shift in function would disclose to the outside the contents of the cell, including potentially harmful pathogens and the products of aberrant genes. This is an attractive hypothesis since it helps explain why modern T lymphocytes are able to respond to not only self-MHC plus antigen, but also allogeneic MHC.

Metazoa *n.*
multicellular animals.

7.4.1 Mapping the MHCs of other species

Allorecognition phenomena have been demonstrated in most vertebrate classes (Figure 7.2) although the presence of an MHC must be confirmed by molecular evidence. Reactivity with alloantisera is contributory evidence. Alloantisera are readily produced in mammals and, as we have seen, the MHCs of human and mouse have been thoroughly mapped as a direct consequence. Alloantisera are less readily produced in non-mammalian species, particularly those that do not breed well in captivity or those with weak MHC-dependent phenomena. However, alloantisera have been produced in the chicken and the amphibian, *Xenopus*. As a consequence, the

BOX 7.2 Kin recognition in colonial tunicates

Invertebrate 'histocompatibility reactions' have been described in several species. Porifera (sponges), for example, are well known for their ability to discriminate between colonies of related and unrelated cells, and will fuse with related colonies. Grafting experiments in other invertebrates, such as cockroaches and earthworms, show that these animals are able to reject allogeneic tissues.

However, much interest has focused on *Botryllus schlosseri* (a tunicate or sea squirt), mainly because this animal is a urochordate. These are grouped together with the vertebrates in the phylum Chordata since they share common 'chordate' characteristics at some stage in their life cycle: a notochord, a dorsal hollow nerve cord, and pharyngeal clefts. Only in the larval stage, which looks like a microscopic tadpole, do urochordates possess these features. It is likely that if the roots of histocompatibility are to be found in any invertebrates, they will more likely be found in the deuterostome branch from which the vertebrates evolved (see Figure in preface). Modern deuterostomes comprise the echinoderms, hemichordates, protochordates and chordates.

Colonies of *Botryllus* are first formed when a tadpole larva settles onto a firm substrate and metamorphoses into a sessile stage called the oozooid. This replicates by a process of asexual reproduction to produce zooids. Figure 1 shows two small colonies. The colony expands by a weekly cycle of asexual reproduction, and mature colonies may contain as many as a 1000 genetically identical zooids. The colony becomes embedded in a translucent tunic, or test. Individual zooids are connected to each other by a vascular network, through which the blood circulates and which terminates at the periphery of the test as bulbous ampullae.

Botryllus colonies are sessile and live in close proximity to each other on the surfaces of rocks. When the edges of two colonies meet, the test dissolves and the ampullae of the opposing zooids make contact. Colonies that are related fuse with each other and the vasculatures of the two systems become connected. However, unrelated colonies are rejected within 30 min. The junction then becomes necrotic and the ampullae involved are severed. This response prevents unrelated, allogeneic zooids from gaining entry to the colony. Why this is important to the colony is not clear but it may prevent the colony from being parasitized. *Botryllus* thus displays a kind of simple allorecognition system.

Breeding experiments have shown that kin recognition in *Botryllus* is controlled by a single locus called the **fusibility/histocompatibility locus** (or Hu/Fc). Several studies have revealed this to be a very polymorphic locus, that has many codominantly

Figure 1 The colonial ascidian *Botryllus schlosseri*. (a) Two colonies, each consisting of six zooids. (b) Section through two zooids of a colony. (c) Fusion between oozooid pairs. Arrows show the connection between the vascular ampullae. (d) Rejection between two oozooids. Arrows show the necrosis and auto-amputation of ampullae. Panels (a) and (b) redrawn from Barnes, R.D. (1974) *Invertebrate Zoology*. 3rd edition, copyright © 1974 by Saunders College Publishing, by permission of the publisher, panels (c) and (d) reproduced with permission from Scofield, V.L. and Nagashima, L.S. (1983) *Biological Bulletin*, **165**, 733.

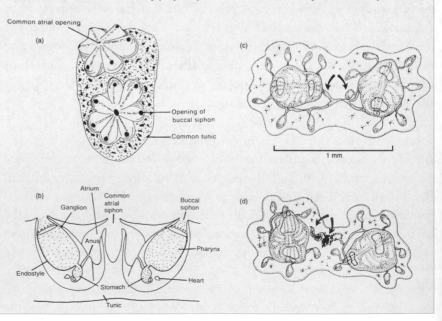

expressed alleles. Some populations in the Mediterranean may have over 300 different alleles. Individuals or colonies sharing one or both alleles at the locus fuse while individuals sharing no alleles do not fuse. This has many similarities with the MHC of vertebrates, although it remains to be determined whether the tunicate genes are related in any way to the vertebrate MHC.

Further reading

Weissman, I.L., Saito, Y. and Rinkevich, B. (1990) Allorecognition and histocompatibility in a protochordate species: is the relationship to MHC somatic or structural? *Immunological Review*, **113**, 227–41.

MHCs of these two species are the best understood of the non-mammalian species.

A strategy to overcome the difficulty of raising alloantisera in non-mammalian vertebrates has been to produce antibodies to their cells in mammals (**xenoantisera**), or to use mammalian alloantisera that cross-react with non-mammalian MHC molecules. There are several examples of antibodies against HLA molecules recognizing MHC molecules from birds, reptiles, amphibians and fish.

A more versatile handle on the MHCs of other mammals exploits the homology between mammalian genes. Probes of human class I MHC genes cross-hybridize with their homologues in several primates, including chimpanzee, gorilla, orang-utan, gibbon, macaque and cotton-top tamarin. Human probes have allowed cloning of some chicken and swine MHC genes, although in the main, mammalian probes generally do not cross-hybridize well with non-mammalian species. Mouse probes have provided the points of entry into the MHCs of rat (RT1) and rabbit (RLA).

For more distantly related vertebrates, mammalian probes seldom work. Instead, probes are designed 'retrospectively' on the basis of amino acid data of MHC molecules. For example, *Xenopus* MHC molecules have been purified with xenoantisera and partially sequenced. From the amino acid sequences, corresponding oligonucleotide probes have been synthesized and used successfully for identifying MHC genes in *Xenopus* cDNA libraries.

A similar strategy has been used to probe cDNA libraries of teleost fish *without* fish amino acid sequence data. On the assumption that fish MHC molecules would also belong to the immunoglobulin gene superfamily, oligonucleotide probes were made corresponding to well-conserved amino acid sequences that flank the intradomain disulphide bonds of different immunoglobulin superfamily members. This has provided a point of entry and mapping for the MHCs of several teleost fishes, including carp, salmon, zebra fish and cichlids. These probes have also identified cDNAs within libraries of several species of shark, which is remarkable considering that allorecognition phenomena are usually undetectable in these animals.

xenoantiserum *n.* antiserum raised in one species of animal against the cells from another species (e.g. in mice against human cells).

7.4.2 Sources of MHC polymorphism

Whatever the origins of MHC molecules, the extensive polymorphism displayed by the classical MHC molecules of most vertebrates is a characteristic feature. Polymorphism may have already existed in the primordial MHC molecules if they were involved previously in recognition of related individuals (see Box 7.2). Polymorphism is also likely to have been generated after the shift in function within non-colonial metazoans.

The rates of mutation in human and murine MHCs are no greater than elsewhere in the genome. It is hypothesized, therefore, that the extensive polymorphism of MHCs in contemporaneous animals is a result of the steady production of new alleles, which have accumulated as each new species evolved. In other words, MHC alleles are inherited in a **trans-species** fashion. Some MHC alleles are remarkably ancient. For example, alleles found in humans and chimpanzees predate the divergence of these animals around 7–8 million years ago. Some alleles may have been around for as long as 10 million years.

Mutations are the original source of any new genetic material. In the case of MHC molecules, gene conversion and homologous recombination have also played a role in mixing up genetic information to create novel alleles. This can occur within the same gene, or between different genes. Figure 7.7

Figure 7.7 Genetic recombination in the MHC. Shown are schematic representations of introns and exons of some human class I MHC α chains. Two basic mechanisms are involved: gene conversion (a) and crossovers between homologous DNA (b). The abbreviations for the exons are: L = leader, TM = transmembrane, C = cytoplasmic, UT = untranslated. Adapted, with permission, from the Annual Review of Immunology, Volume 8, ©1990, by Annual Reviews Inc.

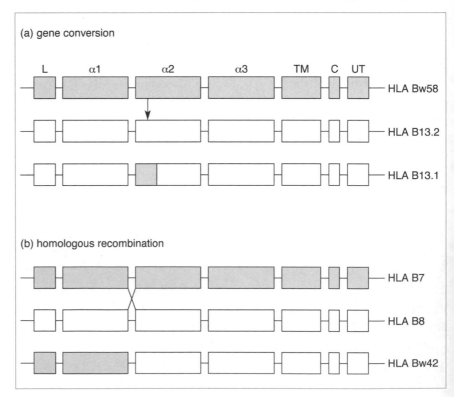

shows a diagram of the human B13.1 allele. This allele arose by a gene conversion event between two other alleles, B13.2 and Bw58. Similarly, the HLA-Bw42 allele is a mixture of exons acquired from two different alleles, B7 and B8, by homologous recombination.

7.4.3 Natural selection has been involved in generating MHC polymorphism

The non-random nature of amino acid substitutions in different alleles shows that natural selection has been involved in shaping MHC polymorphism. As we saw in Chapter 6, these substitutions are clustered in and around the peptide-binding cleft. These confer the cleft of each allele with a unique surface topography. As a consequence, the spectrum of antigenic peptides that a particular allele can present is different from those of other alleles. The remainder of the molecule is comparatively well conserved among different alleles.

Random mutational processes alone, such as gene conversion and homologous recombination, cannot account for the non-random distribution of substitutions, without the effects of subsequent natural selection. We can see evidence of the influence of natural selection if we examine the ratios of **synonymous** and **non-synonymous** nucleotide substitutions within MHC genes. Synonymous (or silent) substitutions do not change the amino acid encoded, and can occur because most amino acids are encoded by more than one different triplet of nucleotides (codon). Non-synonymous (or non-silent) substitutions, in contrast, change the amino acid encoded. These changes are expressed in the phenotype and can therefore be accessible to the forces of natural selection.

> **synonymous mutation**
> *n.* a mutation that does not affect the sequence of amino acids encoded.

If no natural selection is involved, the numbers of silent and non-silent mutations would accumulate in a predictable ratio. A divergence from this ratio would indicate the gene was exposed, through the phenotype it encoded, to the forces of natural selection. For most genes the influence of natural selection is evident because the rate of accumulation of non-silent mutations usually is a fifth of that for silent mutations. This is because, on the whole, non-silent mutations are deleterious to the functions of the protein product.

MHC genes also show a skewing of the ratio of silent to non-silent mutations, showing natural selection is being exerted on their products. However, the selective pressures on the peptide-binding cleft and those on the remaining framework of the molecule operate in different directions. Outside the peptide-binding cleft, there are, as in most genes, more silent than non-silent mutations (Figure 7.8). On the whole therefore, substitutions in these regions are deleterious, presumably because they might alter

Figure 7.8 Mutations in the peptide binding cleft of MHC molecules are predominantly non-synonymous (amino acid changing). The nucleotide sequences of several human and murine class I and class II loci were pooled for the study. On the *y*-axis is the number of non-synonymous mutations minus the number of synonymous mutations: a negative value indicates there are more synonymous mutations, whereas a positive value indicates there are more non-synonymous. For 'average' genes, non-synonymous mutations predominate. In this study, codons encoding different parts of the molecule, the peptide binding cleft (shaded) and the rest of the molecule (diagonal lines) are analysed separately. Reproduced with permission from Potts, W.K. and Wakeland, E.K. (1990) *Trends in Ecology and Evolution*, **5**, 181.

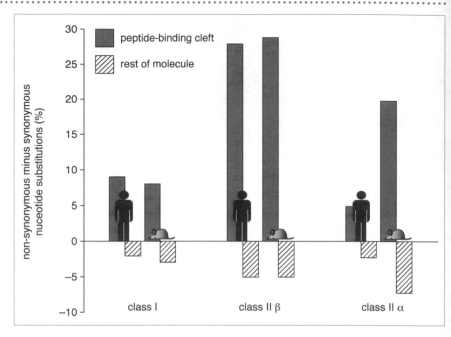

heterozygote advantage a situation in which an individual that is heterozygous for a given gene has an increased fitness (relative capacity to pass its genes to the next generation) over a homozygote.

the interaction with important non-polymorphic ligands such as CD4, CD8 (see Figure 6.10) or β_2-microglobulin. Such mutations would be expected to be deleterious to the host since they would impair the immune system and thus be eliminated by natural selection from pathogens.

In the peptide-binding cleft the situation is reversed, with the proportion of non-silent mutations out-numbering the silent mutations. This indicates that mutations within the peptide-binding cleft are not deleterious. Populations with greater polymorphism may be better able to resist infections by pathogens, whereas less polymorphic populations may be unable to present peptides derived from a novel pathogen and could be wiped out. For the same reason, individuals that are heterozygous at MHC loci are better able to resist the effects of a greater range of pathogens than homozygotes. This phenomenon is called the **heterozygote advantage.**

7.4.4 Natural selection by pathogens may maintain MHC polymorphism

The mechanisms by which MHC genes have become so polymorphic have been the subject of considerable debate. We saw evidence in the previous chapter of the effect of the malarial parasite, *Plasmodium*, on the frequency of a 'protective allele', HLA Bw53, in populations where malaria is endemic. Such studies support the hypothesis that particular MHC alleles, or combination of alleles, are maintained in the population by selective pressure imposed by pathogens. Presumably because each allele increases the fitness of an individual by being able to present an epitope from an indigenous, life-threatening pathogen. Many mammals such as mice and humans are social animals in which infectious diseases are readily transmitted. These animals all have polymorphic MHCs. It is thought that the numerous alleles are a legacy of numerous previous encounters with pathogens which have been collected in the gene pool over the generations.

Pathogens may also have been the driving force behind the *duplication* of class I and class II loci within the MHC. More loci means more alleles can be expressed. However, there seems to be an upper limit to the number of loci an animal can have, possibly because an animal with too many different MHC molecules would delete too many thymocytes during ontogeny in the thymus and severely deplete the T lymphocyte repertoire.

Despite these observations, the role of pathogens in maintaining MHC polymorphism remains a contentious issue. Vertebrates of all classes are exposed to different kind of pathogens, and yet some, such as the Syrian hamster, the cotton-top tamarin, whales and the Florida panther, display very limited MHC polymorphism. This may be because these species may have been exposed to fewer pathogens, or may be the result of genetic drift or inbreeding. Moreover, many other less polymorphic genes, that are not part of the immune system, may play a more important role in survival. For example, resistance to malaria in humans is also associated with several alleles encoding proteins found in erythrocytes, and which influence the resistance of erythrocytes to infection by the merozoite stage in the life cycle (see Box 5.2).

Negative selection in the thymus: Section 9.2.3

genetic drift *n.* random changes in the frequency of alleles in a breeding population by means other than natural selection; usually more pronounced in small populations.

7.4.5 Evolution of pseudogenes and non-classical MHC genes

Intermingled among the classical genes of the MHC are non-classical genes and pseudogenes that have arisen by a process of gene duplication and divergence. These genes are not polymorphic, and although not involved in presentation of peptides to T lymphocytes, non-classical genes may have some other specialized functions in antigen processing and presentation (see Section 7.1.2).

Pseudogenes are unable to produce a protein product, usually because of a deleterious mutation. Because pseudogenes make no product, there are few constraints on their accumulating mutations and these tend to diverge more rapidly than fully functional genes. Pseudogenes may persist in MHCs because of their usefulness as a reservoir of genetic information, from which polymorphism can be generated by gene conversion.

Recombination and gene conversion: Box 2.1

Other genes in the MHC, such as the genes of the TAP transporter proteins and the proteasome components, seem to play important roles alongside classical MHC molecules in the processing and presentation of peptides to T lymphocytes. The significance of the co-localization of these genes together in the MHC is not clear, particularly since several other genes involved in antigen processing and presentation are encoded outside the MHC (e.g. invariant chain and β_2-microglobulin). Precisely why these genes, or indeed any genes, remain in clusters is unclear, but tightly linked genes are less likely to be lost through unequal crossovers.

7.5 Evolution of other molecules

Other proteins of importance in the immune system also illustrate the role of gene duplication and divergence in molecular evolution.

7.5.1 The neonatal Fc receptor

The neonatal Fc receptor (FcRn) is an interesting example of molecular evolution in the immune system. In suckling mice and rats, this receptor is found on the intestinal epithelial cells where is responsible for the uptake of maternal IgG provided in milk. These receptors transport maternal IgG from inside the gut into the blood by **transcytosis** (see Figure 4.22). The FcRn molecule and the classical class I MHC molecule appear to have evolved from a common ancestor. Both are heterodimers of β_2-microglobulin and a heavy chain of three domains, α_1, α_2 and α_3. The heavy

Figure 7.9 Comparison of the α_1 and α_2 domains of the rat neonatal Fc receptor (a) and the mouse class I MHC molecule H-2 K^b complexed with a viral peptide (b). From Ravetch, J.V. and Margulies, D.H. (1994) *Nature,* **372**, 323–4.

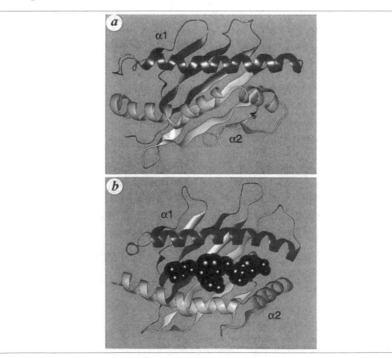

chains of both molecules also have a similar three-dimensional structure of two α-helices surmounting a platform of β-strands (Figure 7.9). However, the cleft of the FcRn molecule is closed and it does not appear to be able to present peptides in the same way as a class I MHC molecule. Instead, FcRn interacts with the C_H2 and C_H3 domains of the IgG Fc region using a surface formed in part by its α_1, α_2 and β_2m domains.

The modern MHC class I molecule, class I-like CD1 molecules (section 7.2.2) and the FcRn molecule are clearly descended from the same common ancestral gene, and have evolved different functions in the immune system. It has been calculated that duplication and divergence of the common ancestor probably occurred around the time of the divergence of mammals and reptiles.

7.5.2 Complement components

Studies in humans and mice show that many components of the complement cascade have arisen by gene duplication and divergence. Consequently, the nucleic acid sequences of such components are similar and usually linked (on the same chromosome). Examples are Factor B and C2, C1r and C1s, and C6 and C7. Similarly, C3 and C4 of the alternative and classical pathways arose by duplication of a common ancestor. These are homologous, thioester bond-containing molecules, whose cleavage products have similar opsonic and inflammatory properties.

Complement components are also examples of the role of exon shuffling in the generation of novel proteins. For example, the lytic components C6–C9 contain regions that are found in other proteins such as epidermal growth factor, low density lipoprotein receptor, perforin and thrombospondin.

Summary

- The T lymphocyte antigen receptor–MHC molecule interaction may have originated as a more conventional heterotypic interaction, possibly involved in allorecognition in primordial colonial protochordates.

- Polymorphism of MHC molecules is generated by random point mutation, gene conversion and crossovers.

- Polymorphism is localized to the peptide-binding cleft and is maintained in populations by the advantage that heterozygosity confers in the resistance to pathogenic microorganisms.

Further reading

Books

Manning, M.J. and Turner, R.J. (1976) *Comparative Immunobiology*. Blackie, Edinburgh.

Turner, R.J. (ed.) (1994) *Immunology. A Comparative Approach*. J. Wiley, London.

Review articles

Farries, T.C. and Atkinson, J.P. (1991) Evolution of the complement system. *Immunology Today,* **12**(9), 295–300.

Lawlor, D.A., Zemmour, J., Ennis, P.D. and Parham, P. (1990) Evolution of class-I MHC genes and proteins: from natural selection to thymic selection. *Annual Review of Immunology,* **8**, 23–63.

Potts, W.K. and Wakeland, E.K. (1990) Evolution and diversity at the major histocompatibility complex. *Trends in Ecology and Evolution,* **5**, 181.

Sperling, A.I. and Bluestone, J.A. (1993) Non-classical MHC molecules: the first line in defence? *Current Biology,* **3**(5), 294–6.

Trowsdale, J. (1995) 'Both man and bird and beast': comparative organization of MHC genes. *Immunogenetics,* **41**, 1–17.

Williams, A.F. and Barclay, A.N. (1988) The immunoglobulin superfamily domains for cell surface recognition. *Annual Review of Immunology,* **6**, 381–4.

Zapata, A.G., Varas, A. and Torroba, M. (1992) Seasonal variations in the immune system of lower vertebrates. *Immunology Today,* **13**(4), 142–7.

Review questions

Fill in the blanks

_____[1] is the oldest immunoglobulin isotype because it is found in all vertebrates except _____[2]. In all vertebrates it exists as a membrane-bound _____[3] form and as _____[4] molecular weight _____[5] forms in the serum. Additional _____[6] molecular weight isotypes are seen in _____[7] and all the vertebrates that evolved afterwards. In amphibians, reptiles and birds the major _____[8] molecular weight isotype is called IgY, whereas _____[9] have at least _____[10] major isotypes. MHC molecules have been found in all vertebrates except _____[11]. Amphibians have a free-living

_____ [12] stage which reaches immunocompetence _____ [13]; in contrast, reptiles, birds and mammals are _____ [14] whose embryos are protected from pathogens. In these animals, immunocompetence is acquired after hatching/birth. During this time, _____ [15] are transferred from the mother. In birds, this occurs during passage of the egg through the _____ [16], whereas in mammals this occurs by transfer across the _____ [17]. Mammals continue to passively transfer _____ [18] to suckling young in secretions of the _____ [19] glands, in the form of _____ [20] followed by _____ [21]. Birds and mammals are also _____ [22] which has enabled these animals to maintain a stable body _____ [23]. In comparison with fishes, amphibians and reptiles, these animals respond more _____ [24] to antigen, have _____ [25] antigen receptor repertoires, more immunoglobulin _____ [26] and more _____ [27] secondary lymphoid tissues. Because the placental mammal embryo bears paternal _____ [28] antigens, several mechanisms have evolved to prevent _____ [29] by the maternal immune system during the lengthy gestation. These include immunological _____ [30] by the trophoblast, and the production by the placenta of a several _____ [31] substances.

Short answer questions

1. True or false.

    (a) The MHC is polymorphic only in eutherian (placental) mammals.

    (b) IgY is the major low molecular weight immunoglobulin in secondary responses by amphibians, reptiles and birds.

    (c) Unequal crossovers at meiosis are a major source of gene duplication.

    (d) The lytic phase of the complement pathway is present in Agnatha.

2. Gene conversion, exon duplication, exon shuffling and gene duplication are genetic events that influence the evolution of proteins. Which most likely accounts for the following?

    (a) The multiple constant domains of immunoglobulin heavy chains.

    (b) The generation of immunoglobulin diversity in the chicken.

    (c) The use by selectins of domains related to domains in other proteins.

    (d) The multiple loci encoding class I MHC molecules in the HLA complex.

(e)   The structural similarities between a class I MHC molecule and the neonatal Fc receptor.

3.   Which of the following pairs of homologous molecules do you think is more likely to have evolved first? (give your reasons).

(a)   IgM or IgG?

(b)   IgG or IgY?

(c)   C3 or C4 of complement?

(d)   The neonatal Fc receptor or the class I MHC molecule?

4.   In what general features of the immune system do endotherms and ectotherms differ?

Signals and control 8

Introduction

The ability to perceive, and respond to, the external environment is fundamental to all cells. Unicellular organisms locate elements essential for their survival, such as nutrients or oxygen. In metazoan organisms, however, cells have foregone an independent existence to become specialized to specific functions. They depend on, and cooperate with, other cells in the body. The extracellular environment to which metazoan cells respond consists of other cells and the substances they secrete. Complex behavioural responses by cells, such as cell division, changing into a new phenotype or undergoing apoptosis, require changes in gene expression. These responses are under strict regulation, as uncontrolled responses may damage the organism. Cancer cells, for example, are the result of uncontrolled cell division.

Many signals instructing a cell to change its behaviour are mediated by soluble signalling molecules – hormones, growth factors and cytokines. Some hormones are soluble in lipid and can pass directly through the plasma membrane of the target cell. In the main, however, extracellular signalling molecules are only soluble in water and cannot pass through membranes on their own. For this reason, the vast majority of signalling molecules are received by specific receptors on the plasma membrane of the target cell.

This chapter is mostly about receptors, specifically those on cells of immune systems. In the first half we will learn how receptors are involved in the activation of lymphocytes into effector cells. In the second half, we will examine cytokines – the extracellular signalling molecules of immune systems – and their receptors.

8.1 Receptors and signalling cascades

The function of a receptor is to communicate the event of binding an extracellular signalling molecule (the input signal) into the interior of the cell (the output signal). Accordingly, a receptor has an extracellular region that binds the signalling molecule, a hydrophobic transmembrane region that passes through the plasma membrane one or more times, and usually an intracellular domain inside the cell.

A few receptors are **signal-transporting** receptors, which operate by allowing the bound extracellular signalling molecule itself to enter the cell. The majority of receptors, however, are **signal transducers**. These convert the input signal into a different, output signal (Figure 8.1). This is mediated

signal transduction the transmission of an extracellular signal, initiated by the interaction of a cell-surface receptor with its ligand, to the interior of the cell.

Figure 8.1 Receptors for extracellular ligands. Receptors communicate extracellular signals into the cell. A signal-transporting receptor allows the ligand to pass directly into the cytoplasm. A signal-transducing receptor converts the binding of the extracellular ligand into a different intracellular signal. This is achieved by a change in the conformation of the cytoplasmic domain, which enables it to trigger a signalling cascade.

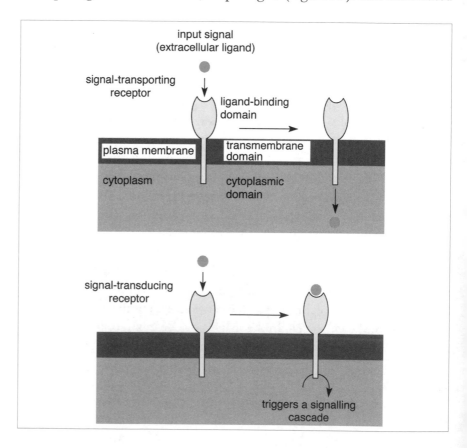

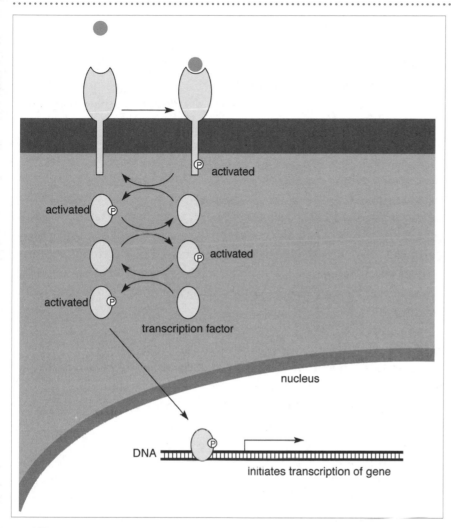

Figure 8.2 Schematic representation of a signalling cascade. A signalling cascade consists of a sequence of components that relay a receptor-mediated signal to the nucleus. Most signalling components are enzymes, switching between active and inactive states. Each is activated by the catalytic activity of the preceding component in the cascade. Activated components return to their inactive states after dissociating from their substrates.

in different ways, but in general depends on a conformational change in the cytoplasmic domain of the receptor. This is then able to interact with intracellular signalling proteins, either by acquiring docking sites for them, or by turning on enzymatic activity that uses other signalling proteins as substrates.

These are the first steps in a cascade of interactions between several signalling proteins that relay the signal to the nucleus (Figure 8.2). The majority of signalling proteins are enzymes that are turned on by the catalytic activity of the preceding component in the cascade. After dissociating from its downstream substrate each signalling enzyme returns to its inactive state.

Switching of a signalling component between active and inactive states is accompanied by the gaining and losing of phosphate. Indeed, the phosphorylation of signalling proteins, cytoskeletal elements and many other substrates in the cytoplasm, is a hallmark of receptor-mediated activation

phosphorylation *n.* the addition of phosphate to a protein or nucleic acid performed by a **kinase**; phosphorylation of a protein often alters its properties.

of eukaryotic cells. The endpoint of a signalling cascade is a transcription factor, a protein that upon activation binds to specific sequences in the control regions of genes and initiates their transcription.

Because a single activated enzyme may activate many molecules of its substrate, a cascade serves to amplify the original extracellular signal. However, activation of the components must be transient to restore function to the receptor. Signalling enzymes are therefore normally maintained in an inactive state and activation is quickly followed by inactivation. A cascade also allows the signal to be modulated in different ways. Many signalling enzymes have more than one substrate, which causes signalling pathways to split. In this way, a single extracellular signal can influence the expression of more than one gene. Moreover, the effect of one particular signal can be modified by another signal received by a different receptor. In reality, cells are subjected to a cocktail of extracellular signals. Each activity by a cell is the net result of a particular combination of multiple input signals.

8.1.1 Three kinds of signal-transducing receptors

There are three major types of receptors, classified according to the way they transduce the signal into a cell. These are **ion channel-linked** receptors, **G-protein-linked** receptors and **enzyme-linked** receptors.

Ion channel-linked receptors

Ion channel-linked receptors convert the extracellular signal into the opening or closing of a channel in the plasma membrane. This controls the flow of ions into or out of the cell. These receptors are involved in rapid signalling, usually independently of changes in gene expression, between electrically excitable cells. This kind of signalling occurs, for example, by receptors for neurotransmitters found at neuromuscular synapses. However, the majority of signal-transducing receptors are G-protein-linked receptors or enzyme-linked receptors. Ion channel-linked receptors will not be discussed further here.

G-protein-linked receptors

G-protein-linked receptors are so called because in response to binding an extracellular ligand, their cytoplasmic domains become associated with a GTP-binding protein (G-protein) (Figure 8.3a). G-protein-linked receptors belong to a phylogenetically ancient gene superfamily, with representatives in eukaryotes and prokaryotes, and almost certainly evolved from a common ancestral gene. Each receptor has a distinctive seven-span transmembrane domain. Over 100 different G-protein-linked receptors

G-proteins *n. plu.* subset of guanine nucleotide-binding regulatory proteins, found at the cytoplasmic face of eukaryotic cell membranes involved in signalling.

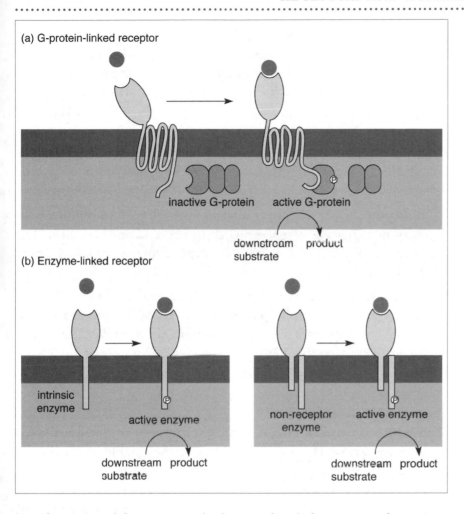

(a) G-protein-linked receptor

inactive G-protein   active G-protein

downstream   product
substrate

(b) Enzyme-linked receptor

intrinsic
enzyme

active enzyme

downstream   product
substrate

non-receptor
enzyme

active enzyme

downstream   product
substrate

Figure 8.3 Signal-transducing receptors. G-protein-linked receptors conduct the conformational change induced by ligand binding to the cytoplasmic domain. This enables it to bind a G-protein, which in turn interacts with signalling enzymes (see Box 8.1). Enzyme-linked receptors have latent enzymatic activity associated with their cytoplasmic domain that is activated by a conformational change induced by the ligand. The activity is either intrinsic to the cytoplasmic domain or comes from a non-covalently attached enzyme.

have been cloned from mammals alone, and include receptors for various hormones, neurotransmitters and chemotactic factors.

Enzyme-linked receptors

The third type are the enzyme-linked receptors. Receptors of this type are heterogeneous and unrelated to each other. These receptors are characterized by a single transmembrane domain, and enzymatic activity associated with their cytoplasmic domain (Figure 8.3b). There are five major families of enzyme-linked receptors, which are classified according to the type of enzymatic activity that is associated with their cytoplasmic domain.

- **Receptor tyrosine kinases**. When activated, these have intrinsic kinase activity that phosphorylates tyrosines on their substrate. Receptors for most growth factors and hormones belong to this family, including

kinase *n.* an enzyme that transfers phosphate to (phosphorylates) its substrate.

epidermal growth factor (EGF) and insulin. Several belong to the immunoglobulin gene superfamily, including platelet-derived growth factor (PDGF), fibroblast growth factor (FGF), nerve growth factor (NGF) and vascular endothelial cell growth factor (VEGF).

- **Non-receptor tyrosine kinases.** Tyrosine kinase activity is provided by a separate enzyme associated non-covalently with the cytoplasmic domain of the receptor. Antigen receptors and most cytokine receptors use this arrangement (discussed below). Receptor and non-receptor tyrosine kinases account for the majority of enzyme-linked receptors.

- **Receptor serine/threonine kinases**. These have intrinsic kinase activity that phosphorylates serine and threonine on their substrates. Members of this family include the receptors for the cytokine TGF-β_1 (see later).

- **Receptor tyrosine phosphatases**. These inactivate the activity of kinases by removing phosphates. A member of this group is CD45, which is an important membrane glycoprotein on mammalian leucocytes involved in signalling.

- **Receptor guanylyl cyclases** which catalyse the production of cGMP.

The role of phosphorylation

G-proteins and signalling enzymes both work by switching between active and inactive states. This is accompanied by the gaining and losing of phosphate (Figure 8.4).

For receptor associated G-proteins, this occurs when it binds to the cytoplasmic domain of the receptor, causing an exchange of bound GDP

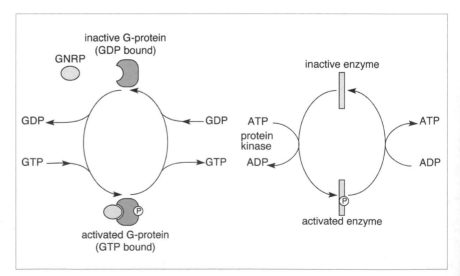

Figure 8.4 Switching of G-proteins and signalling enzymes between active and inactive states. The activation status of signalling proteins depends on the type of protein. G-proteins (GTP-binding proteins) become activated when they bind to a guanidine nucleotide-replacing protein (GNRP) which causes the G-protein to exchange GDP for GTP. Other signalling proteins are controlled by ATP-dependent protein kinases that phosphorylate their substrates and protein phosphatases that remove phosphate.

for GTP and a net gain of phosphate. GTP is exchanged back for GDP as the G-protein dissociates from the receptor. This switching back and forth is governed by alternating conformational changes in the cytoplasmic domain. These changes originate in the extracellular domain when it binds to its ligand. Binding causes a conformational change that is transmitted through the whole molecule, from the extracellular domain to the intracellular domain via the seven-pass transmembrane region.

Enzyme-linked receptors, in contrast, pass through the plasma membrane only once and cannot conduct information into the cell in the same way. Receptor and non-receptor tyrosine kinases achieve this instead by **autophosphorylation** (Figure 8.5). This occurs when two receptor chains assemble into a dimer as a result of binding to their ligand. Dimerization may be promoted by the conformational change that occurs in the extracellular region of the receptor, or when the ligand itself is a dimer and can cross-link the receptor chains. Upon dimerization, the cytoplasmic regions then cross-phosphorylate each other on multiple tyrosines. Phosphorylation of tyrosines within the catalytic site activates kinase activity, whereas phosphorylation of other tyrosines creates anchorage points for several downstream signalling proteins. This process of cross-linking of enzyme-linked receptors by ligands is a recurring theme in the signalling between cells in immune systems.

> **autophosphorylation** *n.* self-phosphorylation of a protein with kinase activity; the first event in the signal transduction pathway of many receptors.

Signalling cascades

Downstream of a receptor are one or more cascades of signalling components (see Box 8.1). Many components are also G-proteins, kinases and phosphatases. Another type of cascade, important notably in lymphocytes,

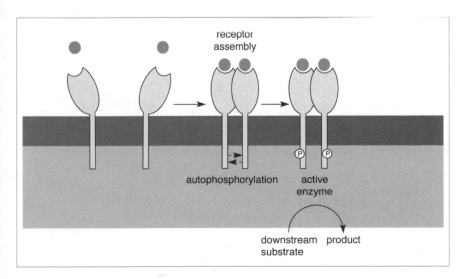

Figure 8.5 Autophosphorylation by receptor associated protein kinases. Enzyme-linked receptors pass through the membrane once and do not conduct the conformational change induced by ligand binding. Instead, ligand binding promotes the assembly of two or more receptor chains, either by cross-linking receptor chains or by causing a change in extracellular conformation that favours assembly, thereby bringing the cytoplasmic domains into close proximity.

BOX 8.1 Signalling cascades

Signalling cascades serve to relay extracellular signals received by signal-transducing receptors to the nucleus. Shown in Figure 1 is a composite of several signalling pathways that have been found to arise from different signal-transducing receptors found on eukaryotic cells. The majority of known hormones, growth factors, cytokines and other extracellular signalling molecules trigger one or more of these pathways through their specific receptors. New receptors are often discovered (usually before their natural ligands or functions have been found) by looking for these signalling events after experimental ligation or cross-linking of membrane proteins on cells using antibodies.

The inositol phospholipid pathway

The major effect of G-protein-linked receptors is the activation of cascades that release Ca^{2+} via the **inositol phospholipid pathway**, or **cyclic AMP** (cAMP). These are both important intracellular second messengers. A G-protein is normally in an inactive, GDP-bound form and resides anchored to the cytoplasmic face of the plasma membrane dissociated from the receptor. A change in the conformation of the receptor caused by binding an extracellular ligand is passed to its cytoplasmic domain. This in effect produces a guanidine nucleotide releasing protein (described in Figure 8.4) to which a G-protein binds. When this happens, the G-protein also undergoes conformational change and displaces GDP for GTP. Activated membrane G-proteins have an affinity for two downstream substrates which are in turn activated; **phospholipase Cβ** and **adenylyl cyclase**. Here the signalling pathway branches.

Activated adenylyl cyclase produces cyclic AMP (cAMP) from ATP. This enzyme is counteracted by cAMP phosphodiesterases which help keep down the concentrations of cAMP in the cytosol. The burst of cAMP produced is used predominantly to phosphorylate and activate A-kinase. This enzyme operates on many different substrates involved in cellular

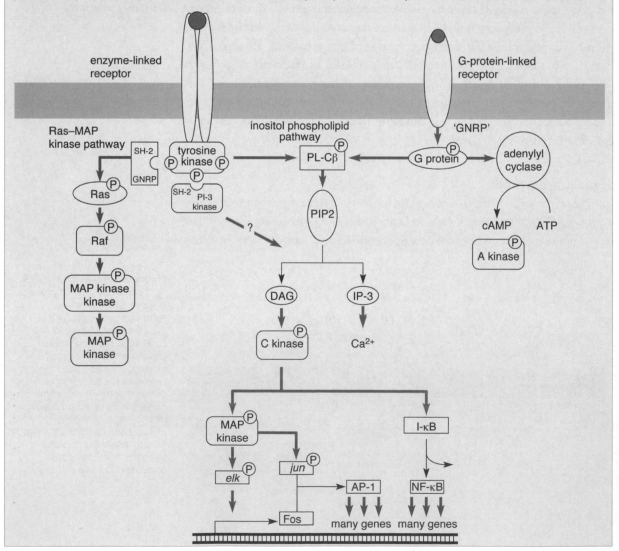

metabolism and, depending on the cell type, has a multitude of effects. Its activity is counteracted by a family of serine/threonine phosphatases. Meanwhile, activated phospholipase Cβ cleaves a phospholipid found in the plasma membrane, called **phosphatidylinositol biphosphate** (or PIP2) into two products: **diacylglycerol** (or DAG) which remains in the plasma membrane, and **inositol triphosphate** (IP3) that diffuses into the cytosol. Here the signalling pathways diverge again. The concentration of Ca^{2+} in the cytosol is normally kept low, either by Ca^{2+}-dependent ATPases that pump Ca^{2+} out through the plasma membrane or into the endoplasmic reticulum, or by a cytosolic protein, called **calmodulin**, that sequesters Ca^{2+}. However, free IP3 opens calcium channels in membranes, causing Ca^{2+} to surge in. IP3 is rapidly dephosphorylated by phosphatases, ensuring the rise in Ca^{2+} is a transient burst. On the other hand, DAG activates **C-kinase**. C-kinase activates downstream components by phosphorylating serine and threonine residues on them. The best characterized targets are **Iκ-B,** and mitogen activated protein kinase or **MAP kinase**. We will encounter MAP kinase again below. Upon phosphorylation, Iκ-B releases a smaller protein called **NF-κB** which is a transcription regulatory protein for many genes, notably cytokines and cytokine receptors. It is through transcription factors such as NF-κB that a single receptor can influence the expression of more than one gene.

Kinases and phosphatases

Many enzyme-linked receptors and downstream signalling components are tyrosine kinases that phosphorylate tyrosines on their substrates. There are several families of tyrosine kinases, the largest of which is the **Src kinase family**. Members include Src, Fyn, Lyn and Lck. The catalytic activity of tyrosine kinases is also dependent on phosphorylation of tyrosines in the catalytic site. Phosphorylation at other sites changes the conformation and creates 'docking' sites for other downstream signalling proteins which bind, either directly or indirectly through adapter proteins. Each docking site can be recognized by a different signalling protein, causing the signalling pathway to branch off in several directions. Signalling proteins that bind to the docking sites are heterogeneous but all have the hallmark of one or more **SH-2** (Src-homology) domains with which they bind to phosphorylated tyrosine. In so doing they in turn become activated. Another family of cytoplasmic tyrosine kinases are the **Janus kinases** (or **Jaks**) which activate a family of proteins called signal transducers and activators of transcription (or **Stats**). This novel signalling pathway is shared by all members of the cytokine receptor superfamily.

A pathway leading from many enzyme-linked receptors is the **Ras–MAP kinase** pathway. This pathway is remarkably well conserved in different eukaryotes. Stimulation of this pathway usually results in the cell moving from a state of quiescence into the cell cycle. Ras is normally in an inactive, GDP-bound state, but becomes activated when it exchanges GDP for GTP. This is catalysed by an activated SH-2-binding protein which is also a guanidine nucleotide releasing protein. Ras-GTP has a short half life and is rapidly inactivated by **GAP** (GTPase activating enzyme). The downstream target for Ras-GTP is a cascade of serine/threonine kinases, starting with **Raf**. The phosphorylation of serine and threonine is longer lived than on tyrosine and serves to sustain the signal initiated by the receptor. Activated Raf in turn phosphorylates MAP (mitogen activated protein) kinase, which in turn phosphorylates MAP kinase. Activated MAP kinase moves into the nucleus where it phosphorylates a protein called **Jun** and DNA-bound transcription factors, including **Elk**. Phosphorylated Elk initiates the transcription of several 'early genes', including the gene for **Fos**, whose product assembles with phosphorylated Jun to create a transcription factor, **AP-1**, that initiates the transcription of many genes required for cellular proliferation.

Many other proteins with SH-2 domains have been found in many cell types that presumably distribute signals to different pathways. A potentially important branch comes from **phosphoinositide 3-kinase**. This enzyme has SH-2 domains and is activated by binding to phosphotyrosines on activated kinases or their substrates. This enzyme phosphorylates phosphoinositides and may in some way connect enzyme-linked receptors to the inositol phospholipid pathway, although precisely how is not known.

Further reading

Nakamura, S.I. and Nishizuka, Y. (1994) Lipid mediators and protein kinase C activation for the intracellular signalling network. *Journal of Biochemistry,* **115,** 1029–34.

Superti-Furga, G. (1995) Regulation of the Src protein tyrosine kinase. *FEBS Letters,* **369**(1), 62–6.

Blumer, K.J. and Johnson, G.L. (1994) Diversity in function and regulation of MAP kinase pathways. *Trends in Biochemical Sciences,* **19**(6), 236–40.

Ihle, J.N. and Kerr, I.M. (1995) Jaks and Stats in signaling by the cytokine receptor superfamily. *Trends in Genetics,* **11**(2), 69–74

endocrine *a.* applied to a system of ductless glands and the hormones they produce.

is the **phosphatidyl inositol pathway** which culminates in the release of intracellular stores of Ca^{2+} and the synthesis of **cyclic AMP**. Signalling components are modular, since the same components feature in different signalling pathways, in the immune, nervous, endocrine and other systems requiring cell-to-cell communication.

8.1.2   Signal transduction by antigen receptors

Lymphocyte activation through the antigen receptor, which we examine first, is probably the most complex example of receptor-mediated cellular responses and not yet fully understood. This is partly because the antigen receptor of T and B lymphocytes is associated with several other receptors that also must be engaged for activation to occur.

It has long been known that artificial cross-linking of the antigen receptors of either T or B lymphocytes, with antibodies or lectins, mimics the effect of antigen binding and causes lymphocytes to proliferate. It is now known that cross-linking brings the cytoplasmic domains of the receptors into close proximity and triggers the signalling cascades. B lymphocyte antigen receptors (surface immunoglobulin) can be engaged by free antigen, but studies in mammals and birds show they are more responsive to antigen when it is collected and immobilized on the surface of follicular dendritic cells within secondary lymphoid tissue. The T lymphocyte receptor is engaged by a specific complex of antigenic peptide and MHC molecule on the surface of antigen-presenting cells. Here too, the ligand is immobilized and better able to cross-link the receptors.

However, only the antigen *specificity* of the interaction is governed by a T or B lymphocyte antigen receptor. Their cytoplasmic domains are short and they possess no intrinsic enzymatic activity of their own. Signal transduction is performed instead by separate signal-transducing molecules that are non-covalently attached to the receptor, forming an antigen–receptor *complex*.

The T lymphocyte receptor

T lymphocyte antigen receptors are non-covalently associated with a cluster of signal-transducing proteins in the plasma membrane (Figure 8.6). CD3γ, CD3δ and CD3ε are members of the immunoglobulin gene superfamily, each having a single extracellular immunoglobulin fold. These chains have very similar amino acid sequences, and in humans and mice their genes are linked (located on the same chromosome) suggesting they arose by gene duplication and divergence.

Also part of the receptor complex is a disulphide-linked homodimer consisting of two ζ (zeta) chains, or a heterodimer of one ζ chain and one

Gene duplication: see Molecular evolution: Box 7.1

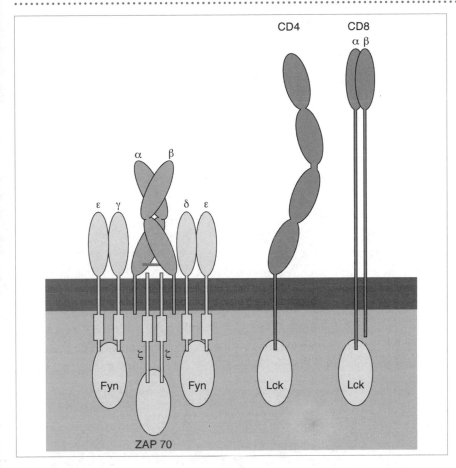

Figure 8.6 The T lymphocyte antigen–receptor complex. The antigen receptor (the αβ or γδ heterodimer) is associated with the CD3 polypeptides (γ, δ and ε) and a ζζ or ζη heterodimer. These molecules operate to transduce signals received by the antigen receptor. Two non-receptor tyrosine kinases, Fyn and ZAP 70 (zeta associated protein), are associated with CD3 and ζζ, respectively. Also involved are the coreceptors, CD4 or CD8 – depending on the type of T lymphocyte. The cytoplasmic domains are associated with another non-receptor tyrosine kinase, Lck. Lck and Fyn are members of the **Src kinase family** (see Box 8.1). Boxes in intracellular domains indicate conserved tyrosine motif (depicted in Figure 8.7).

η (eta) chain. Although genetically unrelated to the CD3 chains, these are also signal-transducing proteins.

Cross-linking of the antigen receptor of T lymphocytes results in the phosphorylation of all three CD3 polypeptides. The precise functions of the CD3 polypeptides in T lymphocyte activation are not fully understood. The ζ and η are also phosphorylated. These molecules have long intracellular domains in which six tyrosine residues are phosphorylated during activation. Studies have shown that the tyrosines that become phosphorylated in ζ and the CD3 chains are contained within a common pattern of amino acids (Figure 8.7). Interestingly, this motif is found in the cytoplasmic domains of many receptors of immunological importance, including the T and B lymphocyte antigen receptors and several Fc receptors.

Also involved in the activation process are at least three non-receptor tyrosine kinases. As shown in Figure 8.6, these are associated with the CD3 chains and with the CD4 and CD8 coreceptor molecules. (Recall that mature T lymphocytes have either CD4 or CD8 in their plasma membranes.)

Figure 8.7 The immunoreceptor tyrosine-based activation motif (ITAM). A motif of amino acids found in the cytoplasmic domains of homologous signal-transducing proteins, including CD3γ, CD3δ, CD3ε, ζ and η of the T lymphocyte receptor complex, Igα and Igβ of the B lymphocyte receptor complex, and several receptors for the Fc region of antibodies (FcεR1, FcγRI, FcγRIIA, FcαR and FcγRIIIA). During lymphocyte activation the tyrosines are phosphorylated by tyrosine kinases, and provide docking sites for additional Src-like kinases.

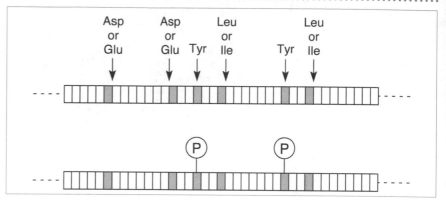

Normally, the CD4 or CD8 molecules are not associated with the antigen–receptor complex, although during activation they are moved within the membrane into the area of contact established between the T lymphocyte and the antigen-presenting cell. CD4 or CD8 participate in the interaction by binding to non-polymorphic surfaces of class II and class I MHC molecules, respectively, on the antigen-presenting cell (Figure 8.8). This has two consequences. One is to strengthen the molecular interaction with the MHC molecule established by the antigen receptor. The second is the introduction of tyrosine kinases into close proximity with other

Figure 8.8 Model of T lymphocyte activation through the antigen receptor. Activation is mediated by cross-linking of the receptor by appropriate peptide–MHC complexes on an antigen-presenting cell. The close proximity of CD3 associated kinase (Fyn) allows autophosphorylation. The tyrosine kinase associated with the cytoplasmic domain of CD4 or CD8 (Lck) is brought into close proximity and is phosphorylated, probably by Fyn. Activated Lck is known to phosphorylate the tyrosines in the CD3 and ζ polypeptides. Zeta (ζ) has six tyrosines in its cytoplasmic domain that are phosphorylated by Lck. These provide docking sites for another tyrosine kinase, ZAP-70 (short for zeta associated protein). This in turn becomes activated. The downstream target of this kinase is not yet known.

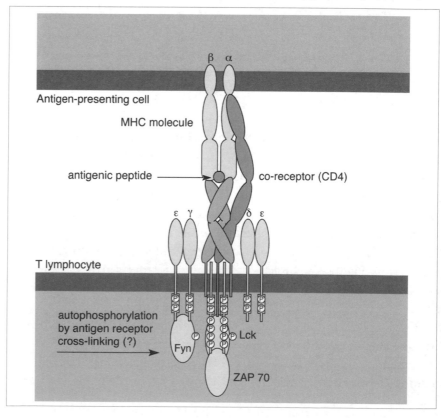

components of the receptor complex, whereupon it becomes phosphorylated and activated.

Precisely which kinase phosphorylates which substrate, and the temporal sequence of events in T lymphocyte activation, is not yet clear. One possible sequence of events is described in Figure 8.7. The result is that the cytoplasmic domains of the CD3 chains, and the tyrosine kinases themselves, become phosphorylated.

The B lymphocyte receptor

The layout of the B lymphocyte receptor is strikingly similar to that of the T lymphocyte. As shown in Figure 8.9, it is associated with a complex consisting of Igα (CD79a) and Igβ (CD79b) heterodimers. There are reasons for believing that this is the B lymphocyte homologue of CD3. First, the chains are structurally similar to each other, and have a single extracellular immunoglobulin-like domain. Second, both chains have the conserved tyrosine-containing motif, whose tyrosines become phosphorylated when the B lymphocyte receptor is cross-linked. Third, the Igα chain is associated with a tyrosine kinase which is thought to become autophosphorylated in response to cross-linking of the B lymphocyte receptor. This in turn phosphorylates the Igα and Igβ chains to create docking sites for several other tyrosine kinases. Here the signalling pathways branch.

A conspicuous downstream event in both T and B lymphocyte activation is the activation of C-kinase and the release of Ca^{2+} via the phosphatidyl inositol pathway (described in Box 8.1). This is known because drugs that activate C-kinase called **phorbol esters** and **calcium ionophores** that make membranes permeable to Ca^{2+} mimic the effects of activation of T and B lymphocytes by antigen. To date, an antigen receptor associated G-protein has not been identified.

8.1.3 CD45

Signalling through the antigen receptor is also influenced by another transmembrane molecule called the **leucocyte common antigen** or **CD45**. In humans and rodents, it is found in abundance on the surfaces of all leucocytes. The cytoplasmic domain of CD45 is a tyrosine phosphatase that counteracts the effects of tyrosine kinases. Src kinases are activated by CD45 as it removes phosphate from a regulatory tyrosine. CD45 is itself activated by phosphorylation, by a kinase called Csk. In so doing, a docking site for Src-like kinases is exposed. CD45 is used in several different receptor complexes, including the antigen receptors of T and B lymphocytes, and the receptors for the Fc region of IgG2a (CD64 and CD32) on monocytes and macrophages.

Figure 8.9 The B lymphocyte antigen–receptor complex. This diagram is drawn to emphasize the homologies with the T lymphocyte receptor. Igα and Igβ are signal-transducing polypeptides that are homologous to the CD3 polypeptides of the T lymphocyte receptor. The tyrosine kinase Syk is homologous to ZAP-70. It is activated during receptor cross-linking, possibly by autophosphorylation, and then phosphorylates Igα and Igβ, creating docking sites for several Src-like kinases. Boxes in intracellular domains indicate conserved tyrosine motif.

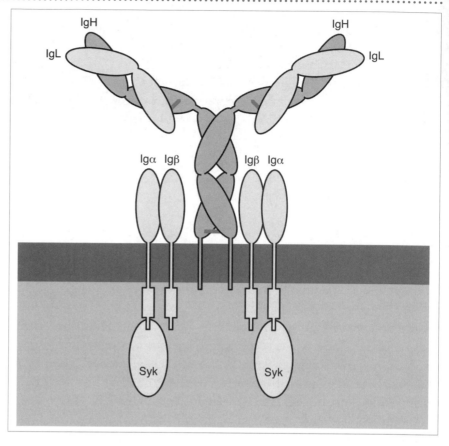

Understanding the role of CD45 in leucocyte activation is complicated because different isoforms of CD45 exist. Its gene has nine exons, of which exons 4, 5 and 6 can be spliced out, either individually or in various combinations, during processing of the primary RNA transcript. There are thus nine potential products of the CD45 gene. These isoforms differ only in their extracellular domain, and it is likely each binds specifically to a different ligand and has a different physiological function. Little is known of the potential spectrum of ligands for CD45. One ligand may be an adhesion molecule on B lymphocytes called CD22, a member of the immunoglobulin superfamily, which is emerging as being another important molecule in activation.

Summary

- Antigen receptors are associated with signal-transducing proteins in the membrane. These are CD3γ, δ, ε and ζζ or ζη in T lymphocytes, and Igα, β and γ in B lymphocytes.

- Non-receptor tyrosine kinases are also associated. These autophosphorylate when receptors are cross-linked extracellularly. These are Fyn and ZAP 70 in T lymphocytes and Lyn in B lymphocytes. Lck, a non-receptor tyrosine kinase associated with CD4 and CD8 in T lymphocytes, is also activated when these molecules move to the site of contact with the antigen-presenting cell.

- Activation of lymphocytes is characterized by the transient phosphorylation of multiple intracellular substrates, and bursts of intracellular calcium and cAMP. This is a sign that the phosphatidyl inositol signalling pathway is triggered.

- CD45 is a receptor tyrosine phosphatase that controls the activation of Src-like kinases in different leucocytes.

> **ZAP 70** an acronym for zeta associated protein (70 kDa); a tyrosine kinase part of the T lymphocyte receptor complex related to Syk in B lymphocytes.

8.1.4 Lymphocytes also require costimulatory signals for activation

Costimulation is a general requirement for the antigen-dependent activation of both T and B lymphocytes. Stimulation through the antigen receptor, or 'signal 1', is usually insufficient for full activation (although there are exceptions – see below), and additional costimulatory signals, or 'signal 2', through a separate receptor must be received (Figure 8.10). Stimulation of T or B lymphocytes through the antigen receptor *in the absence* of costimulation usually leads to the cell becoming unresponsive to subsequent exposure to antigen. This state is called **anergy**. Anergic lymphocytes are unable to proliferate or secrete cytokines, and in some circumstances may undergo cell death by apoptosis.

> **anergy** *n.* a state of antigen-induced paralysis of lymphocytes, sometimes reversible.

This extra level of stringency for lymphocyte activation may have evolved as a way of controlling lymphocytes that recognize 'self' antigens which could consequently cause autoimmune disease. The signalling pathways that lead to anergy are not yet understood, but anergy appears to exist in reversible and irreversible forms. Anergic helper T lymphocytes may not be entirely inert but may be able to perform other immunomodulatory functions such as suppressing effector T lymphocyte activities.

What are the costimulatory signals? For T lymphocytes, costimulation is delivered by members of the **B7 family** of membrane molecules, which are found mainly on the surfaces of dendritic cells and activated B lymphocytes. Currently two B7 molecules have been characterized, called B7.1 (now called CD80) and B7.2 (now CD86). Both are immunoglobulin gene superfamily members (Figure 8.11). Two receptors for these molecules have been found on T lymphocytes of mouse and human. One is **CD28**; the other is the less abundant **CTLA-4**. These are also immunoglobulin gene superfamily members with a single immunoglobulin-like domain, which are

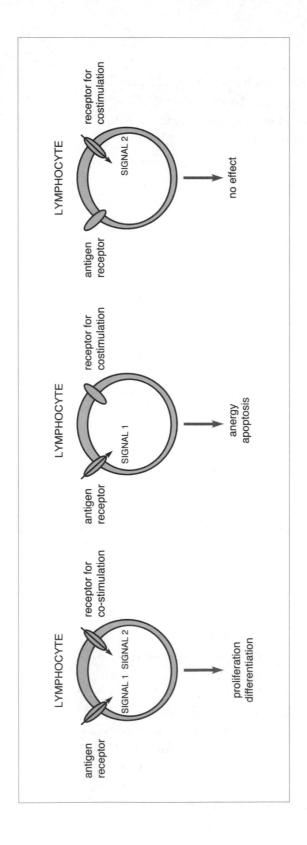

Figure 8.10 Costimulation is required for antigen-dependent activation of T and B lymphocytes.

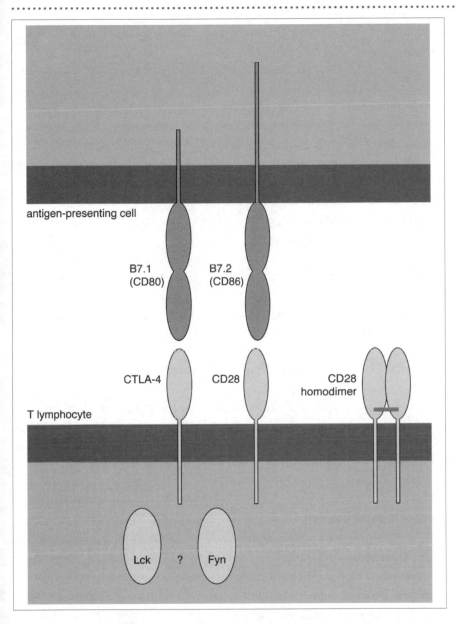

Figure 8.11 Costimulatory molecules for T lymphocytes. B7.1 and B7.2 are costimulatory molecules found on antigen-presenting cells. Both can bind to CD28 or CTLA-4 on T lymphocytes. No intrinsic enzymatic activity is associated with the cytoplasmic domains of CD28 and CTLA-4 although they may interact with Src kinases.

found at the cell surface as single chains (monomers) or disulphide-linked homodimers. The genes for CD28 and CTLA-4 are homologous, suggestive of a gene duplication event. CD28-like molecules with similar functions have also been found in birds.

The signalling function of mammalian CD28 and CTLA-4 is reflected by their cytoplasmic domains, which become phosphorylated during antigen-dependent activation. Many phenotypic changes occur after antigen-dependent activation in the presence of costimulation, although a particular hallmark is the enhanced transcription of the genes encoding interleukin 2 (see below) and its receptor, and the stabilization of the transcripts of several other cytokine genes.

B lymphocytes also require costimulatory signals for activation. Here too, signal 1 is transduced by the antigen receptor. This causes the cell to ingest the complex of antigen and receptor into an endosome wherein it is digested by proteinases. Recall that B lymphocytes are antigen-presenting cells that express class II MHC molecules. Using their membrane immunoglobulin, B lymphocytes capture native antigen. Receptor cross-linking causes endocytosis of the antigen–immunoglobulin complex into endosomes, wherein it is processed into peptides. These are exported back out to the cell surface bound within the peptide-binding cleft of class II MHC molecules and presented to helper T lymphocytes.

The cell biology of antigen processing: Section 6.5

At this point, B lymphocytes require costimulation in order to progress further along the differentiation pathway. The main costimulatory signal is delivered by a molecule called **CD40 ligand** (CD40L) which is only found on the surface of helper T lymphocytes briefly after activation. The receptor on the B lymphocyte is a signal transducer called **CD40**. The interaction between CD40 and CD40L is functionally analogous to the CD80–CD28 interaction in T lymphocytes, although there is no genetic homology.

Stimulation through CD40 is critical for B lymphocyte differentiation. Humans with mutations in the gene for CD40L develop an immunodeficiency syndrome, called X-linked hyper-IgM syndrome, in which class switching by B lymphocytes is blocked owing to defective helper T lymphocytes. The production of IgM, which is constitutive to B lymphocytes and does not require help from T lymphocytes, remains normal or is elevated in these patients.

8.1.5 T lymphocyte help

As mentioned earlier, B lymphocytes are antigen-presenting cells, able to ingest and process antigen captured by surface immunoglobulin and present the processed antigenic peptides in the context of class II MHC molecules to the antigen receptors of $CD4^+$ helper T lymphocytes. In this way, helper T lymphocytes make physical contact with B lymphocytes with antigen receptors for epitopes on the *same* antigenic structure (Figure 8.12). This may prevent T lymphocytes delivering help to irrelevant clones of B lymphocytes with a different antigen specificity.

Naïve T lymphocytes lack surface CD40L and are unable to deliver costimulatory signals to B lymphocytes. For this to occur, naïve helper T lymphocytes must themselves be activated. A naïve B lymphocyte, because it lacks CD80 required for T lymphocyte costimulation, is unable to per-

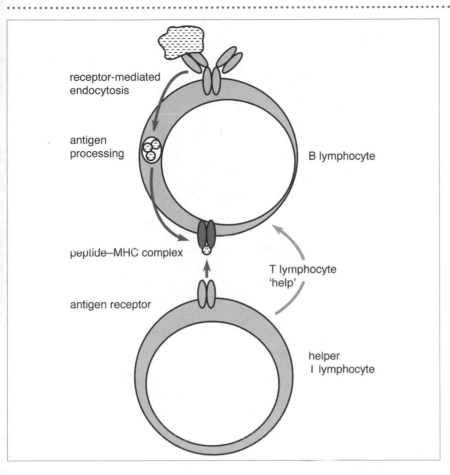

Figure 8.12 Helper T lymphocytes recognize processed antigen on B lymphocytes. B lymphocytes are antigen-presenting cells and present processed peptides to T lymphocytes. If the T lymphocyte is activated it delivers 'help' to the B lymphocyte. This ensures that cooperation between T and B lymphocytes is confined to those that recognize epitopes that are part of the same antigenic structure. In this sense, an antigenic structure may be a single protein or an assembly of proteins such as a virus. Compare this with Figures 5.2 and 5.13, in which different T and B lymphocyte epitopes are shown in a typical protein antigen.

form this role. Instead naïve T lymphocytes are normally activated first by antigen presented on the surface of an interdigitating dendritic cell within secondary lymphoid tissue. Dendritic cells constitutively bear class II MHC molecules and CD80 at their surface and so are able to deliver signal 1 and 2 required for activation of the helper T lymphocyte.

After activation of helper T lymphocytes by dendritic cells, there is an immediate but short-lived increase in the surface expression of CD40L (Figure 8.13). The activated helper T lymphocyte can now respond to the peptide–class II MHC molecule complex presented on the surface of a B lymphocyte and deliver costimulation through the CD40–CD40L interaction. This provides signal 2 required for activation of the B lymphocyte (Figure 8.14).

Several changes occur when the B lymphocyte is activated. One is the expression of CD80, which endows it with both signal 1 (peptide–MHC) and signal 2 (costimulation) required to stimulate helper T lymphocytes. Thus a B lymphocyte receiving costimulation from an activated T lymphocyte simultaneously delivers costimulation in return. This restimulation of

Figure 8.13 Activation of naïve helper T lymphocytes.

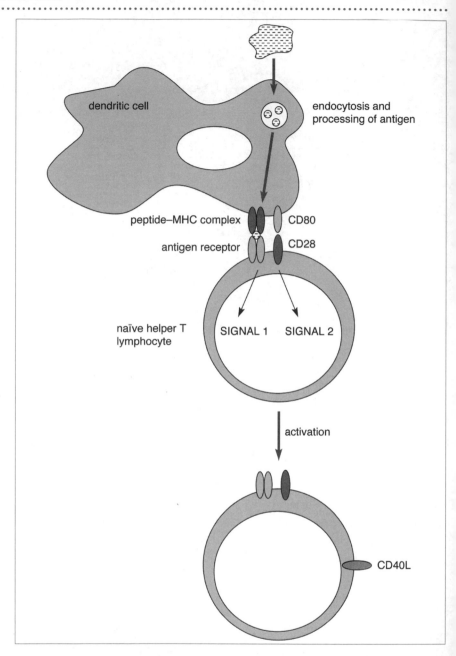

the T lymphocyte stimulates the expression of cytokine genes, in particular IL-3 and IL-4, which cause nearby activated B lymphocytes to proliferate and to undergo immunoglobulin class switching (see below). Unlike a naïve B lymphocyte, an activated B lymphocyte also has the ability to activate *naïve* T lymphocytes and amplify the response.

Some antigens are unusual because they can elicit specific antibodies from B lymphocytes in the absence of costimulation from helper T lympho-

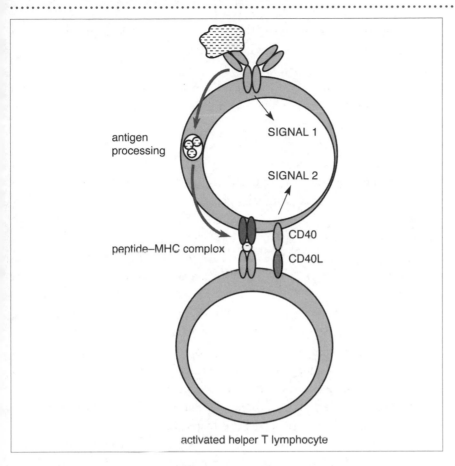

antigen
processing

SIGNAL 1

SIGNAL 2

CD40

CD40L

peptide–MHC complex

activated helper T lymphocyte

Figure 8.14 Activated helper T lymphocytes deliver costimulation to B lymphocytes. Activation of the B lymphocyte induces the expression of CD80 (not shown). CD80, together with the peptide–MHC complex, restimulates the helper T lymphocyte, which responds by releasing cytokines.

cytes. These so called **T-independent antigens** are typically long-chained molecules with repeating subunits, such as the long-chained polysaccharides found in the glycocalyxes surrounding some bacteria. These cause extensive cross-linking of surface immunoglobulin of B lymphocytes which overrides the need for additional help from T lymphocytes. The majority of antigens, however, require the involvement of helper T lymphocytes.

8.1.6 Memory cells

Some of the cells in a clone of antigen-activated T lymphocytes acquire a memory function. These persist long after the effector phase of the immune response has subsided. Often these cells retain an elevated expression of surface adhesion molecules, including LFA-1 (CD11a/CD18), LFA-2 (CD2) and several α_1-integrins.

Cell adhesion molecules: Section 3.7

One particular molecule that may represent a marker for memory, or recently activated, T lymphocytes is an isoform of **CD45**. Recall that the cytoplasmic domain of this molecule is a tyrosine phosphatase that is involved in antigen-dependent activation. Before activation, exons 5 and 6 are spliced from the primary RNA transcript to yield a product called CD45RA, whereas after activation, exon 4 is also spliced out to produce the CD45RO isoform.

As a phenomenon, immunological memory can persist for long periods of time. In humans, for example, life-time resistance to polio virus infection can be elicited from a polio virus vaccine given in childhood. How such memory is maintained is not well understood. The life span of a memory lymphocyte can be measured by marking lymphocytes on a time point, such as with a persistent dye or by creating chromosomal abnormalities with radiation, and then measuring the gradual disappearance of cells with a 'memory phenotype' over time. Such studies in humans, using the CD45RO$^+$ phenotypic marker for memory T lymphocytes, and in sheep, using lymphocytes homing to a specific tissue as a marker for memory, both suggest that memory cells are a dividing, relatively short-lived population of cells.

Memory is likely to be maintained in part by a constant but low level, 'tickover' proliferation of activated cells, presumably driven by periodic re-exposure to antigen. It is known that some antigens, such as replicating live organisms, persist in the body. These elicit longer lasting immunity than a single protein or other non-replicative antigen that is cleared away more rapidly. Follicular dendritic cells retain antigens on their surfaces for many weeks after the effector phase of the response has subsided. These cells could therefore act as an antigen depot and provide the constant restimulation needed to maintain a population of memory B lymphocytes. How this occurs for memory T lymphocytes is more difficult to resolve, although they may receive periodic restimulation from cross-reactive peptides derived from unrelated antigens. Any such theories have to explain how the phenomenon of memory can last for many years.

Summary

- T and B lymphocytes require two extracellular signals for activation. One is antigen, the other is costimulation. Stimulation through the antigen receptor in the absence of costimulation can anergize T and B lymphocytes.

- Costimulatory signals for T lymphocytes are delivered by CD80 on antigen-presenting cells, and transduced by the CD28 and CTLA-4 on the T lymphocyte. The costimulatory signal for B lymphocytes is delivered by CD40L on activated helper T lymphocytes and transduced by CD40 on the B cell.

- To produce antibody, naïve B lymphocytes receive help from helper T lymphocytes. 'Help' refers to a combination of *costimulation* required for activation of the B lymphocyte, and *cytokines* that stimulate proliferation and maturation. Collaborating T and B lymphocytes will have receptors for the same antigenic structure.

- A naïve helper T lymphocyte is unable to deliver help, but must be activated first. This is achieved by antigen presentation on dendritic cells, or on activated (not naïve) B lymphocytes.

- For B lymphocyte activation, antigen is first internalized by immunoglobulin-mediated endocytosis, and processed, and selected peptides are displayed by class II MHC molecules. An activated helper T lymphocyte that recognizes the peptide–MHC complex delivers signal 2 through CD40–CD40L.

- An activated B lymphocyte bears CD80 molecules. The cell is able to reciprocally restimulate the helper T lymphocyte. This causes the expression of cytokines by the helper T lymphocyte required for further B lymphocyte maturation into a plasma cell.

- Memory lymphocytes share phenotypic markers with activated lymphocytes. Particularly useful markers of memory T lymphocytes include certain adhesion molecules and the CD45RO isoform.

8.2 Cytokines

An important medium of extracellular communication comprises soluble signalling molecules called **cytokines**. Cytokines are similar in many ways to hormones of the endocrine system. Both are released by cells. The cytokine or hormone then diffuses away whereupon it will influence the activity of other cells bearing the appropriate receptors (Figure 8.15). Whereas hormones are carried from the source to the target cell some distance away by the circulatory system, the majority of cytokines act locally and are frequently released between cells in physical contact.

Cytokines are not used exclusively in the immune system. These molecules are part of a much more extensive intercellular communication system, used also by the endocrine and nervous systems in particular, and

Figure 8.15 The possible effects of cytokines on cell activity. Cytokines are released by cells in response to a particular stimulus. This in turn stimulates target cells that bear the appropriate cytokine receptor. The receptor is connected to particular genes via an intracellular signalling cascade which causes one or more changes in cell activity. The majority of cytokines influence cellular proliferation or differentiation (including the up- or down-regulation of surface molecules or secretion of other cytokines).

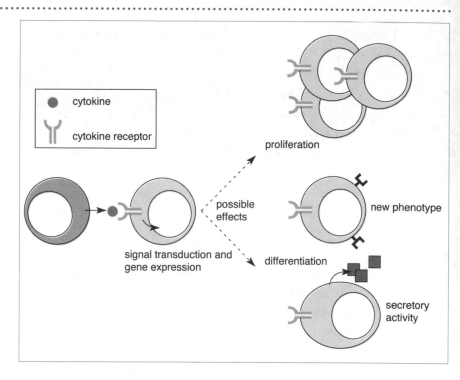

also during developmental processes and embryogenesis. In this section we will focus mainly on the function and evolution of human and murine immunomodulatory cytokines, and their receptors.

8.2.1 Principles of cytokine actions

Cytokines are involved at every stage in the development of leucocytes. Lymphocytes, for example, differentiate from lymphoid stem cells into mature T or B lymphocytes under the influence of cytokines secreted within primary lymphoid tissue. After maturation, their movements around the body are influenced by locally produced cytokines, such as inflammatory cytokines produced at the sites of inflammation. After activation by antigen, the differentiation and proliferation into effector cells is controlled by cytokines released by other cells, notably macrophages and helper T lymphocytes. The death of lymphocytes is also influenced by cytokines, particularly the lack of vital growth factors.

Before we examine individual cytokines, there are a number of general properties that are characteristic of cytokines.

- Cytokines operate over short distances and usually between two different cells. This type of intercellular signalling is described as **paracrine** (Figure 8.16). Short-range signalling helps maintain the fidelity of intercellular interactions, particularly in antigen-specific

paracrine *n.* actions of extracellular signalling molecules (cytokines, etc.) over short distances.

interactions between lymphocytes and antigen-presenting cells.

- Occasionally cytokines bind to receptors on the same cell from which they were released. This is described as **autocrine** signalling.

- A given cytokine can have different effects on different target cells expressing the appropriate receptor. Cytokines are therefore multi-functional, or **pleiotropic**.

- Different cytokines can induce the same effect within a particular target cell type. Some cytokines therefore show **redundancy** in their activities.

- Cytokines rarely operate in isolation. Cells *in vivo* are subjected to a mixture of cytokines and growth factors. The particular activity of a cell is the net result of several cytokines operating simultaneously.

- The effects of some cytokines can be blocked by other cytokines (**antagonism**) or alternatively are amplified (**synergy**).

So far, it has not been possible to organize cytokines into a logical classification scheme. This is because there is little homology between the nucleotide sequences of cytokine genes. Despite this, structural analyses

autocrine *a.* effects caused by a hormone, cytokine, etc., on the same cell from which it was released.

pleiotropy *n.* the multiple effects of a single protein or gene.

synergic *a.* acting together (such as two cytokines) where the combined effect is often greater than the sum of the individual parts.

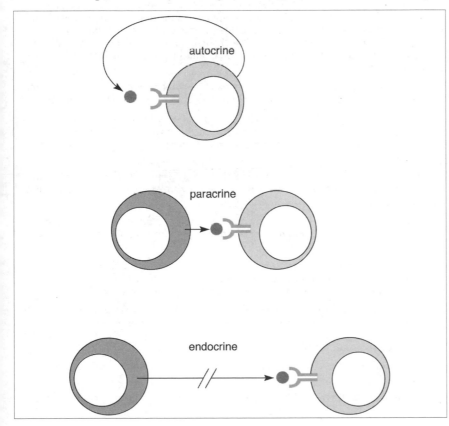

Figure 8.16 Types of cytokine activity. Cytokines with autocrine activity act upon the cell from which they were released. Cytokines with paracrine activity affect cells in close proximity or in physical contact. Cytokines with endocrine activity are carried to target cells via the circulatory system like hormones.

BOX 8.2 Mammalian cytokines

- **Interleukin 1** (IL-1) is produced by activated antigen-presenting cells. Its main role is inflammatory (see Table 1.2), causing increased vascular permeability, and increased expression of intercellular adhesion molecules (ICAMs) by vascular endothelial cells. It is also chemotactic for macrophages and neutrophils. It acts on the hypothalamus to cause fever and on liver hepatocytes to release acute phase proteins. Several of these effects are also caused by IL-6 and TNF-α (a property called redundancy; see text) because the receptors for these cytokines activate the same intracellular signalling pathways. IL-1 also induces the proliferation of activated B lymphocytes, and it is required for the antigen-dependent activation of helper T lymphocytes. The IL-1 receptor exists in two forms: a signal-transducing receptor and a non-signal-transducing, secreted form.

- **Interleukin 2** (IL-2) is produced by activated T helper 1 lymphocytes (Th1; discussed in the next section). IL-2 is essential for $CD4^+$ and $CD8^+$ T lymphocyte proliferation, and it enhances cytotoxicity by NK cells. $CD8^+$ cytotoxic T lymphocytes are dependent on helper cells because they produce little IL-2 of their own. IL-2 can act in autocrine and paracrine fashions. The receptor for IL-2 exists in low, medium and high affinity states (see text).

- **Interleukin 3** (IL-3) is produced by activated helper T lymphocytes and it is necessary for growth of haemopoietic stem cells. IL-3 has similar properties to IL-5 and GM-CSF because the receptors for each of these cytokines share a common signal-transducing β chain and trigger the same intracellular signalling pathways.

- **Interleukin 4** (IL-4). This cytokine causes antigen-activated helper T lymphocytes to differentiate into T helper 2 (Th2) lymphocytes, while preventing them from differentiating into T helper 1 (Th1) cells. Th2 also produce IL-4. This cytokine contributes (along with IL-5, IL-6 and IL-10) to the 'help' provided by helper T lymphocytes to enable activated B lymphocytes to differentiate into antibody-secreting plasma cells. IL-4 causes proliferation, class switching from IgG1 to IgE, and the surface expression of B7 (CD80) costimulatory molecules by B lymphocytes. IL-4 also increases the expression of class II MHC molecules on resting B lymphocytes.

- **Interleukin 5** (IL-5) is produced by activated T helper 2 (Th2) lymphocytes, and stimulates activated B lymphocytes to proliferate and class switching to IgA.

- **Interleukin 6** (IL-6). Like IL-1 and TNF-α, this cytokine is produced by macrophages and monocytes and has inflammatory effects. IL-6 is also produced by activated Th2 lymphocytes for stimulating activated B lymphocyte to differentiate into plasma cells.

- **Interleukin 7** (IL-7) is produced by epithelial cells in primary lymphoid tissue (bone marrow and thymus) and it supports the growth of stem cells into T and B lymphocytes.

- **Interleukin 8** (IL-8) is produced by macrophages and endothelial cells, and is a chemotactic cytokine (a **chemokine**) for neutrophils and enhances their adhesion to, and migration between, vascular endothelial cells (**extravasation**).

- **Interleukin 9** (IL-9) is a product of activated helper T lymphocytes that causes activated helper T lymphocytes to proliferate (i.e. a T lymphocyte **mitogen**). IL-9 also enhances IL-4-induced IgE production and mast cell survival.

- **Interleukin 10** (IL-10). This cytokine, produced by activated T helper 2 (Th2) lymphocytes, has a major role in suppressing Th1 differentiation, mainly by inhibiting the production of cytokines by Th1 cells, as well as by activated macrophages and NK cells.

- **Interleukin 11** (IL-11) is produced by bone marrow and it supports the growth of stem cells.

- **Interleukin 12** (IL-12) is produced by macrophages and B lymphocytes. In the presence of IL-12 a CD4$^+$ T lymphocyte differentiates into a T helper 1 (Th1) lymphocyte, rather than a Th2.

- **Interleukin 13** (IL-13) is produced by activated T helper 1 (Th1) lymphocytes and it is genetically homologous to IL-4. Both are involved in promoting the differentiation of B lymphocytes. This cytokine also down-regulates the release of inflammatory cytokines.

- **Interleukin 15** (IL-15), although unrelated to IL-2, has very similar effects on the activation and proliferation of T lymphocytes. IL-15 uses the same β and γ chains as the IL-2 receptor.

- **Interferons.** Type I interferons (**IFN-α** and **-β**) are released by a variety of mammalian and avian cells in response to viral infection. These bind to receptors on uninfected cells nearby and stimulate the expression of genes that enhance resistance to viral infection, and an increase in surface MHC class I expression. Type II interferon (**IFN-γ**) is produced by activated T helper 1 (Th1) lymphocytes, cytotoxic T lymphocytes and NK cells. In addition to enhancing resistance to viruses, it is now known than IFN-γ has several other very important immunomodulatory roles. Notable is its ability to increase the transcription of genes in the MHC in many different cells, especially class I and class II genes. IFN-γ enhances the cytotoxicity of cytotoxic T lymphocytes and NK cells. IFN-γ also induces class switching in B lymphocytes to IgG2a. It also antagonizes IL-4 thereby promoting the differentiation of Th1 cells.

- **Tumour Necrosis Factor-α and -β.** TNF-α is produced by macrophages and it shares many of the inflammatory properties of IL-1 and IL-6. TNF-α is antagonized by IL-10. TNF-β is produced by Th1 and cytotoxic T lymphocytes and it enhances phagocytosis by macrophages and neutrophils. Both TNF-α and TNF-β are toxic to tumour cells.

- **Transforming Growth Factor-β** (TGF-β). There are five TGF-β isoforms in mammals. TGF-β1 is produced by many different cells, and its major property is to inhibit the proliferation of virtually all cells. The most widely distributed receptors are types I and II, which have intrinsic serine/threonine kinase activity, and which assemble into heterodimers upon binding to TGF-β1. The type II chain stimulates the membrane phosphatidyl inositol pathway via phospholipase C, as described in Box 9.1. The type I chain is dependent on dimerization with the II chain for kinase activity.

- **Leukaemia Inhibitory Factor** (LIF) is produced by the epithelial cells of primary lymphoid tissue, and it supports the growth on stem cells therein.

- **Colony Stimulating Factors** (CSF) are a genetically related family of growth factors that support the growth of myeloid leukocytes (leukocytes other than lymphocytes). **Multi-CSF** supports the earliest, multipotential myeloid progenitor cells. Granulocyte/macrophage-CSF (**GM-CSF**) acts on oligopotential myeloid progenitor cells, and granulocyte-, macrophage- and eosinophil-CSF (**G-CSF, M-CSF** and **E-CSF**) support monopotential progenitors.

have revealed a three-dimensional structure that is common to some cytokines and growth hormones. In contrast, many cytokine receptors are related to each other (discussed later).

8.2.2 Cytokines define multiple subsets of helper T lymphocytes

In humans and in mice it has been observed that the majority of activated CD4$^+$ T lymphocytes appear to belong to one of two subsets, called T helper 1 (Th1) and T helper 2 (Th2) cells, defined according to the

Figure 8.17 Cytokine production by activated CD4 T lymphocytes.

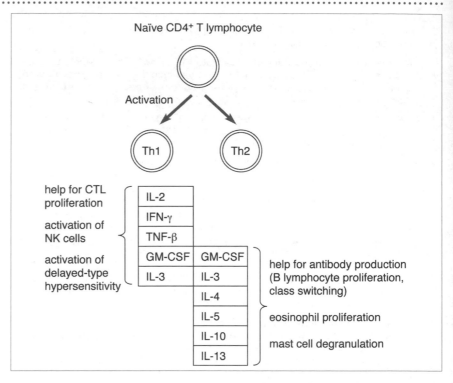

Th1 *n.* a subset of CD4$^+$ helper T lymphocytes that characteristically produce IL-2 and IFN-γ, and which are important for the development of cell-mediated immunity.

Th2 *n.* a subset of CD4$^+$ helper T lymphocytes that characteristically produce IL-4, IL-5, etc., and which are important for the development of humoral immunity.

particular combination of cytokines they secrete (Figure 8.17). The kind of immune responses in which these two types participate are different.

- IL-2 and other cytokines produced by Th1 cells help CD8$^+$ cytotoxic T lymphocytes differentiate into effector cells, and enhance the lytic activity natural killer (NK) cells. Effector CD4$^+$ lymphocytes, which are involved in the inflammatory response in the skin, called delayed-type hypersensitivity (DTH), also differentiate under the influence of Th1-derived cytokines. Th1 lymphocytes are therefore instrumental in promoting cell-mediated immune responses.

- Th2 cytokines assist activated B lymphocytes to differentiate into antibody-secreting plasma cells. Th2 are therefore instrumental in promoting antibody-mediated (humoral) immune responses.

Each is thought to arise from naïve cells that are uncommitted to either pathway. Upon activation by antigen, differentiation is skewed heavily towards the Th1 (cell-mediated) or the Th2 (humoral) direction, and both are rarely seen simultaneously. Individual Th1 and Th2 cells are fixed in their cytokine profiles and do not switch from one type to the other, although their progeny during clonal expansion may differ leading to a switch from one direction to the other during the course of an immune response. In reality, Th1 and Th2 may represent the extremes of a spectrum of cytokine profiles: several intermediate CD4$^+$ cells that produce cytokines characteristic of both Th1 and Th2 have been described, called Th0.

The direction in which a response initially proceeds is influenced by the type of microorganism encountered and perhaps also the antigen-presenting cell involved (Figure 8.18). In general, pathogens that live inside cells usually elicit cell-mediated responses rather than humoral ones. Intracellular microorganisms include some species of bacteria and protozoa, and viruses in the intracellular stage of their life cycle. In contrast, extracellular pathogens, such as most bacteria, protozoan parasites and viruses in the extracellular phase of their life cycle, elicit humoral responses rather than cell-mediated responses. Each response is appropriate for the type of microorganism. An intracellular pathogen is safe from antibodies but 'visible' to cytotoxic T lymphocytes and NK cells, whereas an extracellular pathogen is accessible to antibody.

The mechanisms that push T lymphocytes to differentiate one way or the other after activation by antigen operate early in the response. How this occurs is not clear but exposure to cytokines at this early stage is important. This is summarized in Figure 8.19. Intracellular parasites often stimulate the production of IL-12 by macrophages and monocytes. IL-12

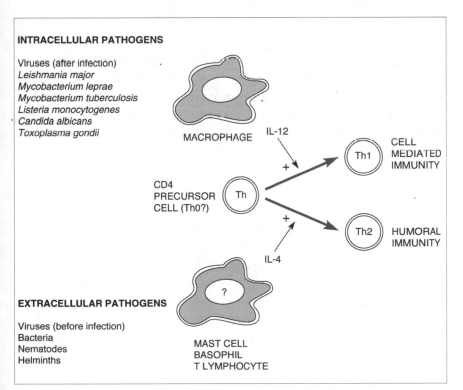

INTRACELLULAR PATHOGENS

Viruses (after infection)
Leishmania major
Mycobacterium leprae
Mycobacterium tuberculosis
Listeria monocytogenes
Candida albicans
Toxoplasma gondii

MACROPHAGE   IL-12

CD4
PRECURSOR
CELL (Th0?)   Th

Th1   CELL
MEDIATED
IMMUNITY

Th2   HUMORAL
IMMUNITY

IL-4

EXTRACELLULAR PATHOGENS

Viruses (before infection)
Bacteria
Nematodes
Helminths

MAST CELL
BASOPHIL
T LYMPHOCYTE

Figure 8.18 Polarization of helper T lymphocyte differentiation according to the type of pathogen. Pathogens with intracellular lifestyles stimulate macrophages to secrete interleukin 12 (IL-12). Several pathogens that are known to elicit predominantly Th1-type responses are listed. A CD4 T lymphocyte responding to antigen in the presence of IL-12 differentiates into a Th1 cell. These are important for the development of cell-mediated responses. Extracellular pathogens, on the other hand, stimulate IL-4 production, possibly from mast cells, basophils and T lymphocytes. This promotes the differentiation of Th2 cells, which in turn help B lymphocytes differentiate into antibody-secreting plasma cells.

Figure 8.19 Cross-regulation by Th1 and Th2 cells. IFN-γ produced by Th1 cells inhibits Th2 cells from proliferating in response to antigen. Reciprocally, IL-4 and IL-10 produced by Th2 antagonize the effect of IL-12 and block its release from macrophages, respectively. In addition, IL-2 secreted by Th1 reinforces the Th1 response by operating in an autocrine fashion on Th1 cells; similarly IL-4 produced by Th2 promotes the differentiation of more Th2 cells.

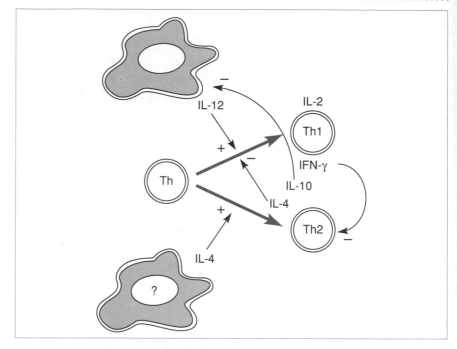

pushes an uncommitted naïve Th into becoming a Th1 cell, resulting in a strong cell-mediated immune response. Extracellular pathogens, on the other hand, elicit the secretion of IL-4 by T lymphocytes and other cells. IL-4 will push a precursor Th into becoming a Th2, resulting in humoral immunity.

Once an activated Th cell is committed to one phenotype or the other, the cytokines it produces reinforce the pathway along which the response proceeds by suppressing the other pathway. Thus, cytokines produced by Th1 suppress Th2 production and *vice versa*. These cross-regulatory mechanisms are shown in Figure 8.19.

Often, the mounting of an inappropriate Th1 or Th2 response may allow the pathogen to replicate with impunity. This is in part because the response elicited cannot clear the pathogen, and partly because the appropriate response is suppressed by the cross-regulatory mechanisms. The *Leishmania* parasite, for example, is a protozoan that lives inside macrophages. In mice, infections are eliminated by a Th1-dependent cell-mediated immune response. In susceptible mice, such as the BALB/c inbred strain, the Th2-dependent humoral response predominates and the mice ultimately die. Similarly in humans, the bacterium that causes leprosy, *Mycobacterium leprae*, also lives inside macrophages. A Th1-mediated response effectively clears the bacteria, although a side effect of the successful response is damage to peripheral nerves causing tuberculoid leprosy (Figure 8.20). Individuals who mount predominantly Th2-dependent humoral responses and weak cell-

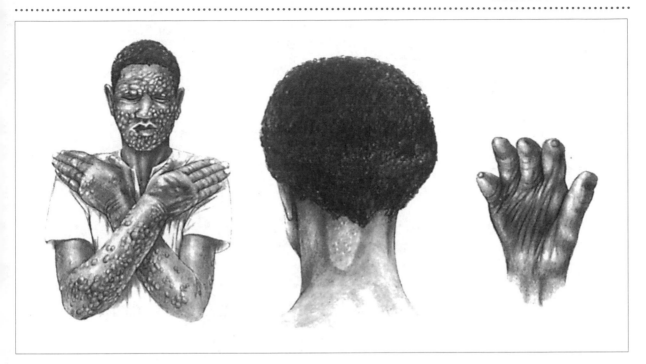

mediated responses cannot clear the bacteria, which persist and multiply, causing severe pathology called lepromatous leprosy.

Figure 8.20 Pathology in leprosy is determined by the balance of Th1 and Th2 responses. Cell-mediated immunity is important in clearing *Mycobacterium leprae*, whereas antibodies have no apparent role. Tuberculoid leprosy (right) develops in individuals who mount Th1-dependent cell-mediated responses, whereas lepromatous leprosy (left) develops in those who mount Th2 responses. Borderline leprosy (centre) occurs in individuals whose response lies in between. From Mitchison, N.A. (1993) *Scientific American*, **269**, 102–8.

Summary

- Cytokines are extracellular signalling molecules that generally operate over short distances in the immune, nervous and endocrine systems.

- Cytokines are bound by specific receptors on target cells. These are linked, via signalling cascades, to cytokine response elements located upstream of genes controlling proliferation and differentiation.

- Antigen-activated helper T lymphocytes develop into Th1 or Th2 lymphocytes according to the pathogen. Th1 cells characteristically produce IL-2 and IFN-γ which are required for cell-mediated immunity. Th2 cells characteristically produce IL-4 which is required for humoral immunity.

- In general Th1 cytokines feed back to promote Th1 activity and inhibit Th2 activity and *vice versa*.

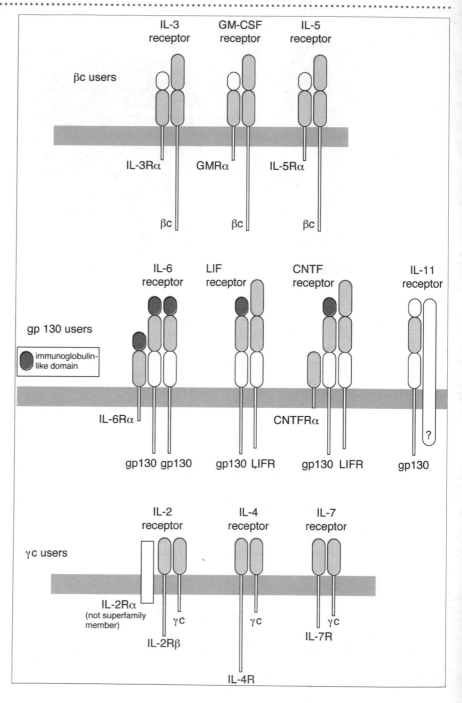

Figure 8.21 Members of the mammalian haemopoeitin (type 1) cytokine receptor superfamily. Each have one or more homologous domains (shaded). Some have immunoglobulin folds and are therefore also members of the immunoglobulin superfamily. Many of the members of this superfamily can bind to specific cytokines as monomers, although with low affinity. Redrawn with permission from Leonard, W.J. (1994) *Current Opinions in Immunology*, **6**, 631.

8.2.3 Cytokine receptors are modular

Unlike cytokines, which are heterogeneous and largely unrelated to each other, many cytokine receptors are homologous and belong to a receptor superfamily (Figure 8.21). A characteristic of this family is the pairing of the same signal-transducing polypeptide chains with different cytokine-binding chains. On this basis, three subfamilies of receptors are recognized.

- The IL-3, IL-5 and GM-CSF receptor group. These cytokines bind to different α chains with low affinity. Upon binding, each α chain assembles with a common β chain (called βc) to produce a high affinity heterodimer (see Figure 8.22). The βc chain is the signal-transducing component of each receptor whereas the α chain accounts for the binding specificity of the receptor. Because these receptors use a common signalling pathway, the biological effects of IL-3, IL-5 and GM-CSF are similar (see Box 8.2)

- The IL-6, IL-11, leukaemia inhibitory factor (LIF), oncostatin M and ciliary neurotrophic factor (a cytokine used in the nervous system) receptors. These share a common signal-transducing chain called gp130.

- The IL-2, IL-4, IL-7, IL-9 and IL-13 receptors. These have in common a γ chain (called γc), although these cytokines have very distinct functions. This is because, in contrast to the other groups, the properties of signal transduction and specificity for cytokine binding are shared by both the γc chain and its partner(s).

The IL-2 receptor is unusual as it can exist in three assembly states. On antigen-activated T lymphocytes it exists as a heterotrimer (three different chains) and a monomer (single chain). The trimer consists of α, β and γ_c and is a high affinity receptor for IL-2. The monomer, which is the α chain alone, has a low affinity for IL-2. The monomer is found only on activated T lymphocytes, which has earned it the name T lymphocyte activation antigen (or **TAC**, now called **CD25**). This molecule is a useful marker of activation. Resting T lymphocytes express a $\beta\gamma$ heterodimer, which has an

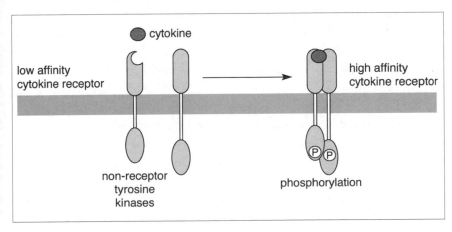

Figure 8.22 Assembly of a typical cytokine receptor. Many cytokine receptors exist as single chains that bind specifically to certain cytokines but with low affinity. This promotes assembly with another chain to create a high affinity receptor and to trigger signalling cascades within the cell.

low affinity
cytokine receptor

cytokine

high affinity
cytokine receptor

non-receptor
tyrosine
kinases

phosphorylation

intermediate affinity for IL-2. This heterodimer is also a receptor for IL-15. Moreover, NK cells have βγc receptors and are thus able to respond to IL-2.

8.2.4 Signal transduction by cytokine receptors: Jaks and Stats

The receptors for growth factors and hormones usually have intrinsic enzymatic activity in their cytoplasmic domains. Those for cytokines and antigen receptors, in contrast, transduce signals via non-receptor protein kinases. Cytokine receptors belonging to the cytokine superfamily are associated with non-receptor tyrosine kinases called **Janus kinases** or **Jaks** (Jak-1 and Jak-2). These phosphorylate cytoplasmic signalling proteins belonging to a family of signalling proteins called the signal transducers and activators of transcription (**Stats**). A few Src kinases are also involved. For example, the β chain of the IL-2 receptor is associated with Lck and Fyn (see the T lymphocyte receptor in Figure 8.7). Details of the signalling cascades downstream of these kinases are yet to be worked out.

8.2.5 Origins?

The common domain structure found in many cytokine receptors and the use of immunoglobulin-like domains in several cytokine and growth factor receptors (see Figure 8.21) indicates that the processes of exon shuffling, gene duplication and divergence have played a major role in the recent evolution of the cytokine receptor superfamily.

However, the origins and evolution of the cytokines themselves are poorly understood. Some human cytokines stimulate murine cells and *vice versa*, suggesting that some human and murine cytokines and receptors have diverged little since the two species departed from a common ancestor some 60 million years ago. Cytokines with activities like those of most human and murine cytokines have been described in several other mammals, including rats, rabbits, horses, guinea pigs, swine and dogs. Avian versions of IL-1, IL-2, IL-6, IL-8, IFN-α, IFN-β, IFN-γ, TNF-α and the receptors for several of these have been described. Even further bridges of evolutionary time can be spanned. For example, macrophages of the rainbow trout can be activated by human TNF-α. Fish cytokines with activities like mammalian IL-1, IL-2, IFN-β and IFN-γ have been described. Cells from tunicates have been reported to proliferate in response to human IL-2.

These cross-reactivities are useful indicators of the presence of human and murine-like cytokines in non-mammalian vertebrates. However, this similarity is at the functional level, and does not necessarily arise from genetic homology that would be expected if they had evolved from common ancestral genes. Genetic homology of cytokines or cytokine receptors from different animals can only be confirmed by cloning and sequencing their genes, although this is a relatively new field and little is known.

8.2.6 Immunomodulatory cytokines outside the immune system

Several immunomodulatory cytokines, such as IL-1, IL-2, IL-4 and IFN-γ, have activities in the nervous and endocrine systems. IL-1 is produced by neurons and glial cells, and it interacts directly with cells in the central nervous system, causing profound physiological effects like fever and changes in appetite and sleep patterns. Cells within major endocrine glands (pituitary, gonads, pancreas, thyroid and adrenals) all have IL-1 receptors, and IL-1 has been shown to modulate the secretion of hormones produced by most of these glands.

Studies of nerve regeneration in lower vertebrates such as fish also suggest other potential functions for immunomodulatory cytokines. In these animals, nerves regenerate spontaneously after they are damaged. This is in contrast to adult mammals in which axons of the central nervous system do not regenerate after injury. This is due to the presence of oligodendrocytes which inhibit axon growth. Regenerating fish optic nerves secrete a fish homologue of IL-2 that is toxic to oligodendrocytes.

IL-4 and IFN-γ are also involved in the nervous system. Their receptors are found on macrophages of the central nervous system (called microglia), and like many other cell types, microglial cells can be induced to express class II MHC molecules in response to IFN-γ, an effect that is antagonized by IL-4. Leukaemia inhibitory factor (LIF) is involved in the differentiation of nerve cells in addition to supporting the growth of haemopoietic stem cells. To this list can be added TGF-β which has been shown to play a role in maintaining the rhythmic contractions of heart muscle cells, and several immunomodulatory cytokines are involved in mesoderm formation during mammalian embryogenesis.

It is not unexpected that the same cytokines are used in different systems in the body, particularly when one considers the extensive pleiotropy they display in the immune system. The extent to which these common factors are used to communicate between the immune, nervous and endocrine systems is not well understood. It is likely that some immunomodulatory cytokines have systemic (whole-body) effects owing to their endocrine activities. The effect of mammalian IL-1 on body temperature is a good example of this. The potential for cross-talk by cytokines blurs the boundaries between the immune, nervous and endocrine systems. Cytokines are coming to be regarded as the intercellular communication language of a much broader 'neuroimmunoendocrine' system.

Fever: see The acute phase of inflammation: Section 1.2.1

Summary

- The majority of cytokine receptors are heterodimers. Often one chain determines binding specificity, and the other transduces the signal. Modularity of receptor chains enables different receptors to be constructed from a relatively small number of chains by combining them in different ways.

- Redundant cytokine actions are in part due to receptors with different cytokine specificities sharing a common signal-transducing chain. Pleiotropic cytokine effects are in part due to receptors with different signal-transducing chains sharing a common binding specificity chain.

- Signalling pathways are also modular, and many of the signalling enzymes are used by different receptors and in different cells.

- There is genetic evidence that some interleukin-like cytokines and interferons have been conserved during evolution.

The future

- It is unknown how many more immunomodulatory cytokines await discovery.

- Of those known, there are many gaps in our full understanding of their range of functions, particularly in combination with other cytokines, and in communicating with the endocrine and nervous systems.

- Much of what we know is from *in vitro* study. Understanding whether cytokines show the same redundancy and pleiotropy *in vivo* will come from the use of transgenic and 'knockout' animal technology.

Further reading

Signalling by antigen receptors

DeFranco, A.L. (1993) Structure and function of the B cell antigen receptor. *Annual Review of Cell Biology,* **9**, 377–410.

Harnett, M. and Rigley, K. (1992) The role of G-proteins versus protein tyrosine kinases in the regulation of lymphocyte activation *Immunology Today,* **13**(12), 482–6.

Janeway, C.A. (1992) The T cell receptor as a multicomponent signalling machine: CD4/CD8 coreceptors and CD45 in T cell activation. *Annual Review of Immunology,* **10**, 645–74.

Pleiman, C.M., D'Ambrosio, D. and Cambier, J.C. (1994) The B-cell antigen receptor complex: structure and signal transduction. *Immunology Today,* **15**, 393–8.

Costimulation

Clark, E.A. and Ledbetter, J.A. (1994) How B and T cells talk to each other. *Nature,* **367**, 425–8.

Durie, F.H., Foy, T.M., Masters, S.R., Laman, J.D. and Noelle, R.J. (1994) The role of CD40 in the regulation of humoral and cell-mediated immunity. *Immunology Today,* **15**, 406–11.

June, C.H., Bluestone, J.A., Nadler, L.M. and Thompson, C.B. (1994) The B7 and CD28 receptor families. *Immunology Today,* **15**, 321–31.

Memory cells

Mackay, C.R. (1993) Immunological memory. *Advances in Immunology,* **53**, 217–65.

Cytokines

O'Garra, A. and Murphy, K. (1994) Role of cytokines in determining T-lymphocyte function. *Current Opinion in Immunology,* **6**, 458–66.

Beck, G. and Habicht, G.S. (1991) Primitive cytokines: harbingers of vertebrate defense. *Immunology Today,* **12**(6), 180–3.

Review questions

Fill in the blanks

A _____-_____[1] receptor relays signals from the extracellular environment to the nucleus via cascades of signalling proteins. Signalling proteins switch between active and inactive states according to their _____[2] state. The majority of receptors involved in modulating lymphocyte behaviour are _____[3]-linked receptors, most of which are receptor- and non-receptor _____ _____[4]. Examples of these enzymes are _____[5] associated with the cytoplasmic domains of CD4 and CD8, and _____[6] associated with the CD_____[7] polypeptides in the T lymphocyte receptor complex. The latter consists of three homologous polypeptides, _____[8], _____[9] and _____[10]. These, and two other signal-transducing polypeptides, _____[11] and/or _____[12] comprise the T lymphocyte receptor complex. The enzymes associated with these chains activate each other by _____[13] when the receptor is cross-linked extracellularly. Immunoglobulin in the B lymphocyte membrane is also associated with signal-transducing polypeptides called _____[14] and _____[15], whose cytoplasmic tails are associated with _____[16].

For T lymphocyte activation through the antigen receptor, _____[17] signals are also required. Antigenic stimulation in the absence of these signals induces a state of antigen-specific non-responsiveness, or _____[18]. There are two _____[19] molecules on antigen-presenting cells that deliver these signals to T lymphocytes: _____[20] and _____[21]. The receptors for these on T lymphocytes are _____[22] and _____[23], which are both members of the _____[24] superfamily. On B lymphocytes, the equivalent signals are received by _____[25] and delivered by _____[26] expressed on the surface of activated helper T lymphocytes. In contrast, _____[27] helper cells are unable to deliver help because they lack surface _____[28].

Short answer questions

1. It is thought that IL-12 might be used in the treatment of cancer, particularly to enhance the efficacy of vaccines against tumour-causing viruses. What is the reasoning behind this?

2. Which mammalian cytokines are being described below?

   (a) The major T lymphocyte growth factor and produced by activated T helper 1 (Th1) lymphocytes; works in an autocrine fashion but also essential for the proliferation of cytotoxic T lymphocytes; also enhances the activity of NK cells.

   (b) Produced by activated Th1 cells, cytotoxic T lymphocytes and NK cells; has many effects, most notable of which is its ability to increase the surface expression of class I and class II MHC molecules on many other cell types.

   (c) Produced by Th2 cells to help activated B lymphocytes undergo class switching; it also causes uncommitted Th precursors to differentiate into more Th2 cells; its effects are antagonized by IFN-γ which causes the progenitors to follow the pathway to Th1 instead.

(d) A chemotactic cytokine for neutrophils.

(e) An inflammatory cytokine that increases the expression of intercellular adhesion molecule-1 (ICAM-1) on endothelial cells; also has cytotoxic effects on tumour cells.

3. What kinds of signals do helper T lymphocytes (Th2) deliver to B lymphocytes to help them produce antibody?

4. Following on from Question 3, how do they recognize which B lymphocytes to deliver this help to?

9 Tolerance to self

In this chapter

- Acquired tolerance to antigen.
- Positive and negative selection of T lymphocytes in the thymus.
- Peripheral tolerance mechanisms: anergy and suppression.
- Tolerance to antigen by B lymphocytes.

Introduction

By its very nature, an immune system is able to inflict damage on cells. Therefore, it must confine this activity to cells that are a potential threat, such as a microorganism or cells in the body that have become infected or neoplastic, while also remaining unresponsive to the normal cells of the body. We have learned that antigen receptors are generated by an essentially random process of gene rearrangement. Consequently, individual lymphocytes have no inherent capacity to make the distinction between what is self and what is foreign, and many clones with self-reactive antigen receptors are generated. In this chapter we will examine a variety of mechanisms that prevent lymphocytes harming normal cells in the body, which thereby endow the lymphocyte repertoire with the capacity to discriminate self from non-self.

9.1 Tolerance to self

9.1.1 Passive mechanisms
Some mechanisms that prevent lymphocytes responding to self-antigens are passive. Self-antigens, such as antigens of the central nervous system (brain or eye), may simply be inaccessible to self-reactive lymphocytes in

the blood. Such an anatomical location is called an **immunologically privileged site**.

Alternatively, antigen may simply be present at concentrations too low to trigger a self-reactive response, or present in the absence of the required costimulatory signals required for lymphocyte activation. This is described as **immunological silence**. We have already encountered several examples of this. The trophoblast that surrounds the mammalian foetus achieves immunological silence in this way. This, together with an assortment of immunosuppressive mechanisms, protects the allogeneic foetus from 'rejection' by the maternal T lymphocytes. Interference with the export of class I MHC molecules by viruses is a mechanism used to achieve immunological silence and escape recognition by cytotoxic T lymphocytes. Similarly, the majority of tumours that successfully avoid elimination by cytotoxic T lymphocytes often have spontaneously lost their expression of class I MHC molecules.

In addition to these passive mechanisms are active mechanisms, in which self-reactive lymphocyte clones are suppressed, paralysed or killed. Active non-responsiveness to particular antigens is the means by which lymphocytes appear to discriminate between self-antigens and non-self-antigens. In reality the effect is achieved, at the population level, by establishing active non-responsiveness among the self-reactive clones in the repertoire and leaving untouched those that, by default, have receptors for non-self (or foreign) antigen (Figure 9.1).

9.1.2 Autoimmune disease is a consequence of failures in tolerance

Immunological phenomena are normally regarded as being based mainly on defence mechanisms to protect the body from pathogens and there is little doubt that a competent immune system is important in eliminating infections. Another view is that an additional function of the immune system is to eliminate unwanted cells and other material produced in the body. This idea has stemmed from observations in healthy animals in which low titres of self-reactive antibodies (**autoantibodies**) are a normal component of the serum. Antibodies that bind to antigens in the body, including those of mitochondria, cytoskeletal proteins, nuclear proteins and nucleic acids, have been found. One possibility is that these are important and function to clear cellular debris.

Similar evidence that autoreactivity is normal comes from self-reactive T lymphocytes that can be readily demonstrated by *in vitro* assays. These probably function *in vivo* to eliminate unwanted cells in the body, particularly

Mammals: Section 7.1.5

Some viruses interfere with antigen processing: see Box 6.5

immunologically privileged site an anatomical site in the body that is inaccessible to the recognition systems of the immune system.

immunological silence non-responsiveness by the immune system owing to insufficiently high levels of antigen or antigen in the absence of essential costimulatory signals; often used by cells as an evasion strategy.

autoantibody *n.* antibody that binds to self-antigen and usually a feature of autoimmune disease.

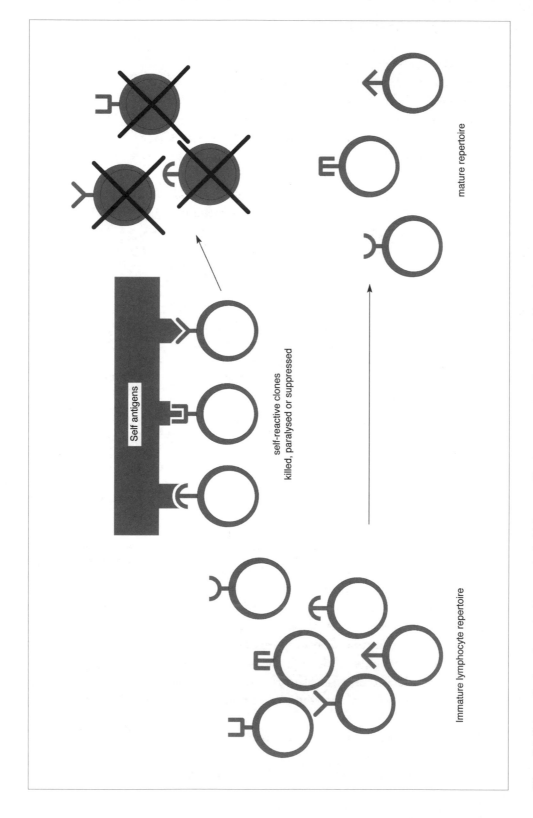

Figure 9.1 Schematic view of acquired immunological tolerance. The immature repertoire consists of lymphocyte clones bearing antigen receptors with randomly generated antigen specificities. Tolerance in the repertoire is established by death, suppression or paralysis (anergy) of self-reactive clones.

Self antigens

self-reactive clones
killed, paralysed or suppressed

mature repertoire

Immature lymphocyte repertoire

those that are unable to stop proliferating (i.e. neoplastic or tumour cells) which may pose a threat to the organism.

This normal autoimmunity is distinct from **autoimmune disease**, which is a pathological condition caused by excessive activation of self-reactive T and B lymphocytes. There are over 40 human autoimmune diseases, in which a variety of self-antigens become the targets of T and B lymphocytes. In a few of these diseases the self-antigen targeted has been identified (Table 9.1). The emergence of autoimmune disease is not caused by the presence of self-reactivity *per se*, but rather by a breakdown of the mechanisms of tolerance that prevent self-reactive lymphocytes from responding. Nevertheless, the existence of such diseases is another indicator that self-reactive lymphocytes are a normal component of the repertoire. How then can lymphocytes and antigens co-exist in peace?

9.1.3 Tolerance to antigen is not genetically predetermined but is acquired

Our understanding of tolerance has its roots in the pioneering studies of graft rejection in the 1950s. As we saw in Chapter 6, tissue grafts exchanged between allogeneic animals (two genetically distinct members of the same species) are rejected by the recipient's immune system. This is not the case, however, if the recipient of the graft was previously exposed to the allogeneic cells during embryonic development. This was first observed occurring naturally in non-identical or **dizygotic twins** of cattle. Non-

autoimmunity *n.* the presence of self-reactive lymphocytes; may lead to autoimmune disease if the mechanisms to control their activation fail.

autoimmune disease cell or tissue damage caused by the body mounting immune responses to its own antigens.

dizygotic *a.* arising from two separately fertilized eggs (e.g. non-identical twins).

Table 9.1 Human autoimmune diseases in which the autoantigen(s) is known

Disease (in man)	Target self-antigen	Pathology
Insulin-dependent diabetes	glutamic acid decarboxylase and insulin receptors in β cells of pancreas	destruction of insulin-secreting β cells
Rheumatoid arthritis	immunoglobulin Fc region	immune complexes formed in joints; damage to synovium
Myasthenia gravis	acetylcholine receptor	destruction of neuromuscular synapses; muscle atrophy
Systemic lupus erythematosus	double-stranded DNA	
Thyroiditis	thyroglobulin	destruction of thyroid glands; hypothyroidism
Multiple sclerosis	myelin (sheaths nerve fibres)	muscle wasting

identical twins develop independently from separate eggs and can be allogeneic. Most dizygotic animals, therefore, will reject tissue grafts exchanged between each other in adult life. Cattle, however, are unusual mammals as the blood supplies of dizygotic cattle twins often become interconnected during embryonic development. This allows blood cells from one animal to circulate around the body of its twin. Because of this mutual exposure to each other's antigens, dizygotic cattle twins do not reject tissue grafts from each other as adults.

This phenomenon was reproduced experimentally in other mammals and birds. For example, two chickens are rendered unable to produce antibodies against each others' cells, or to reject exchanged tissue grafts, if their blood supplies are connected while they are both embryos (**parabiosis**). This is achieved by cutting a window in the eggshell and allowing the blood vessels of the chorioallantoic membrane to fuse. Similarly, tolerance to alloantigens can be produced in inbred strains of mice by infusing newborns with cells from a different, allogeneic strain (Figure 9.2). As adults, the donor mice readily accept grafts from the original allogeneic strain. In each of these cases, the embryonic recipient establishes tolerance to the alloantigens, and is unable to reject grafts bearing the same alloantigens later in life.

While in these models tolerance is induced to allogeneic MHC molecules, administering protein antigen during this early stage of development can also sometimes engender a state of tolerance to it in adult life (Figure 9.3). These studies show that birds and mammals are particularly susceptible to becoming tolerant to antigens to which they are exposed during their embryonic or newborn (**neonate**) stage of development.

> **parabiosis** *n.* joined at birth.

Figure 9.2 Neonatally acquired tolerance to alloantigen. Experimental demonstration that exposure to alloantigen of a newborn (neonatal) mouse renders it tolerant to the same antigens in adult life. Based on the experiments of Billingham, R.E., *et al.* (1956) *Royal Society of London Philosophical Transactions B*, **239**, 357–414.

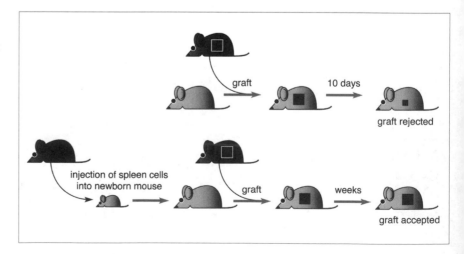

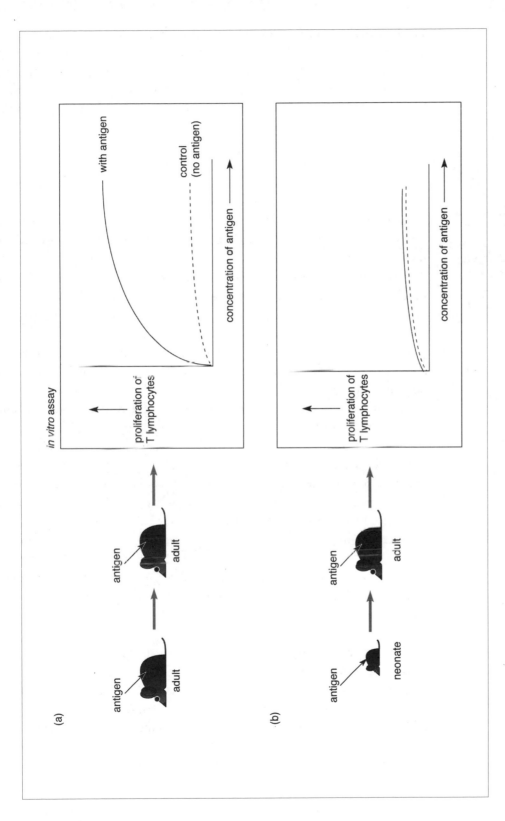

Figure 9.3 Experimental demonstration that mammals become tolerant of antigen experienced in neonatal life. Responsiveness can be measured in different ways. Antigen-induced proliferation of T lymphocytes *in vitro* is shown, although cytotoxicity, antibody production or, in the case of tolerance to alloantigens, mixed lymphocyte reactions or graft rejection can be used. This is active non-responsiveness (or tolerance) rather than a simple failure to prime the immune system, because a second challenge in adulthood that would normally elicit a response fails to do so.

9.1.4 Mature animals are less susceptible to tolerance induction

While immature animals are particularly susceptible to becoming tolerant to antigens, mature animals seem less so. Instead, an antigen gaining entry to the body of a mature animal generally has the opposite effect. Nevertheless, a long-lasting state of unresponsiveness to a particular antigen *can* be induced in certain circumstances.

It has been noticed in adult mammals that protein antigens that enter the body by the oral or nasal routes, or via intravenous injection, induce tolerance. The antigen is described as **tolerogenic**. The same antigen given to the mature mammal by injection into muscle, into the peritoneal cavity, or under the skin, elicits a response. The antigen is now **immunogenic**. These routes are illustrated in Figure 9.4. It is not understood why different routes lead to different responses but it is likely to be due to the nature of the antigen-presenting cells that acquire the antigen. The physiological significance of oral tolerance may be a way of preventing chronic stimulation of the immune system by foreign antigens to which the body is constantly exposed, such as those ingested as food.

There is considerable interest in the deliberate induction of tolerance as it may be useful for the treatment of autoimmune disease in humans. The potential for such therapy has been demonstrated in animals. Certain inbred strains of rats and mice are particularly susceptible to developing autoimmune arthritis if they are immunized with bovine collagen. In this way, T and B lymphocytes in the mouse that have receptors for self-collagen are activated. Arthritis then develops, although precisely why the state of tolerance to self-collagen is breached is not clear. However, if these animals are then given a tolerogenic dose of collagen by the oral route the disease process is halted, presumably because tolerance is re-established.

The other major factor that influences whether an antigen is immunogenic or tolerogenic in an adult animal is the amount of antigen received. At very high or very low doses of certain antigens a state of unresponsiveness is induced and the animal fails to respond to a subsequent immunogenic dose (Figure 9.5).

9.1.5 Helper T lymphocytes are central to maintaining tolerance to self-antigens

Studies in animals, particularly mice and rats, have allowed many questions about tolerance to be addressed by experimentation. One approach is to use inbred strains of rats and mice that spontaneously develop particular autoimmune diseases (Table 9.2). These strains are useful for identifying the inherited (genetic) components that predispose an animal to developing a particular autoimmune disease.

> **tolerogen** *n.* an antigen capable of eliciting **tolerance** (as opposed to an **immunogen**).
>
> **immunogen** an antigen capable of eliciting an immune response, as opposed to a **tolerogen**.

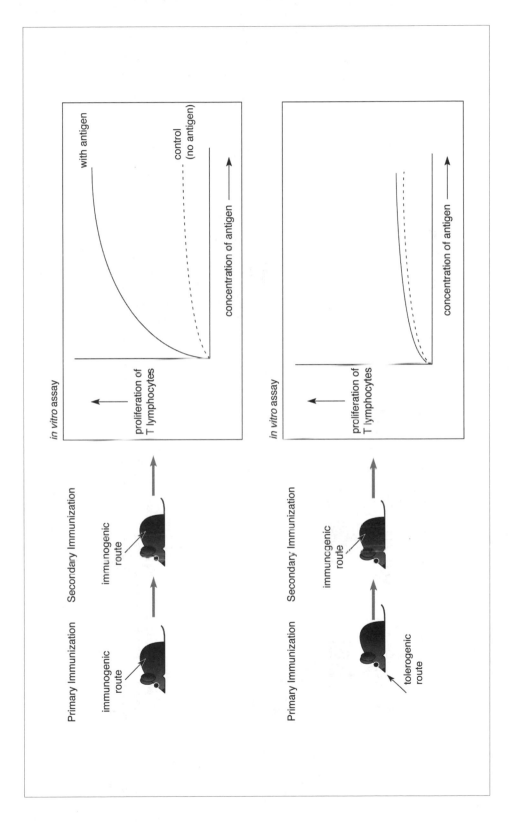

Figure 9.4. Antigens can be immunogenic or tolerogenic depending on the *route* of entry. In adult mice, antigens that are injected through the skin are generally immunogenic, shown here by the antigen-induced proliferation of primed T lymphocytes *in vitro*. Tolerance is induced if the antigen first gains entry through the oral or nasal routes, because subsequent immunization of antigen via an immunogenic route fails to elicit a response.

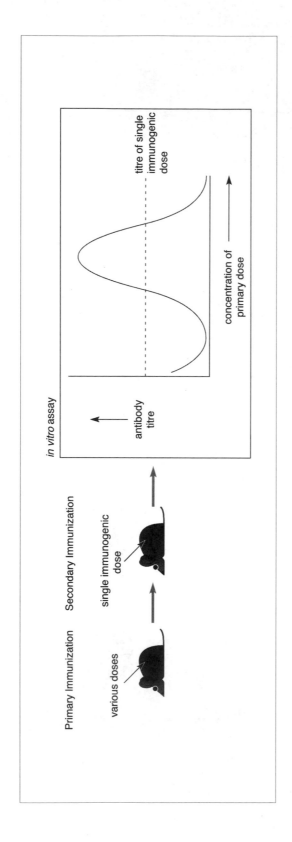

Figure 9.5 Antigens can be immunogenic or tolerogenic depending on *dose*. In adult mice, antigen administered at intermediate dose is immunogenic, seen here by the production of antigen-specific antibodies in response to a second immunization. Very high or very low doses of antigen in the primary immunization are tolerogenic because a subsequent, immunogenic dose elicits titres of antibody that are below normal. Adapted from Mitchison, N.A. (1964) *Proceedings of the Royal Society B*, **161**, 275–92.

Table 9.2 Spontaneous and experimentally induced autoimmune diseases of rats and mice

Inbred strain	Antigen required	Disease induced	Human 'equivalent'
Non-obese diabetic (NOD) mouse	? (spontaneous)	autoimmune diabetes	insulin-dependent diabetes melitus
(NZB × NZW)F$_1$ mouse[a]	?	murine systemic lupus erythematosus	systemic lupus erythematosus
Mouse, most strains	myelin basic protein	experimental allergic encephalomyelitis	multiple sclerosis
DBA-1 mouse (H-2^q)	bovine collagen	autoimmune arthritis	rheumatoid arthritis
Mouse, most strains	acetylcholine receptor (from *Torpedo*, electric ray)	experimental autoimmune myasthenia gravis	myasthenia gravis

[a]NZB, New Zealand black; NZW, New Zealand white.

The other approach involves eliciting autoimmune disease, or conversely, tolerance, by introducing a particular antigen into normal animals. Such studies are useful for defining the environmental components that precipitate autoimmune disease, and possible ways in which tolerance to antigen may be established or breached. The picture that emerges from these different lines of study is one in which the helper T lymphocyte compartment plays a pivotal role in maintaining tolerance to self-antigens.

Let us return to the example of experimental autoimmune arthritis in rodents injected with bovine collagen. Bovine collagen is recognized as foreign and the mouse responds by producing large amounts of anti-bovine collagen antibodies. Because of the similarities between the structure of bovine and murine collagen, some of the antibodies also bind to mouse collagen. These antibodies contribute to the damage of the cartilage in the joints by triggering the complement cascade and the activities of phagocytic cells. Consider a similar example. If a mouse receives an injection of erythrocytes of a rat, it responds by producing high titres of antibodies to rat erythrocyte antigens. Similarities between epitopes on mouse and rat erythrocyte antigens enables some of the anti-rat antibodies to bind to the mouse's own erythrocytes and trigger complement-mediated lysis. In effect, autoimmune haemolytic anaemia has been induced.

Mice do not normally encounter rat erythrocytes or bovine collagen and are unlikely to develop diseases in this way. However, these models do reveal an interesting insight into the cellular basis of tolerance. B lymphocytes that produce autoantibodies to self-antigen and which have the potential to cause autoimmune disease are part of the normal repertoire of B lymphocytes. As we will see below, tolerance can be established in the B lymphocytes. Yet, a significant portion of the responsibility for maintaining tolerance in the immune system seems to fall on the T lymphocyte compartment. It is now necessary to examine the events that occur during the development of T lymphocytes and how this tolerance is established.

Summary

- Passive tolerance to self-antigens is achieved by sequestration of antigens from self-reactive lymphocytes (immunologically privileged sites) or by reducing the levels of available antigens (immunological silence).

- Evidence from mammals and birds suggests tolerance is actively acquired to antigens experienced in the embryonic or early postnatal stages of life. After this time an antigen may be immunogenic or tolerogenic depending on dose or routes of entry.

- Tolerance within the helper T lymphocyte population is thought to be the basis of acquired antigen-specific tolerance, since many effector mechanisms are dependent on helper T lymphocytes.

9.2 Thymocyte ontogeny

T lymphocytes are produced throughout the life of a vertebrate, although the anatomical location in which they develop changes. During the critical embryonic/neonatal stage in which an animal is particularly susceptible to tolerance induction, this site is the thymus. All vertebrates, with the exception of the most primitive of fishes (the agnathans), have a thymus gland.

The thymus is the primary lymphoid tissue for T lymphocytes: Section 3.5.1

The importance of the thymus is clearly revealed in animals born without a thymus, or in athymic strains of inbred animals such as the **nude mouse**. These animals have very few mature T lymphocytes and are consequently incapable of mounting T lymphocyte-dependent immune responses. As a consequence they are severely immunodeficient and succumb easily to infections.

A similar effect is achieved by surgical removal of the thymus (**thymectomy**), although this ablation of T lymphocytes occurs only if the animal is thymectomized in embryonic or early postnatal life. In mammals, such as the mouse, this window extends for a few days after birth. Thymectomy of chickens performed immediately after hatching reduces the size of the T lymphocyte population in the blood of adult birds. However, the lower thymic lobes are difficult to remove so thymectomy is usually supplemented by treatment with anti-T lymphocyte antibodies to deplete the remaining T lymphocytes. As a result, the birds are less able to reject allogeneic skin grafts and tumour cells, and antibody responses to some antigens are also diminished. In the model amphibian, *Xenopus*, the thymus develops around day three after fertilization. Its removal from tadpoles from 4 to 8 days of age abolishes T-dependent responses in the adult frog, including impairment of antibody responses and rejection of allogeneic tissue grafts. Thymectomy of young fishes is less easy to perform because of the small size and accessibility of the thymus. Nevertheless, researchers agree that the role of the thymus in fishes is the same as that in higher vertebrates. In contrast, removal of the thymus from an animal in adult life has a negligible effect on the T lymphocyte repertoire.

Two important events occur in the thymus. Firstly, it is the site in which precursor T lymphocytes (**thymocytes**) rearrange their antigen receptor genes during embryonic and neonatal life. The second major event is the modification of the repertoire by weeding out undesirable thymocyte clones before the remainder are allowed to leave the thymus and enter the **periphery** as mature T lymphocytes.

9.2.1 Positive selection in the thymus

We saw in Chapter 6 that the population of mature T lymphocytes in the periphery recognize peptide antigens presented only by MHC molecules encountered in the thymus. For example, the T lymphocytes in the periphery of a mouse with the H-2^d MHC haplotype recognize only peptide antigens presented by syngeneic (also H-2^d), but not allogeneic, antigen-presenting cells.

MHC restriction: Section 6.1.6

However, we know that in the thymus there are also thymocytes with receptors that are complementary for other MHC molecules. One way of seeing this is to surgically implant thymic lobes into congenitally athymic nude mice (Figure 9.6). Implanted lobes become populated by the recipient's own bone marrow-derived lymphoid stem cells, which mature into T lymphocytes within the lobes. If the thymus is obtained from an

Figure 9.6 Use of athymic nude mice to show the thymus selects thymocytes that are restricted to self-MHC molecules. A nude mouse will produce mature T lymphocytes if it receives a grafted thymus from another mouse (not shown). If T lymphocytes are purified from a nude mouse engrafted with thymic lobes from an allogeneic mouse, and then cultured with antigen in the presence of antigen-presenting cells (APCs) and B lymphocytes, a response by the T lymphocytes is observed if the APCs have the same MHC haplotype as the donor of the thymus and not of the recipient. Based on the experiments of Singer, A., et al. (1982) *Journal of Experimental Medicine*, **155**, 339–44.

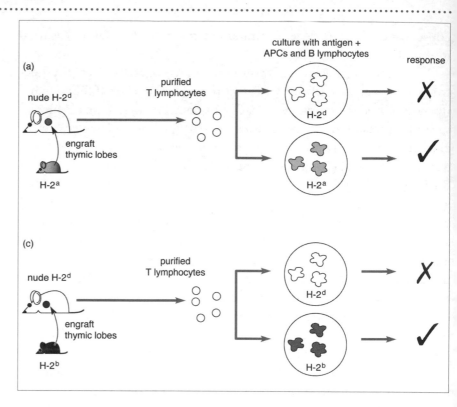

positive selection the process by which thymocytes are stimulated to differentiate into mature T cells in the thymus; thought to be triggered by antigen receptors with an avidity for thymic peptide–MHC complexes that is *below* a certain threshold (see **negative selection**).

allogeneic mouse, the T lymphocytes that mature are restricted by the MHC molecules of the thymus rather than those of the nude mouse itself.

Another (somewhat more complicated) approach is to replace the thymus of a mouse of a particular strain with one from an allogeneic strain. The mouse is then irradiated to kill the existing lymphocytes and then injected with bone marrow-derived stem cells from a syngeneic mouse, which repopulate the thymus. This animal is called a **radiation/bone marrow chimera**. In these kinds of experiments too, the T lymphocytes that were later found in the periphery were restricted by the MHC molecules of the donor of the grafted thymus rather than by the MHC molecules of the recipient.

The reason lies in the 'randomness' of antigen receptor gene rearrangements. Receptors with an inherent affinity for different kinds of MHC molecules are generated, far more in fact than are encountered in the thymus. It is thought that for a thymocyte to survive in the thymus it must be stimulated through its antigen receptor. Consequently, those thymocytes with receptors that are not complementary to the available MHC molecules fail to receive the signal and die in the thymus (see later). Those that engage a complementary MHC molecule survive. This is termed **positive selection**.

9.2.2 The role of CD4 and CD8 in positive selection

We have learned how mature T lymphocytes in the peripheral blood of mice and humans express either CD4 or CD8 molecules. In the thymus, where thymocytes develop into T lymphocytes, the situation is different. Around 80% of thymocytes express both molecules – referred to as 'double-positive' thymocytes. An integral part of the positive selection process is the transition from a double-positive into a 'single-positive' phenotype. The events that determine which developmental pathway a thymocyte takes are not well understood. The key unresolved issue is whether the 'decision' is made randomly (the stochastic model) or whether engagement of the receptor occurs first and this determines which pathway is taken (i.e. according to the class of MHC molecule the receptor binds).

Much of the experimental data obtained on the selection events in the thymus has come from the use of **transgenic** mice. A transgenic animal or plant is one in which a particular gene has been engineered into its genome using the techniques of molecular biology (see Box 9.1).

Consider the example in Figure 9.7. This particular system has provided illuminating insights into thymocyte selection. Here, the genes encoding the α and β chains of a T lymphocyte antigen receptor have been isolated from a clone of $CD8^+$ cytotoxic T lymphocytes. These genes were then inserted into the germline of a mouse to produce a T lymphocyte receptor transgenic mouse. We will digress briefly to examine how this was achieved.

The clone of cytotoxic T lymphocytes was obtained from a female mouse after immunization with cells from a syngeneic male mouse. Despite being the same inbred strain the female responds, albeit weakly, to the male-only antigens on the male cells. Antigenic differences outside the MHC that cause weak histocompatibility reactions between two syngeneic animals, such as the difference between males and females of the same inbred strain, are due to **minor histocompatibility antigens**. In reality 'minors' are conventional peptide antigens, which are derived from processed self-proteins and presented by the appropriate MHC molecules. In the case of the females responding to syngeneic male cells, they are responding to uniquely male peptide antigens which are foreign to females. Such antigens are encoded on the male Y chromosome. The male-only minor histocompatibility antigen to which the female-derived T lymphocyte clone responded was thus called **H-Y**.

The $CD8^+$ cytotoxic T lymphocyte clone that was used as the source of the antigen receptor genes was H-Y antigen-specific, and recognized this antigen in association with the class I MHC molecule, D^b. The next step was to insert the genes encoding the H-Y-specific receptor into the germline of female mice (the significance of the sex of the animal will be explained

transgenic *a.* a plant or animal that has been genetically engineered to contain a stable transgene.

transgene *n.* a foreign gene introduced into the germline of a plant or animal by recombinant DNA technology.

minor histocompatibility antigen *n.* antigen encoded outside the MHC that causes the rejection of **syngeneic** tissue grafts.

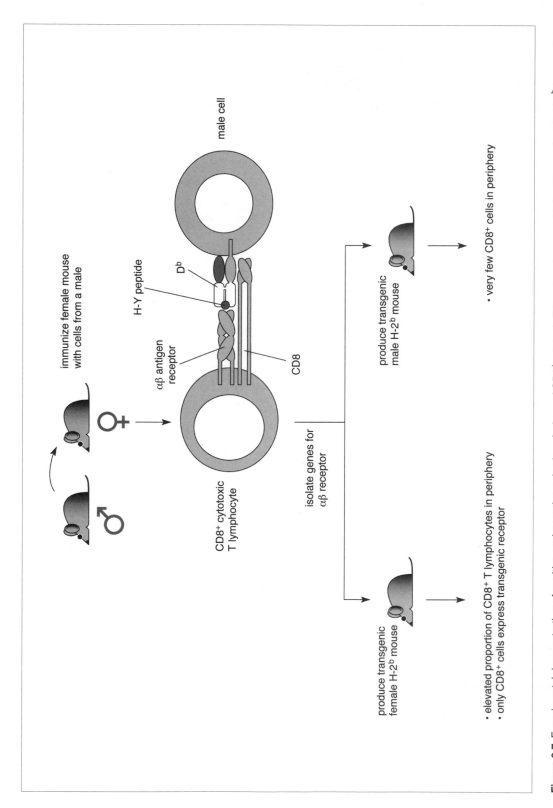

Figure 9.7 Experimental demonstration of positive and negative selection. A clone of CD8$^+$ cytotoxic T lymphocytes was obtained from a female H-2^b mouse after immunization with cells from a syngeneic male mouse. The clone responded to the male-specific antigen, H-Y, in the context of the class I MHC molecule, D^b. The genes for the αβ receptor were then inserted into the germlines of female (left) and male (right) mice (see Box 9.1). The majority of T lymphocytes in transgenic females bear the transgenic receptor and matured mainly as CD8$^+$. In transgenic male mice, which have the H-Y antigen, T lymphocytes with normal levels of CD8 are virtually absent from the peripheral repertoire. See text for details. Based on the experiments of Teh, H.S., *et al.* (1988) *Nature*, **335**, 229–33 and Kisielow, P., *et al.* (1988) *Nature*, **333**, 742–6.

BOX 9.1 Transgenic animals

A transgenic animal or plant is one in which a particular gene has been inserted into its genome using laboratory techniques. Most transgenic animals produced are mice and several of these have been used for studying the immune system. In order that the inserted **transgene** is incorporated into the germline and inherited, the gene is inserted in the early embryonic stage. In mice this is achieved in three ways.

- The original and most widely used method is to inject the gene into the pronucleus of a fertilized mouse egg (Figure 1). Many copies are injected, each contained within a plasmid. Plasmids then integrate or **recombine** into the chromosomes of the host cell, usually at unrelated sequences (**heterologous recombination**). The egg is then implanted into the oviduct of a foster mother that has been made receptive by mating it with a sterile male (a pseudopregnant female). The egg is then allowed to develop and each pup in the resultant litter screened for the presence of the transgene in DNA extracted from a small piece of tail tissue.

- An alternative method of obtaining recombinant embryos is to first insert the desired gene into a mouse retrovirus. Retroviruses characteristically integrate their genomes into the DNA of the cell they infect. Retroviruses of mice have been exploited in the laboratory to deliver foreign genes into the germline of mice embryos. As before, the embryo is implanted into a pseudopregnant female and the offspring that are born screened for the transgene.

Table 1 Knock-out mice

Gene	Description of gene product	Effect of deletion in vivo
CD8 β_2-microglobulin	On cytotoxic T lymphocytes The β chain of class I MHC molecules	• Mice are depleted of mature $CD8^+$ cytotoxic T lymphocytes • Mice are deficient in cell-surface class I MHC molecules, causing arrested development of $CD8^+$ cytotoxic T lymphocytes
TAP-1, TAP-2 (Transporter associated with antigen processing)	Transports peptides from the cytosol into the endoplasmic reticulum	• Class I MHC molecules unable to assemble with peptide in the endoplasmic reticulum • Consequently, mice are deficient in cell-surface class I MHC molecules, causing arrested development of $CD8^+$ cytotoxic T lymphocytes.
CD4	On helper T lymphocytes	• Mice are depleted of mature $CD4^+$ helper T lymphocytes
I-Aβ	The β chain of class II MHC molecules	• Mice are deficient in cell-surface class II MHC molecules, causing arrested development and depletion of $CD4^+$ helper T lymphocytes
RAG-1, RAG-2 (Recombinase associated genes)	Required for rearrangement of antigen receptor genes	• T and B lymphocytes fail to express antigen receptor at cell surface, causing arrested development of T and B lymphocytes
Igμ	Heavy chain of IgM. This class of immunoglobulin is expressed on the membrane of developing B lymphocytes first	• B lymphocytes fail to express membrane IgM, causing arrested development of B lymphocytes
Igκ	Light chain of the majority of immunoglobulin classes	• B lymphocytes fail to rearrange κ, although this is compensated by rearranging λ instead
Perforin	Released by cytotoxic T lymphocytes and natural killer cells; forms pores in target cell membrane	• Cytotoxic T lymphocytes and natural killer cells unable to kill target cells
Interleukin 2	Required for proliferation by T lymphocytes	• T lymphocyte responses normal in vivo, suggesting redundancy, although impared *in vitro* • Mice develop inflammatory bowel disease
Interferon-γ	Antiviral cytokine; also promotes differentiation of Th1 (see Figures 16 and 17 in Chapter 11)	• Activated T lymphocytes differentiate into Th2 • Mice have reduced immunity to intracellular pathogens

BOX 9.1 *continued*

Figure 1 Production of transgenic mice by the microinjection method. A plasmid is constructed containing the foreign gene under control of a promoter that is active in mouse cells. The plasmid also contains an antibiotic resistance gene, under the control of a promoter from *Escherichia coli*. This gene is necessary when the plasmid is grown in bacterial cultures. The plasmids are microinjected into the pronucleus of a fertilized egg and implanted into a pseudopregnant female (one that has been mated with a sterile male). The plasmid then recombines into random sites in the mouse genome. The offspring are screened for the tansgene, often by probing Southern blots of digested mouse DNA with a radioactive probe made from the plasmid itself. Based on the experiments of Gordon, J.W. and Ruddle, F.H. (1981) *Science*, **214**, 1244–1246.

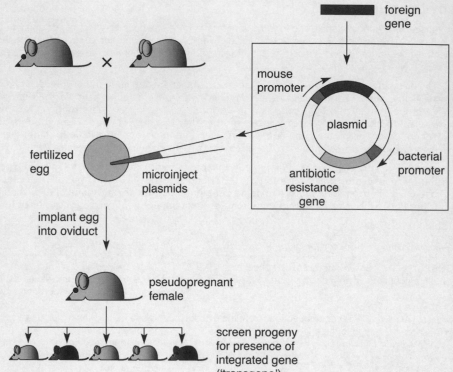

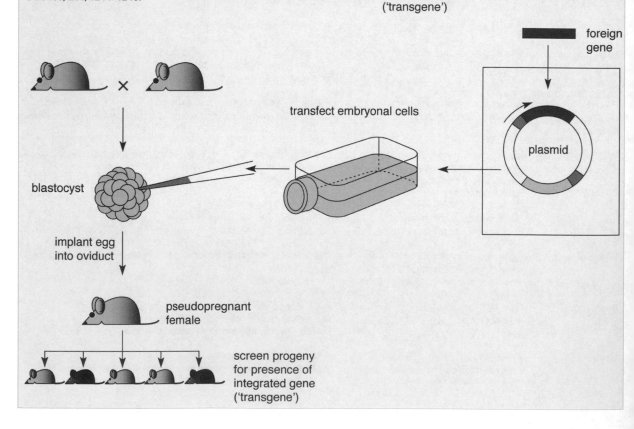

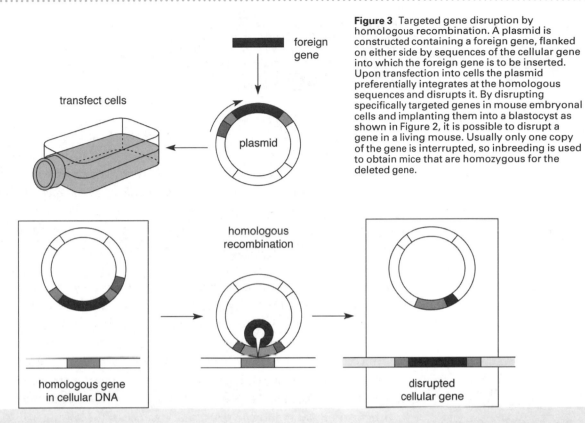

Figure 3 Targeted gene disruption by homologous recombination. A plasmid is constructed containing a foreign gene, flanked on either side by sequences of the cellular gene into which the foreign gene is to be inserted. Upon transfection into cells the plasmid preferentially integrates at the homologous sequences and disrupts it. By disrupting specifically targeted genes in mouse embryonal cells and implanting them into a blastocyst as shown in Figure 2, it is possible to disrupt a gene in a living mouse. Usually only one copy of the gene is interrupted, so inbreeding is used to obtain mice that are homozygous for the deleted gene.

- Heterologous recombination is an essentially random process and the outcome for the viability of the embryo is unpredictable. Sometimes vital host genes are disrupted with lethal consequences. However, a third method has been devised that allows more control over the outcome. This uses **embryonal stem cells**, which are cell lines derived from early mouse embryos (Figure 2). Genes to be incorporated into the genome of the mouse are first introduced by **transfection** into the embryonal cells *in vitro*. Embryonal cells are then injected into a mouse blastocyst wherein they become incorporated into the developing embryo. The advantage of this technique is that the stem cells can be analysed and a suitable clone of cells chosen *before* it is injected into the blastocyst.

As well as coding sequences that specify the protein product, a transgene must also possess the necessary transcription control sequences in order for it to be transcribed in its new cellular environment. A promotor from a mouse gene suffices although, in practice, control sequences from mouse viruses are often chosen (so-called heterologous promotors) because these are often very strong. In this way the transgene is transcribed in those cells in which the viral genome would normally be transcribed. However, by using **tissue-specific promoters** it is possible to confine the expression of the transgene to specific tissues. For example, the insulin promoter can be used to drive the transcription of genes specifically in the insulin-producing β cells of the pancreas. Other promoters are *inducible*, meaning they do not initiate transcription of the (trans)gene until an extracellular signal is received. For example, the regulatory elements of the metallothionein gene respond to heavy metals such as zinc which can be incorporated into the animal's drinking water. Another is the murine mammary tumour virus (MMTV) promoter, which is responsive to glucocorticoids and which can be used to drive the transcription of foreign genes in an inducible fashion.

In recent years a very powerful refinement of this technology has been developed which allows the integration site in the host's genome to be targeted precisely. Unlike the random, heterologous recombination described so far, foreign DNA will also recombine into sites in the host DNA with the same sequence. This is termed **homologous recombination** and it has been exploited as a means of inserting foreign DNA into a specific gene in the host cell genome. In practice this is achieved by constructing a plasmid containing a copy of all or part of the gene to be targeted. A foreign gene is then engineered into the plasmid located within the copy of the targeted gene (Figure 3). The recombinant plasmid is then transfected into a host cell whereupon the matched sequence in the plasmid aligns with and integrates into the homologous region in the cellular DNA. Notice that the targeted gene is disrupted by the insertion of the foreign gene. Disrupting or 'knocking-out' a targeted gene by homologous recombination in this way can reveal what the gene does in the cell. Combining this with transgenic animal technology has enabled knock-out animals to be produced. In mice, for example, the functions of many genes *in vivo* have been studied by disrupting them in embryonal cells by homologous recombination and then implanting them into blastocysts as described. Table 1 lists some of the knock-out mice in which genes of immunological interest have been disrupted.

Successful rearrangement blocks further rearrangement: Section 4.3.3

shortly). In these, and in other mice with transgenic antigen receptors, the most conspicuous effect is that T lymphocytes bearing the transgenic antigen receptor predominate in the mature repertoire. Normally of course, a receptor with a particular antigen specificity is present on only a small fraction of the repertoire. However, in transgenic mice the receptor genes inserted into the germline of the transgenic mouse are *already* in a productively rearranged configuration. Consequently, they block any rearrangement of the mouse's own T lymphocyte antigen receptor genes by allelic exclusion.

Returning to our example, the only mature T lymphocytes that emerged with normal levels of the transgenic $\alpha\beta$ receptor chains were $CD8^+$ rather than $CD4^+$. Notice this is the same phenotype as the original clone of cytotoxic T lymphocytes from which the receptor genes were obtained. This may be a consequence of the inherent affinity of the transgenic receptor for the D^b molecule, an interaction which would also necessitate the involvement of CD8 and not CD4.

Other experiments showed that in the absence of the appropriate MHC molecule, $CD8^+$ cells bearing the transgenic receptor fail to develop. This was discovered when the same antigen receptor genes were inserted into the germlines of mice with $H-2^k$ or $H-2^d$ MHC molecules. It is thought that in these mice, thymocytes bearing the transgenic receptor failed to engage their receptors with the appropriate MHC molecule and consequently did not receive the survival signals needed for positive selection. Without such signals the thymocytes soon die by apoptosis (see below).

More recently, the role of MHC molecules in positive selection has been corroborated by a refinement of transgenic mouse technology, in which the inserted gene can be targeted to a specific gene in the recipient's germline by **homologous recombination** (see Box 9.1). This has the effect of disrupting or 'knocking-out' the targeted gene. Mice with a disrupted β_2-microglobulin gene – which encodes the light chain of class I MHC molecules – produce very low levels of class I MHC molecules in the thymus and elsewhere. Consequently, only thymocytes with an affinity for class II molecules receive adequate stimulation in the thymus and the mature T lymphocyte repertoire in these mice consists almost entirely of $CD4^+$ cells. Conversely, mice with a disrupted I-Aβ gene – which encodes the β chain of a class II MHC molecule – are virtually devoid of class II molecules and

allelic exclusion a mechanism that allows only one of a pair of allelic forms of a gene to be expressed; although T and B lymphocytes rearrange both copies of a receptor gene, allelic exclusion ensures only one is expressed.

homologous recombination recombination between similar sequences of DNA; a phenomenon that is exploited for targeted gene disruption.

only CD8^{+} cells are found in the periphery.

To summarize the model, thymocytes which generate receptors that are not complementary for MHC molecules in the thymus fail to receive stimulation and die. Those that engage MHC molecules are given survival signals. These differentiate into CD4^{+} or CD8^{+} cells according to the class of MHC molecule engaged. This process of allowing self-restricted T lymphocytes to mature is called positive selection. As a consequence, the repertoire of mature T lymphocytes that emerge from the thymus and enter the periphery recognize antigen in the context of self-MHC molecules. This explains the phenomenon of MHC-restricted antigen recognition by T lymphocytes. Let us now return to the issue of how tolerance to self-antigens is established.

9.2.3 Negative selection in the thymus

Cell death plays a major role in shaping the mature T lymphocyte repertoire. It is estimated that 98% of the mouse thymocytes entering the thymus die there. The type of cell death that occurs in the thymus is called **apoptosis** and is the form of cell death used in multicellular organisms to organize and shape developing multicellular tissues.

T lymphocytes with a particular antigen specificity occur in very low numbers in the thymus, so making it virtually impossible to follow their fate. This situation changed, however, when the technology was developed to produce transgenic mice expressing a defined T lymphocyte antigen receptor. As we learned above, the normally low frequency of a particular antigen receptor can be increased by making a transgenic mouse with the receptor genes in its germline. It is then much easier to follow the development of the clone under different antigenic environments.

What happens if the receptor encounters its complementary peptide antigen–MHC molecule in the thymus? Let us return to the transgenic mice carrying a receptor that recognizes the male-only antigen, H-Y (Figure 9.7). As we saw above, transgenic females – which lack the H-Y antigen – have a preponderance of CD8^{+} T lymphocytes bearing the transgenic receptor. In contrast, transgenic males – which have the H-Y antigen – have very few of these cells in the periphery at all. This indicates that a T lymphocyte that encounters its specific peptide antigen during development in the thymus is removed from the repertoire.

This, and similar experiments, support a **clonal deletion** model of tolerance to self-antigen. It is thought that during development in the thymus thymocytes are exposed to a wide array of self-antigenic peptides. These are derived from processed self-antigens and presented by MHC molecules on the surface of thymic epithelial cells, macrophages and

negative selection the process by which thymocytes undergo apoptosis in the thymus; thought to be triggered by antigen receptors with an avidity for thymic peptide–MHC complexes that is *above* a certain threshold (see **positive selection**).

apoptosis *n.* cell death as a result of triggering an internal 'suicide' program, often involving the expression of death-inducing genes; a normal event in most developmental processes (see **necrosis**).

necrosis cell death caused by external trauma (poison, low oxygen, membrane disruption, etc.) causing rupture and local inflammation (see **apoptosis**).

clonal deletion the death, in response to stimulation by antigen, of lymphocytes with the corresponding antigen specificity.

BOX 9.2 Apoptosis

Cells can die by two different pathways.

- **Necrosis** is usually the result of severe physical trauma to the cell such as membrane damage, poison, low oxygen or elevated temperature. A cell undergoing necrosis ruptures, spilling its contents (Figure 1) and causing local inflammation.

- **Apoptosis** [Gr. *apo ptosis* = to fall or descend] is a deliberate and controlled cellular death resulting from the induction of an internal 'suicide' programme.

Apoptosis is triggered by extracellular signals received by receptors on the plasma membrane of the cell, which are often (but not always) relayed to 'suicide genes' by intracellular signalling cascades. A series of distinct morphological changes occur in the cell:

Figure 1 Morphological changes associated with necrosis and apoptosis.

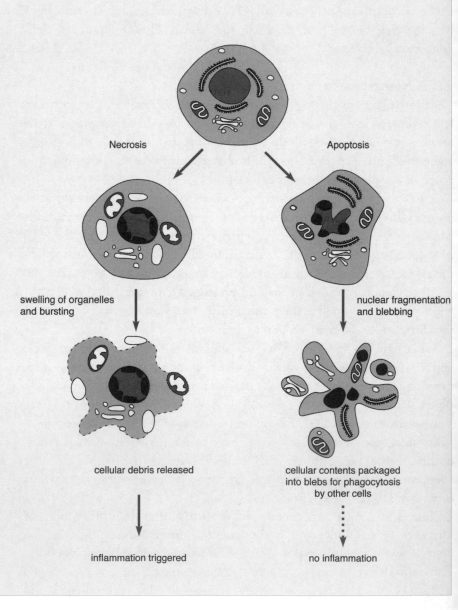

Necrosis

Apoptosis

swelling of organelles and bursting

nuclear fragmentation and blebbing

cellular debris released

cellular contents packaged into blebs for phagocytosis by other cells

inflammation triggered

no inflammation

- The cell first commences to digest its nuclear DNA using endonucleases. In the nucleus, DNA is coiled around proteins called histones, spaced along the DNA at regular intervals. Nuclear DNA looks rather like a string of beads, with each bead or **nucleosome**, occupying around 200 base pairs of DNA. During apoptosis DNA is cleaved between the nucleosomes to produce fragments in multiples of 200 base pairs. This can be observed experimentally as a characteristic 'ladder' of fragments if the DNA of apoptotic cells is resolved by electrophoresis through an agarose gel (Figure 2). The nucleus then breaks into discrete masses of chromatin.

- Meanwhile, the cell synthesizes other enzymes that systematically digest the contents of its cytosol. The leads to a characteristic rounding-up of the cell and violent bubbling of the plasma membrane (or **blebbing**) as the cytoskeletal architecture is lost.

- The cell then breaks up as the blebs pinch off, which are then consumed by phagocytes. This controlled fragmentation into blebs prevents any leakage of cellular components and prevents any inflammation.

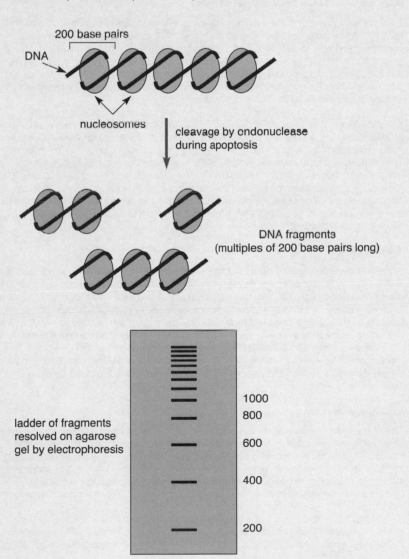

Figure 2 Visualizing DNA fragmentation.

BOX 9.2 *continued*

Apoptosis is a normal event in many developmental processes

Initially, the notion that cells could commit suicide came from observations of tissue sculpting in vertebrate embryogenesis. For example, fingers are created from a webbed hand by the elimination of cells in between. In adult life, apoptosis continues to play vital roles in maintaining constant numbers of cells that are continually proliferating. Apoptosis is also the means by which potentially harmful cells are removed. Cells that sustain severe genetic damage and which may become tumour cells are eliminated by self-induced apoptosis. Indeed many tumours are associated with mutations in apoptosis genes. In the immune system, self-reactive T lymphocytes are purged from the repertoire within the thymus by apoptosis and the affinity maturation of B lymphocytes in lymph node germinal centres involves the apoptotic death of B lymphocytes with low affinity receptors for antigen. Additionally, the 'killing' of tumour cells and virally infected cells by cytotoxic T lymphocytes and natural killer cells is caused by these effector cells triggering apoptotic suicide by the target cell.

Apoptosis – a default pathway of all cells?

One of the simplest experimental models to observe apoptosis is to remove growth factors from immortalized tumour cell lines *in vitro*. Ordinarily, tissue culture medium is supplemented with foetal calf serum, a rich (if undefined) cocktail of growth factors. In serum-free medium the cells round-up and die by apoptosis within a few hours. This has led to the idea that death by apoptosis is the default pathway of all cells and that their very survival is dependent on the continued presence of one or more growth factors or cytokines supplied by other cells.

A similar situation may exist *in vivo*. It is thought that healthy cells may have to compete for limiting amounts of survival factors and many otherwise healthy cells may be sacrificed for the benefit of those that proliferate. While apoptosis may be triggered in the *absence* of survival signals, the *presence* of other signals can induce apoptosis. These include tumour necrosis factor (TNF), a cytokine that induces apoptosis in some tumour cells. The receptor for TNF belongs to a family of receptors that includes CD95 (previously called **Fas** in mouse and **Apo-1** in humans) that transduce extracellular apoptotic signals. For example, apoptosis of thymocytes during deletion in the thymus, and the apoptosis of target cells by cytotoxic T lymphocytes and NK cells are due to signals received through Fas/Apo-1 molecules.

Suicide genes

Much has been discovered from studying an apoptosis-like process in the nematode worm *Caenorhabditis elegans*. During development, 1090 cells are produced and 131 are removed by apoptosis. Two genes, called *ced-3* and *ced-4*, are together necessary for these deaths. A third gene, *ced-9*, represses the activity of *ced-3*/*ced-4* and is needed for the survival of cells. Mammals have genes that are homologous to the worm *ced-3* and *ced-9* genes. The nematode *ced-3* gene encodes a proteinase. One mammalian counterpart is called **ICE** (interleukin-1β converting enzyme) which is a proteinase that cleaves a precursor of IL-1 into an active form. The *ced-3* gene may trigger apoptosis by activating signalling proteins required for apoptosis by cleaving their inactive precursors.

Anti-apoptosis genes

There are several mammalian homologues of the nematode *ced-9* gene, although the best characterized is *bcl-2*. The Bcl-2 protein was first discovered in B lymphocyte tumours in which it was found in abundance. This is caused by the over-expression of the *bcl-2* gene, which in turn prevented the B lymphocyte from undergoing apoptosis. If such cells also lose control over cell proliferation, which may happen in many ways – particularly if it becomes infected with Epstein–Barr virus (EBV), a lymphoma may develop.

Aberrant over-expression of a gene can occur when chromosomes break during mitosis and genetic information is exchanged. For example, some B cell lymphomas are caused when the *bcl-2* gene is translocated into the immunoglobulin heavy chain gene. Because the immunoglobulin gene is transcribed specifically in B lymphocytes, the translocated *bcl-2* gene is also expressed.

Bcl-2 is now known to block apoptosis in many different cell types. B lymphocytes, for example, die very quickly by apoptosis *in vitro* because of a lack of appropriate stimulation. However, they can be rescued if they receive signals through the IL-4 receptor and the CD40 molecule (achieved by culturing B lymphocytes in the presence of IL-4 and soluble CD40 ligand). These together trigger the appropriate signalling pathways for expression of the *bcl-2* gene.

The balance between cell proliferation and cell death is reflected in the use of several common components. One key component is a transcription factor called **c-Myc**, encoded by the c-*myc* gene. This protein binds to DNA and initiates transcription of several genes, in particular those that are required for cellular proliferation. Aberrant over-expression of c-*myc*, which causes uncontrolled proliferation, often occurs in tumour cells. Paradoxically, c-*myc* also regulates genes required for apoptosis. It is thought that the central position occupied by c-Myc in the proliferation and apoptosis pathways enables a cell to commit to one pathway while simultaneously aborting the other.

Two other important genes that control proliferation are **p53** and **Rb**. Normally, the protein products of these genes halt cellular proliferation, although their importance was first revealed in tumours in which their genes were mutated. Mutated *p53* is found in many kinds of human tumours and mutations in *Rb*, described first as underlying cancer of the retina (retinoblastoma) in humans, have now also been detected in several other tumours. Also, some viruses synthesize proteins during infection of a cell that interfere with the functions of these two cellular proteins, mimicking the effects of mutation. This is the way the virus creates a cellular environment suitable for its own replication. However, these viruses are associated with tumours: examples include adenovirus and SV40 (simian virus 40) which can cause tumours in rodents, and papillomaviruses which are associated with tumours in many mammals.

Because tumours are often associated with mutations in *p53* or *Rb*, these genes are often referred to as **tumour suppressor genes**. Normally, *p53* is expressed in response to DNA damage, such as that incurred after exposure to radiation. By stalling cellular proliferation it allows the cell to undertake the necessary repairs. If the damage to the DNA is beyond repair, *p53* then also triggers apoptosis. This additional property was revealed in *p53* 'knock-out' mice (see Box 9.1), whose thymocytes are extraordinarily resistant to apoptosis induced by radiation.

Further reading

Collins, M.K.L. and Lopez Rivas, A. (1993) The control of apoptosis in mammalian cells. *Trends in Biochemical Sciences*, **18**, 307–9.

Evan, I.G. (1994) Better dead than red. *Therapeutic Immunology*, 1, 343–8.

Lane, D.P. (1993) A death in the life of p53. *Nature*, **362**, 786–7.

Nagata, S. (1994) Fas and Fas ligand: a death factor and its receptor. *Advances in Immunology*, **57**, 129–44.

Raff, M.C. (1992) Social controls on cell survival and cell death. *Nature*, **356**, 397–400.

Schwarz, L.M. and Osborne, B.A. (1993) Programmed cell death, apoptosis and killer genes. *Immunology Today*, **14**, 582–90.

Wyllie, A.H. (1994) Death gets a break. *Nature*, **369**, 272–3.

dendritic cells. Those thymocyte clones with receptors that recognize self-peptide–MHC molecule complexes receive negative signals and are eliminated by apoptosis. It is also thought that antigens outside the thymus can gain access in the blood supply, and be picked up by the antigen-presenting cells in the thymus. This would account for the capacity to induce tolerance to foreign antigens or alloantigens that are experienced in early neonatal life. Some researchers have likened negative selection to holes being punched in the T lymphocyte repertoire, where individual clones have been deleted. The process of selection in the thymus is very complex and not yet fully understood. The descriptions above give the impression that positive and negative selection occur as two discrete events. In reality, these events are almost certainly closely intertwined processes.

9.2.4 The differential avidity model of thymocyte selection

Clearly a paradox of thymic selection concerns the rules that determine whether a thymocyte that has an affinity for a self-MHC molecule lives or dies. Although the answer is not clear, the strength of the interaction between T antigen receptor and the peptide–MHC complex (affinity) seems to play an important part in determining the fate of a thymocyte.

At one end of the spectrum are those thymocytes with no affinity – that is, those that generate receptors that do not bind to any of the available MHC molecules on the thymic epithelial cells. These fail to receive survival signals (i.e. no positive selection) and die by apoptosis. At the other end of the spectrum are those with receptors that recognize self-peptide–MHC complexes with high affinity. This high affinity interaction is thought to deliver a negative signal, and these cells also die by apoptosis. In between are receptors with an intermediate affinity for self-peptide–MHC molecule complexes. By the processes of positive selection and clonal deletion, the population of mature T lymphocytes that emerges from the thymus has receptors restricted for self-MHC molecules, and a responsiveness for peptide antigens of low/intermediate affinity.

avidity *n.* the overall stability of the interaction between receptors and ligands; a combination of several influential factors, including affinity, number of receptors and ligands involved, and their spatial density.

Precisely how a T lymphocyte receptor can measure the affinity of the interaction is not known. Moreover, there is now evidence that the number of peptide–MHC complexes engaged is also important. The currently favoured model is the avidity model of selection. The avidity of an antigen receptor for antigen, as we learned in Box 5.1, is a measure of the overall stability of the interaction. In the case of the T lymphocyte receptor, this is a combination of both the strength of the interaction (affinity) and the concentration of appropriate peptide–MHC molecule complexes (ligand density) on the antigen-presenting cell.

Evidence for this model has come from experiments with foetal mouse thymi cultured *in vitro*. In Figure 9.8 are summarized some experiments with thymi obtained from gene knock-out mice (see Box 9.1). As a result of the disrupted gene, these mice produce 'empty' class I MHC molecules (i.e. not loaded with self-peptide). These molecules are unstable. Very few are successfully exported from the endoplasmic reticulum to the surface of antigen-presenting cells and these have a very short life span. Consequently, class I-restricted thymocytes fail to mature within their cultured thymi.

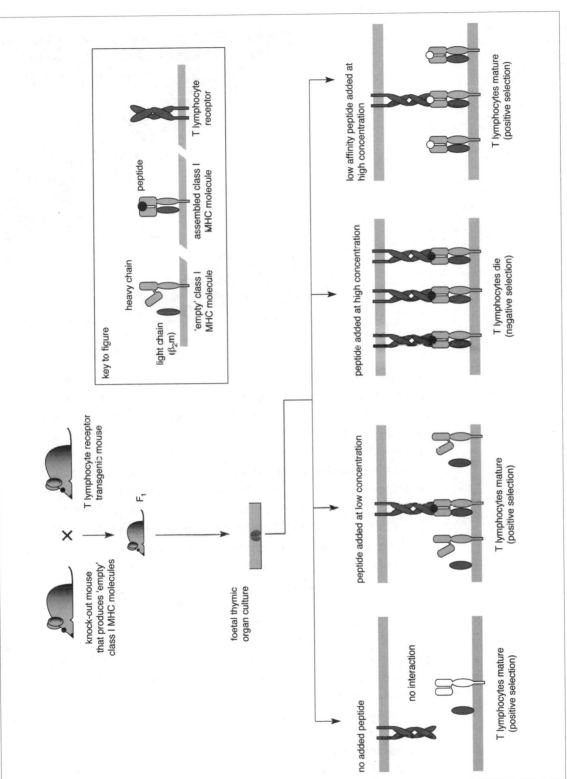

Figure 9.8 The fate of thymocytes is determined by a combination of affinity and the density of peptide–MHC complexes recognized by its antigen receptor. In the absence of appropriate antigenic peptide, T lymphocytes fail to emerge from thymic organ cultures owing to a poor interaction between receptor and MHC molecule. In the presence of the peptide at low concentrations the lymphocytes are positively selected, whereas increasing the concentration above a threshold causes them to die. A different peptide, that also allows the T lymphocyte receptor and the peptide–MHC complex to interact, but at a lower affinity, causes positive selection at high concentration. Based on the experiments of Hogquist, K.A., et al. (1994) Cell, **76**, 17–27 and Ashton-Rickardt, P.G., et al. (1994) Cell, **76**, 651–63.

Figure 9.9 The fate of developing thymocytes is determined by the avidity of its receptor for peptide–MHC molecules. Avidity is a combination of the affinity of the molecular interaction and the density of the peptide–MHC complexes on the antigen-presenting cell. In the shaded portions a thymocyte will die by apoptosis; elsewhere, a thymocyte will survive. Redrawn from Marrak, P. and Parker, D.C. (1994) *Nature,* **368**, 397–8.

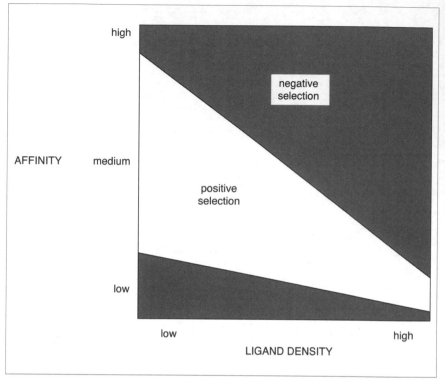

However, supplying synthetic peptides that are known to be able to bind to the empty class I molecules stabilizes them. When such peptides are added to the thymic organ cultures, class I-restricted thymocytes mature successfully.

In order to follow the development of a single clone of T lymphocytes, the knock-out mice were crossed with a mouse with a transgenic αβ T lymphocyte receptor obtained from a cytotoxic T lymphocyte clone. As we saw earlier, nearly all the thymocytes from receptor transgenic mice bear the same transgenic receptor, making their fate easier to follow. The F_1 hybrid mice used for the experiment had both attributes – a transgenic αβ receptor and empty class I molecules. Foetal thymi were then cultured in the presence of different concentrations of the peptide epitope recognized by the transgenic receptor. At high concentration the thymocytes died (negative selection), whereas at low concentration they matured normally (positive selection). This suggests the *density* of peptide–MHC complex on the thymic antigen-presenting cell influences whether the thymocyte is positively or negatively selected.

In further experiments, close relatives of the peptide epitope were made, differing from the original by a single amino acid change. These variant peptides were chosen because they also bound to the appropriate MHC molecule, but the complex produced resulted in a lower affinity interaction with the transgenic receptor. Whereas the original high affinity peptide caused deletion of the transgenic thymocytes, the variant peptides allowed the thymocytes to develop normally. This suggests the *affinity* of peptide–MHC complex on the thymic antigen-presenting cell influences whether the thymocyte is positively or negatively selected. These data are summarized in Figure 9.9.

Summary

- The thymus is the site in which precursor T lymphocytes (thymocytes) rearrange their antigen receptor genes. This endows each thymocyte clone with a predetermined affinity for a particular peptide–MHC complex.

- Thymocyte survival is dependent on receiving stimulation through the receptor.

- The avidity of the interaction between the receptor and a peptide–MHC complex (a combination of affinity and ligand density) is thought to determine the fate of the thymocyte. Positive selection seems to occur within a range of avidities; the lower the ligand density the higher the permitted affinity and *vice versa*. Avidity above this range causes apoptosis (negative selection). Apoptosis is also triggered by interactions below the avidity necessary for survival.

9.3 Post-thymic mechanisms of T lymphocyte tolerance

The deletion of self-reactive clones of T lymphocytes in the thymus seems to be the major means by which tolerance is established in the embryonic and early postnatal stages, when animals seem particularly susceptible to tolerance induction. However, there are many reasons for believing that additional 'post-thymic' mechanisms must exist.

- Firstly, the thymus involutes with age and ceases to exert the same influence on the repertoire. Yet lymphocytes continue to be produced throughout adult life without any obvious increase in the incidence of autoimmune disease.

- Secondly, some self-antigens are only produced after this neonatal period and will have not been available to modify the repertoire emerging from the thymus. Examples include products of genes responsive to sex hormones which are not expressed until sexual maturity, the products of genes that undergo somatic mutation (such as immunoglobulin V genes or mitochondrial DNA) and self-antigens in immunologically privileged sites which may be released by tissue damage. In amphibians, it is thought that several novel antigens are expressed after metamorphosis to which the animal needs to establish tolerance.

• Finally, animals that bear live young (viviparous animals) need a means for becoming tolerant of the paternal alloantigens expressed on the foetus to prevent 'rejection'. This is particularly relevant to the eutherian mammals in which there is comparatively long and intimate relationship during gestation.

These observations indicate that other 'post-thymic' mechanisms exist for establishing tolerance in self-reactive cells that escape thymic deletion or in those generated after the neonatal period of life during which the thymus is important. This so-called 'peripheral' tolerance may be achieved by clonal deletion although the anatomical site where this happens is not known. One hypothesis, based on the embryological origins of the thymus, is that deletion could occur in the epithelium of the skin, although there is no hard evidence for this or the cellular interactions involved.

On the other hand, there is considerably more *in vitro* evidence for two non-deletional mechanisms responsible for establishing peripheral tolerance in T lymphocytes. These mechanisms are **anergy** – a kind of antigen-induced paralysis – and **suppression** by T lymphocytes. Unlike deletion, the latter two mechanisms are reversible. Consequently, failures in these mechanisms may account for autoimmune disease. Suppression and anergy may also account for the experimental induction of tolerance to particular antigens in adult life via specific tolerogenic dosages or routes of entry (Section 9.2.3). In this section we will examine these reversible post-thymic mechanisms, starting with anergy.

9.3.1 Anergy

We saw in the previous chapter that the activation of helper T lymphocytes to antigen requires two signals from the antigen-presenting cell. One is the peptide–MHC complex itself, that is received by the antigen receptor. The second is costimulation received by separate receptors. In mouse and humans examples of receptors for costimulatory signals are CD28 and CTLA-4. Recognition of the appropriate peptide–MHC complex by the

Lymphocytes also require costimulatory signals for activation: Section 8.1.4

antigen receptor in the *absence* of simultaneous costimulation renders the T lymphocyte unresponsive, or anergic, to subsequent antigenic stimulation.

This double requirement of the activation of T lymphocytes helps maintain a high stringency for activation and may have evolved to prevent inappropriate activation of self-reactive T cells. Moreover, costimulatory molecules are, for the most part, found only on the surfaces of professional antigen-presenting cells (i.e. those that express class II MHC molecules) such as dendritic cells and activated B lymphocytes, which presumably helps maintain this stringency in some way.

It has been noticed that the state of anergy in T lymphocytes sometimes can be reversed *in vitro* by stimulating them to proliferate with interleukin 2 (IL-2). A similar situation may occur *in vivo*: the administration of IL-2 to humans with cancer, intended to boost their own tumour-specific cytotoxic T lymphocytes, was also accompanied by the onset of autoimmune diseases. Whether such breaches of tolerance account for spontaneous autoimmune disease is not known, but this remains a possibility.

9.3.2 Immune suppression

It has long been argued that a distinct subset of T lymphocytes, called suppressor cells, exists in the immune system to turn down antigen-driven responses, and as a mechanism to maintain tolerance to self-antigens in the periphery. There have been several different demonstrations of T and B lymphocyte responses being reduced *in vitro* and *in vivo* which could be explained by the activity of suppressor cells. However, the suppressor cell itself has been rather elusive to define, and its phenotype, MHC restriction and mechanism of suppression have been difficult to define consistently.

Although suppression phenomena are genuine, they may be achieved in ways without invoking an additional T lymphocyte subset. For example, Th1 lymphocytes release IFN-γ which inhibits the proliferation of Th2 cells. Because B lymphocytes require the cytokines released by Th2 cells for differentiation into plasma cells, Th1 have the effect of suppressing antibody responses. Reciprocally, Th2 'suppress' Th1 (see Figure 8.19). Similarly, class II MHC-restricted, CD4$^+$ cytotoxic T lymphocytes, which could kill antigen-presenting cells, could account for some suppressor phenomena. Alternatively, anergic helper T lymphocytes competing with other effector T lymphocytes for cytokines like IL-2 could also turn down responses by the effector cells.

Cytokines define multiple subsets of helper T lymphocytes: Section 8.2.2

9.4 B lymphocyte tolerance

Despite the importance of tolerance within the T lymphocyte compartment, experiments have shown that B lymphocytes can also be made tolerant to certain antigens *in vivo*. As with T lymphocytes, the rarity of B lymphocyte clones of a defined antigenic specificity makes it difficult to follow their development in different antigenic environments *in vivo*. This problem was also resolved by the use of transgenic mice. When, as is depicted in Figure 9.10, the genes of the heavy and light chains of rearranged immunoglobulin are obtained from a B lymphocyte clone and then engineered into the germline of a mouse, rearrangement of the endogenous immunoglobulin genes is blocked by allelic exclusion. In non-transgenic mice, the proportion of B lymphocytes that have this specificity is too small to detect, whereas in the immunoglobulin transgenic mice this proportion increases to as much as 90% of the repertoire.

In the experiment illustrated, genes encoding an anti-hen egg lysozyme antibody were used. Another transgenic mouse was made, this time containing the gene for hen egg lysozyme. These mice produced lysozyme which was present at high levels in the blood. It was then possible to see what effect the presence of the antigen (lysozyme) had on the development of the B lymphocytes expressing complementary (anti-lysozyme) receptors. This was achieved by mating the two transgenic mice to produce 'double-transgenic' offspring carrying both the lysozyme gene and the anti-lysozyme immunoglobulin gene. Unlike T lymphocytes, which would be deleted had they developed in the presence of their specific antigen, B lymphocytes expressing the transgenic immunoglobulin *were* found in the spleen and lymph nodes of these transgenic mice. However, they were unable to secrete anti-lysozyme antibody after stimulation with lysozyme, even after transferring to a normal mouse which had lysozyme-specific helper T lymphocytes. This suggests that B lymphocytes can also be rendered tolerant to self-antigen although by becoming anergic rather than being deleted from the repertoire.

> **allelic exclusion** a mechanism that allows only one of a pair of allelic forms of a gene to be expressed; although T and B lymphocytes rearrange both copies of a receptor gene, allelic exclusion ensures only one is expressed.

Summary

- Tolerance may be established by non-thymic or 'peripheral' mechanisms, including anergy and suppression. These mechanisms are

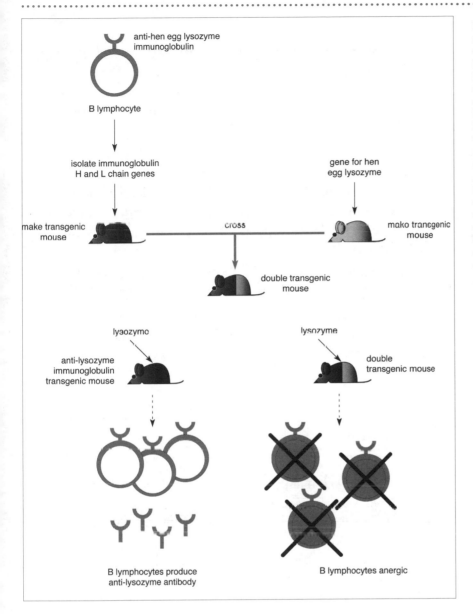

Figure 9.10 Evidence for a non-deletional mechanism for establishing tolerance in B lymphocytes. A transgenic mouse containing the heavy and light chain genes of an anti-hen egg lysozyme antibody has a high proportion of the B lymphocytes expressing the transgene. This mouse responds to an injection of lysozyme by releasing anti-lysozyme antibodies in the blood. The mouse was crossed with a mouse with lysozyme transgenes. These 'double transgenic' mice do not delete the B lymphocytes bearing the transgenic receptor although they are unable to produce antibody in response to challenge with antigen. Based on the experiments of Goodnow, C.C., *et al.* (1988) *Nature*, **334**, 676–82.

reversible and breaches in them may lead to autoimmune disease.

- Anergy of T lymphocytes may be achieved by stimulation through the antigen receptor in the absence of costimulatory signals.

- Discrimination between self and foreign antigens by the adaptive immune system is not genetically predetermined, but is achieved by acquiring central and peripheral tolerance to self-antigens.

The future

- Although breaches of tolerance are used to explain the existence of autoimmune disease, in no case of a naturally occurring autoimmune disease is the cause of the breach known.

- Transgenic animals have been useful in understanding the potential role of deletion in the thymus as a mechanism of tolerance. However, these systems use artificially high numbers of responder lymphocytes. It is not yet clear whether self-peptides have the same effect on the natural repertoire in which the responder frequency to a particular antigen is much lower.

- It is also not clear how or whether thymic tolerance is established to antigens expressed outside the thymus.

- There are many reasons why peripheral tolerance must exist *in vivo*. There is good evidence from *in vitro* studies for antigen-specific non-responsiveness (anergy) but it is still unclear how relevant this phenomenon is *in vivo*.

Further reading

Goodnow, C.C. (1992) Transgenic mice and analysis of B cell tolerance. *Annual Review of Immunology,* **10**, 489.

Goodnow, C.C. (1989) Cellular mechanisms of self-tolerance. *Current Opinion in Immunology,* **2**, 226–36.

Miller, J.F.A.P and Morahan, G. (1992) Peripheral T cell tolerance. *Annual Review of Immunology,* **10**, 51–69.

Nossal, G.V.A. (1983) Cellular mechanisms of immunological tolerance. *Annual Review of Immunology,* **1**, 33–62.

Rennie, J. (1990) The body against self. *Scientific American,* December, 76–85.

Schwartz, R.H. (1989) Acquisition of immunologic self-tolerance. *Cell,* **57**, 1073–81.

Von Boehmer, H. (1990) Developmental biology of T cells in T cell receptor transgenic mice. *Annual Review of Immunology,* **8**, 531–6.

Review questions

Fill in the gaps

Tolerance to self-antigens is achieved at many levels. Physical sequestration of antigens from self-reactive lymphocytes in _____ _____ _____[1] is one means, or immunological silence can be achieved by reducing the levels of surface _____[2] molecules. In addition, tolerance is actively acquired to antigens experienced in the _____[3] or early _____[4] stages of life. After this time an antigen may be _____[5] or _____[6] depending on dose or routes of entry. A key to many forms of tolerance lies in the _____[7] T lymphocyte compartment because many effector mechanisms are dependent on these cells. Tolerance is established during ontogeny in the _____[8]. Survival of _____[9] is dependent on receiving stimulation through the antigen receptor. It is thought that those with receptors having an intermediate affinity for a _____/_____[10] complex survive (_____[11] selection) and those that fail to bind die by _____[12]. Those that bind with _____[13] affinity also die by _____[14] (called _____[15] selection). Tolerance may also be established by 'peripheral' mechanisms, by mechanisms such as _____[16] and immune _____[17]. These mechanisms are reversible and breaches in them may lead to _____[18] disease. Studies *in vitro* suggest _____[19] of T lymphocytes may be achieved by stimulation through the antigen receptor in the absence of _____[20] signals.

Short answer questions

1. Define the following terms: (a) tolerance; (b) anergy; (c) thymic ontogeny; (d) positive selection; (e) negative selection; (f) oral tolerance; (g) apoptosis; (h) necrosis.
2. True or false; if you think the answer is false, give your reasons.
   (a) Tolerance to antigen is learned or acquired.
   (b) Tolerance to antigen can only be established during the embryonic/neonatal period of life.
   (c) Anergy of T lymphocytes can be caused by costimulatory signals in the absence of stimulation through the antigen receptor.
   (d) Th2 cells can suppress cell-mediated immune responses via the effect of IL-4.
   (e) The experimental induction of tolerance can be used as a means of treating autoimmune disease.

(f)   In mammals, introduction of antigen by subcutaneous (beneath the skin) injection usually renders the animal tolerant to the antigen.

3.   In 1962, Triplett reported studies of the tree frog, *Hyla regilla* (Triplett, E.L., *Journal of Immunology,* **89**, 505–10, 1962). The pituitary gland was removed from the animal in the larval (tadpole) stage and the gland kept alive by implanting it under the skin of a recipient frog. Pituitary glands are the source of a hormone that causes the skin to darken. Consequently, glandless larvae develop into adult frogs that are light in colour. If, however, the gland that was removed from a tadpole was transplanted back to the original donor after metamorphosis, the colour of the frog immediately darkened but then gradually lightened again over the course of several days. What explanations are there for this?

4.   You find that when you immunize a particular inbred strain of mouse against hen egg lysozyme, a response is elicited, which you can demonstrate by the proliferation of activated helper T lymphocytes in response to the same antigen *in vitro*. When you perform the same experiment on mice of the same strain expressing a hen egg lysozyme transgene, there is no proliferative response *in vitro*. Why do you think this is and how could test your hypothesis?

5.   Consider another transgenic mouse (Adams, T.E., *et al.*, *Nature,* **325**, 223, 1987). This animal has the DNA encoding for a virus protein that has been engineered into the mouse germline. In addition, the promoter for insulin has been placed upstream of the virus sequence which drives the expression of the virus gene. As the animals mature, it was found that the stage of mouse development in which the virus protein first appeared was varied. In some mice, the protein did not appear for several weeks after birth and many of these developed diabetes. How do you explain this and what does this suggest about the cause of diabetes in humans?

Answers to review questions

Chapter 1

Missing words

parasitic[1]; disease[2]; pathogens[3]; chemical[4]; lysozyme[5]; haemorrhage[6]; damage/trauma[7]; mast[8]; damage/trauma[9]; vasoactive[10]; vascular permeability[11]; infiltration[12]; neutrophils[13]; chemotactic[14]; macrophages[15]; phagocytosis[16]; IL-1/IL-6 and TNF-α[17–19]; effector[20]; complement[21]; chemotactic[22]; cytotoxicity[23]; natural killer[24]; cytotoxic[25]; granzymes[26]; perforins[27]; perforin[28].

Short answer questions

1. C(a) Lysozyme; (b) encapsulation; (c) phagocytosis; (d) interleukin-1; (e) C-reactive protein; (f) phagolysosome; (g) the respiratory burst; (h) chemotaxis; (i) interferon-γ; (j) histamine.

2. TNF-α, IL-1 and IL-6; all act locally to increase vascular permeability and (except IL-6) increase expression of adhesion molecules on vascular epithelium for extravasation of lymphocytes. All three induce systemic effects: fever and the production of acute phase proteins by hepatocytes.

3. (a) Both; (b) invertebrates; (c) vertebrates; (d) vertebrates; (e) invertebrates; (f) invertebrates; (g) both; (h) invertebrates; (i) vertebrates.

4. (a) True; (b) false, it is an anaphylatoxin; (c) true; (d) true; (e) true; (f) false; a more valuable function is the generation of opsonins and chemotactic factors; (g) true.

5. (a) Cytotoxic T cell or natural killer cell response. (b) Phagocytosis and activation of complement. (c) Cytotoxic T cell or natural killer cell response.

Chapter 2

Missing words

adaptive[1]; lymphocytes[2]; gene rearrangement[3]; specificity[4]; repertoire[5]; low[6]; proliferation (or expansion)[7]; slow[8]; memory[9]; auto (or self-) reactive[10].

Short answer questions

1. A lectin is a protein that binds to specific sugar residues, particularly on cell-surface glycoproteins, enabling them to agglutinate prokaryotic and eukaryotic cells. C-reactive protein, serum amyloid protein and several collectins (Table 2.1) are used in mammalian innate defence.

2. Alpha-2 macroglobulin is a serum proteinase inhibitor which is thought to act as a recognition molecule in *Limulus*. In vertebrates, the C3 and C4 components of complement also bind to proteinaceous surfaces, although discrimination between self and non-self is achieved by inhibitors (factor I, decay accelerating factor) on self surfaces.

3. Directly by using the 'NK receptor' (possibly a C-type lectin), and indirectly by using Fc receptor that bind to the Fc domains of antibodies bound to foreign surfaces (antibody-dependent cell-mediated cytotoxicity).

4. The activation and proliferation, in response to antigen, of clones of lymphocytes with the corresponding antigen specificity.

5. The immunoglobulin superfamily is a set of evolutionary related proteins that have one or more immunoglobulin folds as their basic unit.

6. Tolerance is a state of non-responsiveness to particular antigens achieved by removing lymphocytes from the repertoire. Without tolerance, the immune system is not prevented from triggering effector mechanisms in response to self-antigens and causing self-inflicted damage.

7. Recombination is the process during meiosis in which breaks in homologous chromosomes lead to crossover of DNA; general recombination is any exchange or insertion of one DNA molecule in another; rearrangement refers to the type of general recombination that occurs in antigen receptor genes.

Chapter 3

Missing words

red[1]; white[2]; thrombocytes (platelets in humans)[3]; granulocytes[4]; mononuclear[5]; basophils, neutrophils, eosinophils[6–8]; monocytes[9]; lymphocytes[10]; haemopoiesis[11]; pluripotent (or haemopoietic)[12]; lymphocytes[13]; myeloid[14]; tissues[15]; extravasation (or diapedesis)[16]; helper[17]; cytotoxic[18]; CD4[19]; CD8[20]; antigen presenting[21]; naïve[22]; secondary[23]; high endothelial venule[24]; inflammation[25].

Short answer questions

1. (a) True; (b) false; it is generally accepted that fishes do not have a true lymphatic system, at least not of the same level of complexity as that found in tetrapods; (c) false; the thymus is not a site in which pluripotent stem cells proliferate, but is the site in which the lymphoid progenitors undergo maturation into T lymphocytes; (d) false; interdigitating dendritic cells are, like the leucocytes, bone marrow-derived cells that derive from dendritic cells that traffic between the tissues and draining lymph nodes. Follicular dendritic cells are not bone marrow-derived and are sessile cells that reside in the follicles of lymphoid tissues.

2. Primary lymphoid tissues are the sites in which lymphoid stem cells produced by the haemopoietic tissue rearrange their antigen receptor genes and undergo differentiation into mature lymphocytes. Secondary lymphoid tissues are sites in which antigens and immune complexes (aggregates of antigen and antibody) are filtered from the blood and draining lymph, or by transportation by circulating dendritic cells, and through which naïve lymphocytes circulate to encounter antigen; these tissues also provide the necessary cellular microenvironment for activation of naïve lymphocytes by antigen and differentiation into effector cells.

3. Naïve T lymphocytes preferentially recirculate between the secondary lymphoid tissues and leave via the efferent lymphatics to be returned to the blood. Entry into the lymph nodes and mucosa associated lymphoid tissues occurs at high endothelial venules, whereas lymphocytes enter the white pulp of the spleen from the arterioles located in the marginal zone. Activated T lymphocytes preferentially home to inflamed tissue because of the localized induction of a variety of adhesion molecules.

4. Blood-borne antigen and immune complexes are removed by the filtering action of the spleen; antigen and immune complexes in the tissues are carried away by the lymphatic system into the nearest draining lymph nodes; antigens gaining access through mucosal linings are entrapped by mucosal secondary lymphoid tissues – in the case of the gut these include the Peyer's patches, the appendix and diffuse lymphoid tissues (gut associated lymphoid tissues or GALT)

5. Lacking the ectodermal component, the thymus fails to develop properly. Consequently, these mice are unable to generate a competent repertoire of functional T lymphocytes. T lymphocyte-dependent responses are impaired and these animals are susceptible to many opportunistic infections. As we see in Chapter 6, T lymphocytes are responsible also for rejecting grafts of tissue from genetically unrelated donors; nude mice accept such grafts.

6. The function of the spleen is the same in all jawed vertebrates. It is a haemopoietic tissue in which haemopoietic stem cells self renew and differentiate into different blood cells. It is also the site where damaged and worn out erythrocytes are removed from the circulation and consumed by macrophages. It also serves as a secondary lymphoid tissue for trapping blood-borne antigen for presentation to trafficking lymphocytes.

7. (a) P-selectin and S-selectin on the endothelium and glycoproteins on the neutrophil. (b) ICAM on inflamed vascular endothelium and LFA-1 on the neutrophil. (c) MAdCAM on the endothelium and L-selectin on the lymphocyte.

Chapter 4

Missing words

γ-globulin[1]; papain[2]; Fab[3]; Fc[4]; heterogeneous[5]; myeloma[6]; light[7]; variable[8]; constant[9]; restriction endonuclease[10]; germline[11]; B lymphocyte[12]; germline[13]; rearrangement[14]; clone[15]; subtractive hybridization[16]; beta[17].

Short answer questions
1.

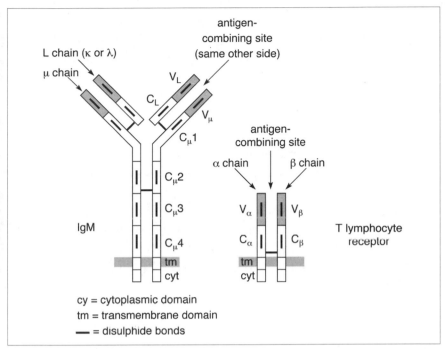

cy = cytoplasmic domain
tm = transmembrane domain
— = disulphide bonds

2. Combinational diversity in antigen receptors is diversity created between V domains by the different combinations in which V(D)J gene segments can be juxtaposed by the rearrangement process. Junctional diversity is additional diversity created by imprecise joining at the junction of the V(D)J segments. Junctional diversity is manifested in the V domain as an external hypervariable loop - the third complementarity-determining region (CDR3). Birds and rabbits generate diversity in their immunoglobulins during the proliferation of immature B lymphocytes by gene conversion.

3. Allelic exclusion is a mechanism that allows only one of a pair of allelic forms of a gene to be expressed. Allelic exclusion ensures that while T and B lymphocytes rearrange both copes of a receptor gene, only one (the one that achieves a productive rearrangement first) is expressed. Therefore, only one version of a functional receptor is found at the lymphocyte surface .

4. (i) Activated B lymphocytes manufacture soluble immunoglobulin molecules (antibodies) which is achieved by differential splicing of the primary RNA transcript of the immunoglobulin gene. (ii) Immunoglobulin genes undergo class switching so that antibodies with the same antigen specificity but of several different classes are produced by different cells in the expanding clone. This is caused by further DNA rearrangements that allow different constant gene segments to be transcribed after the same rearranged V(D)J gene. (iii) As the clone

proliferates, somatic mutation in the V region occurs with each cell division. Immunoglobulins with widely different affinities for the same antigen are produced. Plasma cells producing the highest affinity antibodies are stimulated in germinal centres, causing affinity maturation of the response.

5. (i) Antibody-dependent cell-mediated cytotoxicity (ADCC) of antibody-tagged cells by cytotoxic cells (neutrophils, NK cells, macrophages). (ii) Transcytosis of IgA through mucosa into exocrine secretions (mucus, tears, saliva, milk, colostrum) via the poly-Ig receptor, or of IgG in milk/colostrum through the gut wall of neonatal mammals via the neonatal Fc receptor. (iii) Fixation of complement and activation of the classical pathway. (iv) Opsonization of antigens prior to Fc receptor-mediated endocytosis by phagocytes. (v) Degranulation of mast cells and basophils mediated by binding of IgE by the FcR$_\epsilon$I receptor.

Chapter 5

Missing words

X-ray crystallography[1]; homogeneous[2]; Bence-Jones[3]; myeloma[4]; monoclonal[5]; lysozyme[6]; neuraminidase[7]; complementary[8]; non-covalent[9]; complementarity-determining regions[10]; epitope[11]; surface[12]; discontinuous[13]; affinity[14]; valency[15]; epitopes[16]; avidity[17]; cross-link[18]; immune[19]; antigen presenting[20]; 4[21]; proliferation[22]; interleukin 2[23]; 8[24]; lysis[25]; conformation[26]; processed[27]; major histocompatibility complex[28].

Short answer questions

1. CD4$^+$ T lymphocytes recognize peptide antigens in the context of class II MHC molecules on the surface of antigen-presenting cells. Helper T lymphocytes and T lymphocytes that mediate delayed-type hypersensitivity reactions are usually CD4$^+$ and their responsiveness to antigen *in vitro* is usually measured by proliferation or IL-2 release. CD8$^+$ T lymphocytes recognize peptide antigens in the context of class I MHC molecules on the surface of antigen-presenting cells. Most cytotoxic T lymphocytes are CD8$^+$ and their activity is measured by lysis of antigen-presenting cells (called targets) *in vitro*.

2. Monoclonal antibodies are antibodies produced by a single clone of B lymphocytes with a defined antigen specificity. These are produced by

taking antibody-producing plasma cells (usually from the spleens of rats or mice immunized against the antigen) and immortalizing them in the laboratory by fusing them with myeloma cells to produce hybridomas. The hybridomas are then cloned by limiting dilution.

3. Antibodies generated against the virus capsid during infection are less likely to recognize denatured capsid in immunoassays than capsid in its native configuration. Therefore the ELISA is likely to be the technique of choice, as this uses native antigen to detect the antibody. In Western blotting, the antigen has to be first denatured by boiling in SDS and resolved by polyacrylamide gel electrophoresis.

4. (a): (2), (3), (4). (b): (2), (3), (4). (c): (5).

Chapter 6

Missing words

inbred[1]; cancer[2]; allogeneic[3]; syngeneic[4]; congenic[5]; H-2[6]; allogeneic[7]; allo- (or histocompatibility)[8]; alloantisera[9]; synthetic peptides[10]; allo- (or histocompatibility)[11]; allo- (or histocompatibility)[12]; in vitro assays[13]; syngeneic[14]; restriction[15]; T lymphocytes[16]; thymus[17]; antigen[18]; MHC molecule[19]; peptide[20]; cleft[21].

Short answer questions

1. Classical MHC molecules are polymorphic proteins that present peptides at the surface of cells to the antigen receptors of T lymphocytes. Non-classical MHC molecules have similar structures to classical MHC molecules although they are much less polymorphic and several remain inside the cell rather than on the surface. Their functions are largely unknown.

2. Both have four extracellular domains; a peptide-binding cleft is created by the two membrane-distal domains from a platform of β-strands surmounted by two α-helices. The two membrane-proximal domains are immunoglobulin-like domains. Allelic differences localize to the peptide-binding cleft in both class I and II molecules. Differences lie in the size of the chains. Class I molecules are constructed from a heavy (or α) chain having both membrane-distal domains (α_1 and α_2) and one membrane-proximal domain (α_3). The smaller, membrane-proximal chain is β_2-microglobulin. Class II molecules have two roughly equally sized chains (α and β), each with one proximal and one distal domain. Only the heavy chain of class I molecules is encoded in the

MHC and is polymorphic, whereas both class II chains are MHC encoded and polymorphic (an exception is HLA DRα, which is not polymorphic).

3. Class I molecules present peptide antigens to $CD8^+$ T lymphocytes (mainly cytotoxic T lymphocytes), whereas class II molecules present antigens to $CD4^+$ T lymphocytes (mainly helper T lymphocytes). Class I molecules present peptides from predominantly endogenously derived antigens (i.e. from proteins synthesized on free ribosomes in the cytosol), whereas class II molecules present peptides encountered in the endocytic pathway (derived from extracellular sources or from ingested plasma membrane proteins).

4. A polymorphic gene is one that exists as two or more different alleles in a population at frequencies that are too high to be due to spontaneous mutation alone. This implies that the allelic variants are or have been beneficial to the organism and have been maintained in the population by natural selection. MHC polymorphism refers to the number of different alleles of genes coding for classical MHC molecules. At the protein level, the polymorphisms are confined mainly to the peptide-binding cleft of the MHC molecule.

5. The extensive polymorphism of MHC molecules in most outbred mammalian populations means that two unrelated individuals are unlikely to have the same combination of MHC alleles. This may contribute to differences in immune responsiveness between individuals through the role of these molecules in antigen presentation to T lymphocytes. The surface topographies of different allelic forms of a given MHC molecule are subtly different. It is likely that each allelic form is able to present a different spectrum of peptide antigens; therefore the particular combination of alleles an individual has may influence the quality of peptide presentation.

6. (i) Yes, because they differ at the MHC (are allogeneic). (ii) No, because they share the same MHC alleles (are syngeneic). However, among the 'background' genes will be differences between the two strains, such as other polymorphisms. When the processed peptides of these are presented by MHC molecules they will be recognized as foreign by T lymphocytes of the other strain. Thus despite sharing the same MHC alleles, the background differences (or so called 'minor' histocompatibility antigens) would lead to the development of a gradual response in the mixed lymphocyte reaction. (iii) Yes, because cells from the F1 hybrid express both $H\text{-}2^k$ and $H\text{-}2^d$ haplotypes.

7. (i) No (remember that 'normal' recognition by antigen-specific T lymphocytes is restricted by a 'self' MHC molecule). (ii) Yes. (iii) Yes. (iv) No.

8. To be recognized by CD8$^+$ cytotoxic T lymphocytes, peptides derived from the antigen must be presented by a class I MHC molecule. The antigen must therefore originate in the cytosol – as an 'endogenous' antigen – in order to enter the class I processing pathway (see Figure 6.14). Viruses express their genes in the cytosol of the infected cell and their gene products produced during infection therefore enter the class I processing pathway. Viruses in the extracellular phase of their life cycle may also gain entry to the class II MHC pathway, by being ingested as 'exogenous antigen' by antigen processing, cells such as macrophages, or B lymphocytes bearing capsid-specific immunoglobulins. This pathway is necessary for the activation of CD4$^+$ helper T lymphocytes, and in turn, for B lymphocytes to produce antibody. Hence many viruses elicit responses by both the humoral (antibody-mediated) and cellular (cytotoxic T lymphocyte) arms of the immune system. Inactivation of a virus prevents its ability to normally infect cells and generate endogenous antigens, while having little effect on its capacity to be ingested as an exogenous antigen and processed by antigen-presenting cells.

Chapter 7

Missing words

IgM[1]; Agnatha[2]; monomeric[3]; high[4]; polymeric[5]; low[6]; amphibians[7]; low[8]; mammals[9]; four (IgG, IgA, IgD, IgE)[10]; Agnatha[11]; larval[12]; early[13]; amniotes[14]; antibodies/immunoglobulins[15]; oviduct[16]; placenta[17]; antibodies/immunoglobulins[18]; mammary[19]; colostrum[20]; milk[21]; endothermic[22]; temperature[23]; quickly[24]; larger[25]; isotypes/classes[26]; complex[27]; histocompatibility/MHC/allo-[28]; rejection[29]; silence[30]; immunosuppressive[31].

Short answer questions

1. (a) False; a polymorphic MHC is seen in fishes, amphibians (*Xenopus*) and birds (chicken); (b) true; (c) true; (d) false.
2. (a) Exon duplication; (b) gene conversion; (c) exon shuffling; (d) gene duplication; (e) gene duplication.
3. (a) Probably IgM as this is present in all jawed vertebrates whereas IgG is found only in mammals; (b) probably IgY, as this is present in amphibians, reptiles and birds whereas IgG is only present in mammals; (c) probably C3 as it is present in modern Agnatha whereas C4 and the classical pathway is absent; (d) probably the MHC class I molecules

because they are present in all jawed vertebrates whereas the neonatal Fc receptor is found only in mammals.

4. Generally speaking, the immune response by endotherms is more rapid, they have large antigen receptor repertoires, they have more immunoglobulin isotypes and they have more complex secondary lymphoid tissues.

Chapter 8

Missing words

signal-transducing[1]; phosphorylation[2]; enzyme[3]; tyrosine kinases[4]; Lck[5]; Fyn[6]; CD3[7]; gamma (γ)[8]; delta (δ)[9]; epsilon (ε)[10]; zeta (ζ)[11]; eta (η)[12]; autophosphorylation[13]; Ig(α) (or CD79a)[14]; Ig(β) (or CD79b)[15]; Syk[16]; costimulatory[17]; anergy[18]; costimulatory[19]; B7.1 (or CD80)[20]; B7.2 (or CD86)[21]; CD28[22]; CTLA-4[23]; immunoglobulin[24]; CD40[25]; CD40L[26]; naïve[27]; CD40L[28].

Short answer questions

1. IL-12 helps to promote the response to a particular antigen in favour of a cell-mediated (cytotoxic T lymphocyte) response rather than an antibody-mediated (humoral) response. A cell-mediated response is a more appropriate response to eliminate virus-infected cells or tumour cells than a humoral response. IL-12 achieves this by promoting the differentiation of uncommitted helper T lymphocytes into T helper 1 (Th1) lymphocytes, which produce the cytokines required for the differentiation of cytotoxic T lymphocytes.

2. (a) IL-2; (b) IFN-γ; (c) IL-4; (d) IL-8 (e); TNF-α.

3. In general, two kinds. (a) Costimulatory signals, delivered to the CD40 molecule on the B lymphocyte by CD40L on the surface of activated helper T lymphocytes. (b) Cytokines, notably IL-4, IL-5, IL-6 and IL-10.

4. This is possible because B lymphocytes are antigen-presenting cells. These are able to take up antigen by receptor-mediated endocytosis and process the antigen in endosomes; peptide fragments of the antigen are then transported to the cell surface and displayed by class II MHC molecules to CD4$^+$ (helper) T lymphocytes. The costimulatory signals and cytokines required for full activation and differentiation into a plasma cell are solicited from helper T lymphocytes, activated previously by the same peptide antigen on a professional antigen-

presenting cell (dendritic cell or fully activated B lymphocyte). Cytokines are then released into the area enclosed by cellular contact between the helper cell and B lymphocyte. Together these precautions ensure help is delivered to the appropriate B lymphocytes.

Chapter 9

Missing words

immunologically privileged sites[1]; MHC[2]; embryonic[3]; neonatal[4]; immunogenic[5]; tolerogenic[6]; helper[7]; thymus[8]; thymocytes[9]; peptide–MHC[10]; positive[11]; apoptosis[12]; high[13]; apoptosis[14]; negative[15]; anergy[16]; suppression[17]; autoimmune[18]; anergy[19]; costimulatory[20].

Short answer questions

1. (a) A state of antigen-induced non-responsiveness of lymphocytes achieved by clonal anergy, suppression or deletion; (b) a state of antigen-induced paralysis of lymphocytes, which is often reversible; can be induced in T lymphocytes by stimulation through the antigen receptor in the absence of costimulation; (c) the developmental sequence of events that occur during the differentiation of thymocytes into mature T lymphocytes in the thymus; (d) the process by which thymocytes are stimulated to undergo maturation into mature T lymphocytes, thought to be stimulated through antigen receptors with an avidity for thymic peptide–MHC complexes that is *below* a certain threshold; (e) the process by which thymocytes are stimulated to undergo cell death by apoptosis, thought to be stimulated through antigen receptors with an avidity for thymic peptide–MHC complexes that is *above* a certain threshold; (f) tolerance to antigen induced by oral administration of the antigen; (g) cell death as a result of triggering an internal suicide program, often involving the expression of death-inducing genes; (h) cell death cause by physical damage to cells resulting in lysis and local inflammation.

2. (a) True. (b) False; tolerance can be acquired after this time if antigen enters the body in tolerogenic doses or by tolerogenic routes of entry. (c) False; nothing would happen. Anergy of T lymphocytes *can* be caused, however, by stimulation through the antigen receptor in the absence of costimulatory signals. (d) True. (e) True. (f) False. This is usually an immunogenic route of entry.

3. One explanation is that removal of the pituitary from the tadpole removes pituitary antigens before tolerance to them is established during early development. An animal lacking a pituitary thus reaches adulthood able to mount immune responses to pituitary antigens. The transplanted gland initially produces hormone (hence the transient darkening) but this is gradually rejected by the self-reactive T lymphocytes still present in the repertoire. Another explanation is that the gland has become 'contaminated' by foreign antigens from the foster frog in which it was held while the tadpole matured. (However, when *half* a gland was removed, kept under the skin of a foster animal as before, and transplanted back, the graft was accepted. This indicated that in the first experiment, the rejection was not due to contamination with antigen from the foster frog. More recent studies with frogs, in which the removal of other organs from tadpoles had little effect on their subsequent acceptance, suggests self-tolerance in frogs can also be established later.)

4. The likeliest explanation is because tolerance to the hen egg lysozyme by deletion of the lysozyme-specific T lymphocytes has been established in the thymus, in much the same way as it is to other self-antigens present in the embryonic/neonatal period. Because the number of T lymphocytes bearing lysozyme-specific receptors is too small to see if this had actually occurred, it would be necessary to substitute the normally diverse receptor repertoire with receptors consisting entirely of lysozyme-specific receptors. You could achieve this by making a transgenic mouse expressing such a receptor, which could then be crossed with the lysozyme-transgenic mouse. If your hypothesis is true, there should be no T lymphocytes (bearing the transgene) in the periphery of the F_1 mice.

5. The use of the insulin promoter confines the expression of the virus antigen mainly to the insulin-producing β cells of the pancreas. The delayed onset of expression of the viral gene in some animals is thought to have caused the antigen to appear *after* the embryonic/neonatal period during which tolerance by thymic deletion occurs. Consequently, the animals were not tolerant to the virus antigen, and mounted an immune response against their own β cells when the viral gene became active in these cells and the viral antigen appeared. This resulted in damage to the β cells and disruption of insulin synthesis which caused diabetes. In humans, a failure to establish tolerance to normal β-cell antigens might explain human autoimmune diabetes.

Glossary

adjuvant *n.* a substance, usually an inducer of mild inflammation (e.g. mineral oil, aluminium hydroxide, bacterial cell walls), injected with an antigen to enhance its immunogenicity.

affinity *n.* a measure of the strength of the interaction between two macromolecules, such as hormone and receptor or antibody and antigen.

affinity maturation the gradual increase in overall binding strength (affinity) of antibodies produced during the course of an immune response to antigen.

affinity purification *n.* the use of antibodies immobilized on a column of beads to capture antigen from a complex mixture.

Agnatha *n.* class of primitive vertebrates lacking a jaw; represented today by hagfish and lampreys.

albumins *n. plu.* soluble proteins of serum, milk, synovial fluid and other mammalian fluid secretions remaining after insoluble protein (**globulins**) have been precipitated.

alternative (or **differential**) **splicing** removal of the introns in different ways from the same primary RNA transcript (or pre-mRNA) to produce different translated polypeptides; the process by which mammalian and avian immunoglobulin genes produce membrane-bound or soluble H chains.

allele *n.* one of a number of alternative forms of a gene that can be found at a particular locus on a chromosome.

allelic exclusion a mechanism that allows only one of a pair of allelic forms of a gene to be expressed; although T and B lymphocytes rearrange both copies of a receptor gene, allelic exclusion ensures only one is expressed.

alloantiserum *n.* antiserum raised in one animal against the cells of an unrelated member of the same species; such sera contain antibodies to cell-surface antigens that differ between the two animals, especially alleles of genes in the major histocompatibility complex.

alloantigen *n.* molecules encoded in the major histocompatibility complex recognized by alloantisera.

allogeneic *a.* two genetically dissimilar individuals of the same species, such as would be found in an outbred population; allogeneic mammals reject exchanged tissue grafts (see **syngeneic**).

allograft *n.* a graft of cells, tissue or organs between two allogeneic individuals.

allorecognition *n.* recognition by T lymphocytes of the MHC molecules on an allogeneic individual's antigen-presenting cells; results in cell-mediated immune responses as manifested by allograft rejection *in vivo* or the mixed lymphocyte reaction *in vitro*.

amniotes *n. plu.* vertebrates whose embryos are surrounded by an amniotic membrane; reptiles, birds and mammals.

anamnestic response see **secondary response**.

anergy *n.* a state of antigen-induced paralysis of lymphocytes, sometimes reversible.

antagonism *n.* the counteracting effect of a cytokine, hormone, etc., on the effect of another.

antibody *n.* soluble form of immunoglobulin; membrane-bound immunoglobulin is the antigen-binding component of the B lymphocyte receptor complex.

antibody-dependent cell-mediated cytotoxicity (ADCC) a form of killing of foreign cells coated by bound antibody and mediated by cytotoxic cells bearing Fc receptors (neutrophils, NK cells, macrophages).

antigen *n.* a substance that is bound specifically by immunoglobulin or T lymphocyte (antigen) receptor.

antigen presentation the display of processed antigenic peptides at the surface of an antigen-presenting cell to the antigen receptors of T lymphocytes.

antigen processing intracellular events involving the enzymatic degradation of antigens and the formation and transport to the cell surface of complexes between processed peptides and class I or class II MHC molecules (see **endogenous antigen** and **exogenous antigen**).

antigen receptor cell-surface T lymphocyte receptor ($\alpha\beta$ or $\gamma\delta$) or immunoglobulin on B lymphocytes (IgM, IgD, IgG, IgE or IgA).

antigen specificity the limited recognition of a particular epitope, or group of structurally related epitopes, by a lymphocyte clone.

antigenic a substance that elicits a response by T or B lymphocytes (synonymous with **immunogenic**).

antigenic drift a gradual change in an antigen, typified by the haemagglutinin molecule of influenza virus, as a result of the accumulation of small genetic changes.

antigenic shift a sudden and substantial change in an antigen; typified by changes in influenza virus neuraminidase and haemagglutinin molecules arising from genetic recombination between different strains of the virus.

antigenic variation deliberate changes in surface antigens carried by some pathogens, especially African trypanosomes, to escape recognition by the host immune system.

Anura *n.* one of three orders of extant amphibians containing the frogs and toads (see **Apoda** and **Urodela**).

Apoda *n.* an order of limbless, burrowing amphibians (see **Anura** and **Urodela**).

apoptosis *n.* cell death as a result of triggering an internal 'suicide' program, often involving the expression of death-inducing genes; a normal event in most developmental processes (see **necrosis**).

autoantibody *n.* antibody that binds to self-antigen and usually a feature of autoimmune disease.

autoantigen *n* a self-antigen that is the target of self-reactive lymphocytes in autoimmune disease.

autocrine *a.* effects caused hormone, cytokine, etc., on the same cell from which it was released.

autoimmune disease cell or tissue damage caused by the body mounting immune responses to its own antigens.

autoimmunity *n.* the presence of self-reactive lymphocytes; may lead to autoimmune disease if the mechanisms to control their activation fail.

autophosphorylation *n.* self-phosphorylation of a protein with kinase activity; the first event in the signal transduction pathway of many receptors.

avidity *n.* the overall stability of the interaction between receptors and ligands; a combination of several influential factors, including affinity, number of receptors and ligands involved, and their spatial density.

axolotl *n.* an aquatic, newt-like salamander.

B complex *n.* the major histocompatibility complex of the chicken.

β₂-microglobulin *n.* the non-polymorphic light chain of class I MHC molecules, CD1 and the neonatal Fc receptor.

basophil *n.* class of granulocyte that is non-phagocytic; has **Fc receptors** for **IgE** which trigger degranulation and release of histamine and serotonin.

Bence-Jones protein free immunoglobulin light chains present in the urine of some patients with multiple myeloma; produced by a malignant clone of plasma cells (plasmacytoma).

blood coagulation (**clotting**) a response to haemorrhaging from damaged blood vessels in which blood is converted to a solid plug usually via an enzymatic cascade.

bone marrow connective tissue inside hollow bones; often contains haemopoietic tissue.

bursa of Fabricius *n.* a primary lymphoid tissue for B lymphocytes found only in birds; a pouch of lymphoid tissue connected to the dorsal surface of the cloaca.

bursectomy surgical operation in birds to remove the **bursa of Fabricius**.

cAMP cyclic AMP.

CD *abbrev.* cluster of differentiation; a nomenclature for phenotypic markers on leucocytes detected by specific monoclonal antibodies (see Appendix 1 for details).

CD40L CD40 ligand; expressed on activated T lymphocytes.

cell-mediated immunity *n.* originally used to describe immunity that could be adoptively transferred to irradiated recipient mice by cells and not antibodies; now used to describe responses by effector T lymphocytes (i.e. cell-mediated cytotoxicity or delayed-type hypersensitivity), and sometimes includes NK cells and macrophages.

cDNA library a set of complementary DNA (cDNA) clones derived by reverse transcription of different mRNAs in a cell.

chemokine *n.* a cytokine with chemotactic properties, e.g. IL-9 for neutrophils.

chemotaxis *n.* movement of motile cells to or from the source of a chemical stimulus.

Chordata *n.* a phylum of animals with a notochord, a dorsal hollow nerve cord and pharyngeal clefts at some point in their development; comprises the vertebrates, the cephalochordates and the urochordates (tunicates).

chromosome walking a technique for mapping genes from a collection of partially overlapping cloned restriction fragments.

class I MHC molecule vertebrate cell-surface heterodimers of polymorphic heavy (or α) chains encoded in the MHC and monomorphic β_2-microglobulin encoded elsewhere; function to present peptide antigens to $CD8^+$ T lymphocytes; expressed in varying levels on all nucleated cells.

class II MHC molecule vertebrate cell-surface heterodimers of polymorphic α and β chains encoded in the MHC; function to present peptide antigens to $CD4^+$ T lymphocytes; constitutively expressed only on specialized antigen-presenting cells (dendritic cells, macrophages and B lymphocytes).

class III MHC molecule heterogeneous proteins encoded in the MHC that are distinct from class I and II proteins, and include tumour necrosis factor and some complement components.

classical MHC molecule *n.* polymorphic class I or class II MHC molecules involved in antigen presentation to T lymphocytes (see **non-classical MHC gene**).

class switching change from one class (or isotype) of antibody to another by the rearrangement of C region genes.

clone *n.* a population of cells derived from a single cells; *v.* to produce a set of identical cells or DNA molecules from a single cell or copy.

clonal deletion the death, in response to stimulation by antigen, of lymphocytes with the corresponding antigen specificity.

clonal selection the proliferation in response to stimulation by antigen, of lymphocytes with the corresponding antigen specificity.

codominance *n.* a state in which both alleles of a heterozygous gene are expressed to give a phenotype that is different from that produced by either allele in a homozygous state.

coelom *n.* the body cavity of animals, lined with epithelium and usually containing the gonads and excretory organs; animals possessing a true coelom belong to the phyla Mollusca, Arthropoda, Annelida, Phoronida, Bryozoa, Brachiopoda, Echinodermata, Chaetognatha, Hemichordata and Chordata.

colony forming unit types of haemopoietic stem cells committed to differentiation along different lineages, so called for their ability to populate haemopoietic tissues, esp. the spleen, to form colonies.

colony stimulating factors a group of growth factors that stimulate the differentiation of different colony forming units.

colostrum *n.* a clear fluid secreted by mammary glands before milk is produced; contains a higher concentration of antibodies than milk, which are passively transferred to newborns immediately after birth.

combinational diversity *n.* diversity of antigen receptor V domains that arises from the different ways that single V, D and J gene segments can be combined (see **junctional diversity**).

commensalism *n.* a symbiotic relationship in which one partner (the commensal) benefits without benefit or detriment to the other (the host). See **mutualism** and **parasitism**.

complement *n.* a system of 30 or more serum and membrane-bound proteins connected in an enzymatic cascade, and triggered by foreign surfaces directly (**alternative pathway**) or via antibody binding to antigen (**classical pathway**). The cascade generates opsonins, anaphylatoxins, and chemotactic factors. A **lytic pathway** culminates in the self-assembly of membrane attack complexes.

complementarity-determining regions (CDR) hypervariable loops in the polypeptide chains of antigen receptor V regions that create the antigen-binding site.

congenic *a.* inbred strains of animals that are genetically identical except at a particular locus; produced by superimposing an allele or cluster of alleles from one inbred strain onto the genetic background of another strain by controlled inbreeding.

coreceptor *n.* CD4 and CD8 molecules of T lymphocytes that bind to conserved regions of class II and class I MHC molecules, respectively, and strengthen the interaction of the T lymphocyte antigen receptor with the MHC molecule; also bring tyrosine kinase into the receptor complex to trigger T lymphocyte activation.

cortex *n.* the outer zone of an organ or gland, such as the **thymus** or **lymph node**.

cosmid a cloning vector, combining the elements of a bacterial plasmid for cloning, with the *cos* sequences of λ phage to allow packaging of the DNA into phage particles *in vitro*.

costimulation a second signal (or 'signal 2') in addition to that received through the antigen receptor (or 'signal') required for full activation of lymphocytes; lymphocyte may be paralysed by antigen in the absence of costimulation (**anergy**).

crossing-over the process of exchange (or recombination) of genetic material during meiosis; caused by the breakage of homologous chromatids in the four chromatid stage and exchange of homologous regions of DNA.

Cyclostomata *n.* cyclostomes; the extant Agnatha, hagfish and lampreys (jawless vertebrates with a round mouth).

cytokine *n.* a short-range extracellular signalling molecule that influences the growth, differentiation or behaviour of cells.

cytopathic effect cell abnormality or death caused by viral infection.

cytoplasm *n.* the contents of a eukaryotic cell excluding the nucleus.

cytosol *n.* the cytoplasm of a cell excluding membrane-bound organelles (endosomes, endoplasmic reticulum, Golgi apparatus, mitochondria, etc.).

cytotoxic T lymphocyte (CTL) *n.* a subset of T lymphocytes that kill target cells, usually tumour cells or cells infected with viruses; classical CTL are $CD8^+$ and kill target cells bearing antigenic peptides in the context of class I MHC molecules; a minority of CTL are $CD4^+$ and restricted by class II MHC molecules.

cytotoxic *n.* an action that destroys cells.

delayed-type hypersensitivity (DTH) a reaction by $CD4^+$ T lymphocytes in response to stimulation by antigen in the cutaneous epithelium; manifested as an inflammatory reaction that peaks 2–3 days after application of the antigen to the skin.

dendritic cell *n.* antigen-presenting cells of haemopoietic origin with potent capacity to activate naïve T lymphocytes; these cells are circulatory and are found in the blood, skin (**Langerhans cell**), lymph (**veiled cell**) and secondary lymphoid tissues (**interdigitating dendritic cell**).

deuterostomes *n. plu.* animals with a true coelom in which, during radial cleavage of the egg, the blastopore becomes the anus; includes the phyla Pogonophora, Hemichordata, Echinodermata, Urochordata and Chordata (see **protostomes**).

diapedesis see **extravasation**.

differentiation *n.* development of cells with specialized functions from unspecialized precursor or stem cells.

dimer *n.* a protein made of two polypeptide chains (or subunits) (see **homodimer** and **heterodimer**).

disease *n.* illness or a state of reduced competence to perform normal activities, usually manifested by characteristic symptoms.

disulphide bond S–S bond formed between two cysteines within a polypeptide or between two polypeptides, important in maintaining three-dimensional structure.

dizygotic *a.* arising from two separately fertilized eggs (e.g. non-identical twins).

domain *n.* a structurally defined section of a protein, often compact and globular.

ectotherm *n.* an animal whose source of body heat content is mainly from the external environment, e.g. fish, amphibians and reptiles (see **endotherm**).

endotherm *n.* an animal that derives all or most of its heat content from endogenous metabolic heat rather than from the external environment (see **exotherm**).

electrophoresis *n.* a technique for the separation of proteins or fragments of DNA or RNA on the basis of charge or molecular mass by migration through a gel matrix in an electric field.

endocytosis *n.* the process by which eukaryotic cells ingest extracellular components by the invagination and pinching off of the plasma membrane to form an endocytic vesicle (**endosome**); can be divided according to the size of the endosome into **phagocytosis** and **pinocytosis**.

eosinophil *n.* a granulocyte that participates in antibody-dependent cell-mediated cytotoxicity (ADCC); thought to be important in immunity to helminths.

elasmobranchs *n. plu.* cartilaginous fish including sharks, skates and rays.

embryonal stem cell *n.* stem cell of (mouse) embryo that can be cultured *in vitro* and will develop into an embryo when implanted into a surrogate mother.

encapsulation *n.* a defence reaction of many invertebrates in which macroscopic foreign bodies gaining entry to the coelom are enclosed and sequestered in a multicellular sheath of haemocytes.

endocrine *a.* applied to a system of ductless glands and the hormones they produce.

endogenous antigen *n.* an antigen that originates from within a eukaryotic cell by translation of self or foreign (e.g. viral) RNA in the cytosol; peptides derived from the processing of endogenous antigens are normally presented at the cell surface by class I MHC molecules.

endosome (**endocytic vesicle**) *n.* an intracellular vesicle produced by invagination of the plasma membrane (see **endocytosis**).

epitope *n.* the part of an antigen recognized by an antigen receptor; sometimes also called an **antigenic determinant**.

Epstein–Barr virus (**EBV**) a herpes virus that causes infectious mononucleosis (glandular fever) in humans, and is implicated in the development of several malignancies of B lymphocytes; can immortalize B lymphocytes *in vitro*.

eukaryotes *n. plu.* organisms whose cells possess membrane-bound organelles (such as mitochondria and the Golgi apparatus, a cytoskeleton and a nuclear membrane surrounding the chromosomes); comprise the protozoans, slime moulds, fungi, algae, plants and animals (see prokaryotes).

Eutheria *n.* placental mammals; mammals other than the monotremes and marsupials.

exocrine *n.* glands that secrete fluids into ducts.

exogenous antigen *n.* an antigen that is ingested into endosomes; originate extracellularly, or may be derived from cellular membrane-bound proteins; peptides derived from the processing of exogenous antigens are normally presented at the cell surface by class II MHC molecules.

exon *n.* block of DNA that forms part of the amino acid-encoding part of a eukaryotic gene, and separated from the next exon by a non-coding sequence of DNA called an **intron**.

exon shuffling the evolutionary process of creating novel genes by the recombination of exons from pre-existing genes.

extracellular *a.* the environment outside of a cell.

extravasation *n.* the passage of leucocytes through a blood vessel wall into surrounding tissues by migration between the vascular endothelial cells (also known as **diapedesis** or **transendothelial migration**).

F_1 **hybrid** first filial (son or daughter) hybrids arising from a first cross, with subsequent generation denoted by F_2, F_3, etc.

Fab fragment an antigen-binding fragment of immunoglobulin consisting of one light chain and a similarly sized piece of the heavy chain; produced by mild digestion with **papain**.

Fc fragment a crystallizable fragment of immunoglobulin consisting of the C-terminal halves of both heavy chains; produced by mild digestion with **papain**.

Fc receptor (FcR) *n.* cell-surface receptor that binds to the Fc region of immunoglobulin.

flow cytometer *n.* a machine for quantitating the level of fluorescence on individual cells in suspension stained by immunofluorescence (see **fluorescence activated cell sorter**).

fluorescence activated cell sorter (FACS) a flow cytometer adapted for separating cells into subpopulations according to fluorescence levels.

follicle see lymphoid follice.

follicular dendritic cell *n.* antigen-presenting cell that resides in lymphoid follices in secondary lymphoid tissue; has characteristic long processes that retain antigen.

frameshift mutation deletion or insertion of nucleotides into a coding sequence of DNA that causes a change in the reading frame.

GDP guanosine diphosphate.

gene conversion a phenomenon caused by the repair of **heteroduplex DNA** in which the incorrect strand is used as the synthesis template, leading to an apparent 'conversion' of one allele to another.

gene family set of homologous genes created by duplication and divergence of a single ancestral gene; some families with large numbers are superfamilies (see **immunoglobulin gene superfamily**).

genetic drift *n.* random changes in the frequency of alleles in a breeding population by means other than natural selection; usually more pronounced in small populations.

germline *n.* gametes (sperm and ova) or the cells that give rise to them.

germinal centre a focus of actively proliferating and differentiating B lymphocytes in the centre of follicles in lymph nodes and spleen.

globulin *n.* the serum protein fraction that is precipitated by the addition of salt.

Gnathostomata *n.* jawed vertebrates; all vertebrates other than the Agnatha (lampreys and hagfish).

G-proteins *n. plu.* subset of guanine nucleotide-binding regulatory proteins, found at the cytoplasmic face of eukaryotic cell membranes involved in signalling.

GTP guanosine diphosphate.

granulocytes *n. plu.* a class of vertebrate leucocytes with distinctive cytoplasmic granules (see **neutrophils, basophils** and **eosinophils**).

granular cells *n.* a class of invertebrate (esp. insect) leucocytes with distinctive cytoplasmic granules.

gut associated lymphoid tissue (**GALT**) a system of secondary lymphoid tissues draining the gastrointestinal tract of mammals, consisting of Peyer's patches, the appendix, and the diffuse lymphoid tissues.

H-2 *n.* the major histocompatibility complex of the mouse.

haemagglutinin *n.* a substance that agglutinates erythrocytes; an envelope protein of influenza virus.

haemocyte *n.* a blood cell of insects and other invertebrates.

haemolymph *n.* a fluid in the coelom of some invertebrates containing respiratory pigment (e.g. haemocyanin), haemocytes, etc.

haemopoiesis *n.* the formation of blood and red and white blood cells from stem cells.

haemopoietic stem cell cell in haemopoietic tissue with capacity for self-renewal and to differentiate into any of the mature blood cells.

haplotype *n.* the particular combination of alleles in a linked group, such as the major histocompatibility complex, present on one chromosome of a homologous pair.

helminth *n.* a parasitic flatworm (phylum Platyhelminthes: flukes and tapeworms) or roundworm (phylum Nematoda).

helper T lymphocyte *n.* a subset of T lymphocytes that elaborate cytokines and provide costimulatory signals for B lymphocytes; helper T lymphocytes are CD4$^+$ and response to antigenic peptides in the context of class II MHC molecules on antigen-presenting cells.

heterodimer *n.* a protein consisting of two dissimilar polypeptide chains.

heteroduplex (or **hybrid**) **DNA** *n.* double-stranded DNA (duplex) in which the strands are not complementary; may be caused by crossovers and branch elongation during meiosis; repair of heteroduplex DNA is a source of **gene conversion**.

heterogeneous *a.* consisting of dissimilar parts.

heterologous recombination recombination between dissimilar sequences of DNA .

heterophilic *a.* the binding between two dissimilar molecules (e.g. cell-surface adhesion molecules).

heterozygote advantage a situation in which an individual that is heterozygous for a given gene has an increased fitness (relative capacity to pass its genes to the next generation) over a homozygote.

heterozygous *a.* a gene in which the maternally and paternally derived alleles are different.

high endothelial venule an area of capillary venule consisting of plump endothelial cells where leucocytes can pass through into the tissues (see **extravasation**).

histochemistry the staining of cells and tissues, particularly for examination by microscopy.

HLA (human leucocyte antigen) *n.* the human major histocompatibility complex.

homing receptor receptor on lymphocytes that contributes to determining into which tissues a lymphocyte enters from the blood circulation.

homodimer *n.* a protein consisting of two identical polypeptide chains.

homology *n.* a similarity between DNA sequences that originates from common ancestry. Can also refer to similarities between protein structures. Two such DNA or protein sequences are **homologous**.

homologous recombination recombination between similar sequences of DNA; a phenomenon that is exploited for targeted gene disruption.

homophilic *a.* the binding between two similar molecules (e.g. cell-surface adhesion molecules). See **heterophilic**.

homozygous *a.* a gene in which the paternal and maternal alleles are the same.

humoral immunity *n.* originally used to describe immunity that could be adoptively transferred to irradiated recipient mice by antibodies rather than by cells; antibody-mediated immunity.

hybridoma *n.* a **clone** of hybrid cells formed by the fusion of a plasma cell and a myeloma cell (bone marrow-derived cancer cell); used in the production of **monoclonal antibodies**.

hypervariable region one of three loops in the variable domain of each chain of an immunoglobulin or T lymphocyte receptor that form the antigen-combining site of the receptor.

IgA immunoglobulin A; the predominant antibody found in extravascular secretions in mammals (mucus, tears, milk, colostrum); exists as monomer or dimer.

IgD immunoglobulin D; mammalian B lymphocyte receptor with IgM.

IgE immunoglobulin E; mammalian antibody involved in inflammatory responses via Fc receptors on mast cells and basophils.

IgG immunoglobulin G; the predominant serum antibody produced in secondary responses in mammals.

IgM immunoglobulin M; found in all jawed vertebrates; exists as monomer in membrane of B lymphocytes; secreted forms polymeric (pentameric in mammals).

IgX immunoglobulin X; an isotype found in some amphibians.

IgY immunoglobulin Y; the predominant antibody found in birds, reptiles and amphibians.

IgY(ΔFc) lower molecular weight species of IgY lacking the third and fourth C domains that predominates in secondary responses.

immune complex *n.* aggregations of antigen and antibody.

immune system *n.* the cells and tissues that collectively recognize and eliminate invading microorganisms, parasites, tumour cells, etc., from the bodies of vertebrates and invertebrates.

immunogen an antigen capable of eliciting an immune response, as opposed to a **tolerogen**.

immunoglobulin (**Ig**) *n.* antigen receptor of B lymphocytes; forms the antigen-binding component of the **B lymphocyte receptor complex** and **antibody** when released as soluble molecules.

immunoglobulin fold *n.* a globular **domain** structure of approximately 110 amino acids in length folded into two β-pleated sheets of three or four β-strands each held together by an intradomain disulphide bond.

immunoglobulin gene superfamily a **gene family** whose members all possess one or more **immunoglobulin folds**.

immunology *n.* the study of the immune system of vertebrates and invertebrates.

immunological silence non-responsiveness by the immune system owing to insufficiently high levels of antigen or antigen in the absence of essential costimulatory signals; often used by cells as an evasion strategy.

immunologically privileged site an anatomical site in the body that is inaccessible to the recognition systems of the immune system.

inbreeding *n.* mating between related individuals.

infection *n.* invasion of a host by microorganisms and subsequent multiplication; infection may or may not cause disease (see **latent infection**).

inflammation *n.* localized response to tissue injury and invading microorganisms, characterized by swelling, heat, pain and redness; features increased vascular permeability, infiltration by phagocytes and lymphocytes, removal of foreign cells and debris and tissue repair.

integrin *n.* a member of a family of cell-surface adhesion molecules that bind mainly to extracellular matrix.

interdigitating dendritic cell antigen-presenting cell of lymphoid tissue that present antigen to T lymphocytes.

in vitro experiments conducted with living cells outside the organism.

in vivo within the living organism.

invariant chain *n.* a third chain of class II MHC molecules that promotes correct assembly of the α and β chains in the endoplasmic reticulum (ER), prevents binding of peptides while in the ER, and serves to direct exported class II molecules to the endocytic pathway.

involution *n.* a reduction in size or functional activity, such as with age.

isoforms *n. plu.* different forms of a protein encoded in the same gene produced by alternative RNA splicing.

isotypes *n. plu.* different classes of immunoglobulin determined by the particular heavy chain constant regions expressed.

J chain *n.* a polypeptide chain that joins IgA or IgM into multimers.

junctional diversity *n.* diversity of antigen receptor V domains that arises at the junction of V(D)J gene segments, in part as a result of the number of D segments but also because of the loss or addition of variable numbers of nucleotides at the junction(s). See also **combinational diversity**.

kinase *n.* an enzyme that transfers phosphate to (phosphorylates) its substrate.

latent infection (latency) a virus infection that has ceased replication and is in an apparent dormant state that precedes reactivation.

lectin *n.* proteins found in plants and animals that bind to sugar residues and can agglutinate cells; several animal lectins perform innate recognition of microbes.

leucocyte, leukocyte *n.* white blood cell; in mammals comprise **granulocytes** and **mononuclear cells**.

ligand *n.* the complementary structure bound by a receptor (also called a **counter-receptor**).

Limulus a genus of horseshoe crab (more closely related to arachnids than crustaceans).

linkage *n.* the tendency of two or more parental alleles to be inherited together because of their localization on the same chromosome.

linkage map *n.* a map of the relative positions of genetic loci on the same chromosome; produced by analysing recombination frequencies (the frequency with which linked alleles are segregated into different gametes owing to crossing over).

lipopolysaccharide (LPS) *n.* one of the components of Gram-negative bacteria.

lymph *n.* the clear fluid that collects between tissue cells and which is drained away by a network of vessels called the **lymphatic system**.

lymph node *n.* a small secondary lymphoid organ that collects antigens drained by the lymph; contains antigen-presenting cells and trafficking lymphocytes.

lymphatic system (see **lymph**).

lymph hearts *n.* small contractile areas in the lymphatic systems of non-mammalian vertebrates that push lymph in one direction.

lymphocyte *n.* a mononuclear leucocyte bearing antigen receptors encoded in rearranging genes (see **helper T lymphocyte, cytotoxic T lymphocyte** and **B lymphocyte**).

lymphoid *a.* tissues or cells pertaining to lymphocytes.

lymphoid follicle *n.* a cluster of antigen-presenting cells and lymphocytes within secondary lymphoid tissue where antigen-stimulated B lymphocyte activation occurs (**primary lymphoid follicle** and **secondary lymphoid follicle**).

lymphoid tissue *n.* tissues in which immature lymphocytes develop (**primary lymphoid tissue**) or in which mature lymphocytes undergo antigen-dependent activation (**secondary lymphoid tissue**).

lysate *n.* the mixture of soluble and insoluble material left when cells are disrupted.

lysis *n.* bursting or disruption of cells.

lytic *a.* a toxin, virus or other agent that causes cell lysis (haemolytic, cytolytic).

lysosome *n.* a membrane-bound organelle of eukaryotic cells that contains digestive enzymes at acidic pH; the endpoint of the endocytic pathway in which extracellular material is ingested and broken down.

lysosomotropic agent *n.* substances (chloroquine, etc.) that raise the pH of lysosomes and other compartments of the endocytic pathway; used in experiments to block processing of exogenous antigens.

lysozyme *n.* a ubiquitous enzyme found in animal secretions that breaks the glycosidic bond within **peptidoglycan** in bacterial cell walls.

macrophage *n.* a large phagocytic mononuclear leucocyte found in the tissues of vertebrates.

major histocompatibility complex (**MHC**) a cluster of genes encoding **class I MHC molecules** and **class II MHC molecules**, some **complement** components and other proteins.

marginal zone area of the **spleen** between the red and white pulp that is rich in lymphoid follicles.

medulla *n.* the central part of an organ or tissue such as the **thymus** or **lymph node**.

memory *n.* in immunological parlance, the enhanced immune reactivity mounted against an antigen encountered previously (see **secondary response**).

memory cells *n.* differentiated T and B lymphocytes that remain after antigen-dependent clonal selection and which maintain immunological **memory**.

Mendelian inheritance inheritance of traits, now known to be mediated by nuclear genes, according to the laws defined by Gregor Mendel.

Metazoa *n.* multicellular animals.

MHC *abbrev.* major histocompatibility complex.

MHC restriction the dependence by T lymphocytes on MHC molecules for the recognition of antigen.

microbiology *n.* the study of **microorganisms**.

microlymphocytotoxicity assay *n.* an assay used in the typing of HLA antigens.

microorganism *n.* an organism too small to be seen clearly with the naked eye; often used as a collective term for organisms that colonize host organisms and which may or may not be pathogenic (viruses, bacteria, fungi, protozoa and worms).

minor histocompatibility antigen *n.* antigen encoded outside the MHC that causes the rejection of **syngeneic** tissue grafts.

mixed lymphocyte reaction (MLR) the activation of T lymphocytes by **allogeneic** MHC molecules *in vitro*.

monoclonal antibody *n.* antibodies produced by a single clone of B lymphocytes and having the same antigen specificity; produced in the laboratory by the immortalization and cloning of plasma cells.

monocyte *n.* a leucocyte in the blood that is the precursor of tissue macrophages.

mononuclear leucocyte *n.* leucocytes other than **granulocytes** which comprise the lymphocytes and monocytes.

morphology *n.* the physical shape or form of an organism; also applied to cells.

multiparous *a.* birth of multiple offspring.

mutation *n.* a change in the sequence of DNA caused by errors in DNA replication or chemicals, radiation, etc. (see **somatic mutation**).

mutualism *n.* a type of symbiosis in which both partners gain from the relationship (see **commensalism** and **parasitism**).

myeloid pertaining to the bone marrow; tissue in which haemopoiesis occurs in vertebrates; cells of haemopoietic origin other than lymphoid cells.

myeloma *n.* a cancer of bone marrow cells; also used to describe the disease cause by a malignant clone of plasma cells (**plasmacytoma**) that produce homogeneous immunoglobulin.

naïve lymphocyte a mature lymphocyte that has yet to encounter and be activated by specific antigen (also known as 'unprimed' or 'virgin' lymphocytes).

nascent *n.* newly synthesized nucleic acid or protein.

natural killer (NK) cell *n.* a type of lymphocyte that lacks antigen receptors and displays cell-mediated cytotoxicity (used to be called null cells).

necrosis cell death caused by external trauma (poison, low oxygen, membrane disruption, etc.) causing rupture and local inflammation (see **apoptosis**).

negative selection the process by which thymocytes undergo apoptosis in the thymus; thought to be triggered by antigen receptors with an avidity for thymic peptide–MHC complexes that is *above* a certain threshold (see **positive selection**).

neoplasm *n.* new growth, as in a tumour.

neutrophil *n.* a class of **granulocyte** in vertebrates.

nucleosome *n.* a repeating structure in the chromosomes of eukaryotic cells caused by the winding of DNA around proteins called histones.

oligonucleotide *n.* a short, synthetic DNA molecule used as a probe or a primer of DNA synthesis in the laboratory.

oncogene *n.* a mutated cellular gene, or its viral homologue, that contributes to the loss of control over cellular proliferation and development of cancer; many of these genes normally encode intracellular signalling components.

oncogenic *n*. capable of causing cancer, such as an oncogenic virus.

ontogeny *n*. the history of the development of a cell or organism.

opsonin *n*. a substance that attaches to microorganisms and promotes their phagocytosis by cells bearing receptors for the opsonin.

opsonization *n*. the process in which a microorganism or antigen is coated with opsonins prior to ingestion by phagocytic cells.

organelles *n. plu.* structures found in the cytoplasm of eukaryotic cells in which specialized processes occur.

ovalbumin *n*. main protein component of hen egg white.

papain *n*. an endopeptidase.

parabiosis *n*. joined at birth.

paracrine *n*. actions of extracellular signalling molecules (cytokines, etc.) over short distances.

parasite *n*. an organism that lives on or in another (host) organism and which benefits from the relationship at the expense of the host.

parasitism *n*. a kind of symbiosis in which one partner (the parasite) lives at the expense of the other (the host). See **commensalism** and **mutualism**.

parasitoid *n*. Hymenopteran wasps that are partly free-living and partly parasitic. Adults lay their eggs on or in the larvae of other (host) insects which are consumed by the parasitoid larvae.

parasitology *n*. the study of parasites (especially parasitic protozoa and helminths).

passive immunity immunity acquired by the transfer of preformed antibodies, as in maternal antibodies passed on to newborn mammals in milk.

pathogen *n*. a microorganism that causes disease in its host.

pathogenicity *n*. the relative ability of different species of pathogens to cause disease.

pathology *n*. the science of disease and disease symptoms; cell and tissue damage caused by the disease process.

PCR (polymerase chain reaction) a technique that replicates a specific sequence of DNA .

peptidoglycan *n.* a gel-like polymer in the cell walls of bacteria made from chains of alternating NAM (*N*-acetylmuramic acid) and NAG (*N*-acetylglucosamine) subunits cross-linked by short, peptide side chains. **Lysozyme** digests peptidoglycan by the cleaving the bond connecting NAM and NAG.

perforin *n.* a protein released by **cytotoxic T lymphocytes** and NK cells that polymerizes with other perforin molecules to produce pores in the membranes of target cells.

periarteriolar lymphoid sheath the T lymphocyte-rich **white pulp** of the **spleen** that surrounds the arteries.

Peyer's patch *n.* a cluster of lymphoid follicles that are found in the inner lining of the small intestine in mammals; secondary lymphoid tissue where antigens from the contents of the digestive tract collect.

phagocyte *n.* a cell capable of **phagocytosis**.

phagocytosis *n.* a type of **endocytosis** that refers to the uptake of large particles (microorganisms, other cells, etc.) into a cell.

pharynx *n.* in the vertebrates the tube behind the mouth shared by the alimentary and respiratory tubes; the throat.

phenotype *n.* a collection of visible or measurable characteristics of a cell or organism.

phosphatase *n.* an enzyme that removes phosphate from its substrate.

phosphorylation *n.* the addition of phosphate to a protein or nucleic acid performed by a **kinase**; phosphorylation of a protein often alters its properties.

photoperiod *n.* duration of exposure to daylight.

phylogeny *n.* the evolutionary line of descent of a species.

phylogenetic tree a branching diagram showing the evolutionary history of a group of related species.

plasma *n.* the liquid part of body fluids (blood, lymph, milk).

plasma cell *n.* a fully differentiated B lymphocyte engaging in antibody synthesis.

plasmacytoma *n.* a cancer of plasma cells.

Plasmodium *n.* a genus of parasitic protozoa, including *P. falciparum*, the causative agent of malaria in humans.

platelet *n.* the **thrombocytes** of mammals; non-nucleated cells derived from the fragmentation of megakarocytes that are involved in blood clotting.

pleiotropy *n.* the multiple effects of a single protein or gene.

pluripotent stem cell a stem cell able to give rise to the greatest number of different 'committed' **stem cells**.

poly-Ig receptor receptor on the vascular epithelial cells captures polymeric IgA prior to **transcytosis**.

polymorphism *n.* a gene that exists as two or more different alleles that are present in a population at frequencies too high to be due to spontaneous mutation alone.

Porifera *n.* the phylum containing the sponges.

positive selection the process by which thymocytes are stimulated to differentiate into mature T cells in the thymus; thought to be triggered by antigen receptors with an avidity for thymic peptide–MHC complexes that is *below* a certain threshold (see **negative selection**).

primary lymphoid follice an undifferentiated follicle of B lymphocytes and accessory cells within secondary lymphoid tissue, prior to B lymphocyte activation and development into a **secondary lymphoid follicle**.

primary lymphoid tissue *n.* anatomical site where antigen receptor gene rearrangement occurs (see also **thymus**, **bone marrow** and **bursa of Fabricius**).

professional antigen-presenting cell antigen-presenting cells that can activate naïve T lymphocytes; constitutively express class I and class II MHC molecules, and T lymphocyte costimulatory molecules; especially cells of the dendritic cell family and activated B lymphocytes.

prokaryotes *n. plu.* the bacteria; unicellular organisms lacking a membrane-bound nucleus or membrane-bound organelles (see **eukaryotes**).

proteasome *n.* a large, multisubunit complex of proteinases found in the cytoplasm of eukaryotic cells.

protostomes *n. plu.* invertebrates with a true coelom in which, during spiral cleavage of the egg, the blastopore becomes the mouth; includes the phyla Mollusca, Annelida, Arthropoda, Phoronida, Bryozoa and Brachiopoda (see deuterostomes).

pseudogene *n.* a non-functional gene derived from a once functional gene by the accumulation of mutations.

RAG-1 and **-2** (**recombinase associated genes**) genes whose products are required for rearrangement of T and B lymphocyte antigen receptor genes; so far found only in vertebrates.

rearrangement *n.* a type of DNA **recombination** seen in antigen receptor genes.

receptor-mediated endocytosis uptake by cells of extracellular substances which first bind to receptors on the plasma membrane.

recombination *n.* any exchange or integration of one DNA molecule into another.

red pulp *n.* area of the spleen rich in erythrocytes and macrophages where effete erythrocytes are destroyed.

repertoire *n.* the combined assortment of antigen receptors present in the peripheral pool of mature T and/or B lymphocytes.

reticulocyte *n.* a non-nucleated cell containing few organelles which is the penultimate stage in the differentiation pathway of erythrocytes.

retrovirus *n.* type of virus with a genome of RNA which is reversed transcribed into cDNA in the infected cell prior to integration into the host cell genome.

RFLP (**restriction fragment length polymorphism**) genetic polymorphism as revealed by the sizes of fragments generated with a particular restriction endonuclease.

rhinovirus *n.* common cold viruses.

secondary antibody *n.* an antibody that is coupled to a fluorescent dye, radioisotope, or chromogenic enzyme, which used in immunoassays to visualize the primary antibody bound to antigen.

secondary lymphoid follice an active lymphoid follicle with a **germinal centre** in which activated B lymphocytes are proliferating and differentiating into plasma cells.

secondary lymphoid tissue *n.* tissue in which lymphocytes encounter antigen and undergo antigen-dependent activation; **lymph node**, **spleen**, **Peyer's patch**, tonsil.

secondary response a rapid humoral or cell-mediated response mediated by memory B or T lymphocytes, that is stimulated upon the second exposure to a particular antigen (also known as **memory** or **anamnestic** responses).

secretory component *n.* a fragment of the **poly-Ig receptor** that remains with secreted IgA and IgM.

selectins *n. plu.* a family of carbohydrate-binding cell-surface adhesion molecules.

serological typing the identification of histocompatibility antigens (class I and II MHC molecules) on cells by using antibodies.

serum *n.* the cell-free fluid component of blood that remains after blood has been allowed to clot.

signal transduction the transmission of an extracellular signal, initiated by the interaction of a cell-surface receptor with its ligand, to the interior of the cell.

somatic cell *n.* any cell in the body other than a germline cell.

somatic cell hybrid a hybrid cell formed by the fusion of cells, often from different species of animals.

somatic mutation mutation of DNA in somatic cells and not inherited.

spleen *n.* a multifunctional organ in vertebrates that is the site of haemopoiesis (blood cell production), and removal of effete erythrocytes, and acts as a secondary lymphoid tissue for collecting blood-borne antigens.

splice *v.* the processing of pre-mRNA by the removal of introns and the joining of the exons into a contiguous sequence.

Srk kinases *n. plu.* a family of tyrosine kinases involved in signalling that are related to the prototype, Src. Other members include Fyn, Lyn, Lck, Blk, Hck, Fgr, Yrk and Yrs.

stem cell *n.* a cell with the capacity for constant self-renewal and to undergo differentiation.

symbiosis *n.* the living together in close association of two dissimilar organisms (symbionts), one or both of which are dependent on the relationship for survival (see **commensalism**, **mutualism** and **parasitism**).

synergic *a.* acting together (such as two cytokines) where the combined effect is often greater than the sum of the individual parts.

syngeneic *a.* two genetically identical individuals of the same species, such as would be found in an inbred strain of mice; syngeneic mammals accept exchanged tissue grafts because of identity at the MHC (see **allogeneic**).

synonymous mutation *n.* a mutation that does not affect the sequence of amino acids encoded.

target cell *n.* the cell lysed by cytotoxic T lymphocytes, NK cells, etc.

Teleostei *n.* a group of modern bony fishes (carp, salmon, zebrafish, etc.).

tetrapods *n. plu.* vertebrates with four limbs (includes birds).

Th1 *n.* a subset of CD4$^+$ helper T lymphocytes that characteristically produce IL-2 and IFN-γ, and which are important for the development of cell-mediated immunity.

Th2 *n.* a subset of CD4$^+$ helper T lymphocytes that characteristically produce IL-4, IL-5, etc., and which are important for the development of humoral immunity.

thioester *n.* (or thiol ester) an ester with sulphur instead of oxygen.

thrombocyte *n.* the non-mammalian platelet, involved in blood clotting.

thymectomy *n.* surgical operation to remove the thymus.

thymocyte *n.* an immature T lymphocyte within the thymus.

thymus *n.* a primary lymphoid tissue that hosts the maturation of T lymphocytes from lymphoid stem cells.

tolerance *n.* a state of antigen-induced non-responsiveness of lymphocytes achieved by **clonal deletion**, suppression or **anergy**.

tolerogen *n.* an antigen capable of eliciting **tolerance** (as opposed to an **immunogen**).

transcytosis *n.* the transport of material through a layer of epithelial cells by endocytosis.

transfection *n.* the introduction of foreign DNA into eukaryotic cells *in vitro*, which may become stably integrated into the cell genome.

transgene *n.* a foreign gene introduced into the germline of a plant or animal by recombinant DNA technology.

transgenic *a.* a plant or animal that has been genetically engineered to contain a stable transgene.

transposable elements (or **transposons**) *n.* a sequence of DNA that moves by translocational or replicative means to new locations within eukaryotic or prokaryotic genomes.

trophoblast *n.* the outer layer of cells surrounding the embryo.

Trypanosoma a genus containing parasitic flagellate protozoans transmitted by blood-sucking flies, including *T. cruzi* (American trypanosomiasis or Chagas' disease) and *T. brucei* (African trypanosomiasis or sleeping sickness).

tumour suppressor genes genes that normally play a role in preventing a cell from proliferating but which, when they become inactivated by mutation or by the activities of certain viruses, become an important component in the pathway leading to uncontrolled cellular proliferation and cancer.

tunicates *n. plu.* sub-phylum of chordates (urochordates) consisting of sea squirts.

typhlosole *n.* an invagination of the stomach of larval lampreys that contains haemopoietic and lymphoid cells, together called a protospleen.

ubiquitin *n.* a protein found in prokaryotic and eukaryotic cells that is attached to abnormal proteins and which targets them for degradation by proteinases.

Urodela *n.* one of three orders of extant amphibians containing the newts and salamanders (see **Anura** and **Apoda**).

vaccine *n.* a preparation, usually derived from an infectious pathogen, administered to provide protective immunity without causing disease.

vascular addressins *n. plu.* a family of adhesion molecules on vascular endothelial cells that regulate the homing of lymphocytes to specific tissues in the body.

vascular endothelium *n.* a layer of endothelial cells that form the blood vessels.

vasodilation enlargement of blood vessels.

virulence *n.* the relative pathogenicity of different strains of the same species of pathogenic microorganism.

virus *n.* small, infectious particles that are totally dependent on a living host cell for replication; viruses consist of a genome of DNA or RNA (not both) surrounded by a protective protein shell (capsid) and sometimes surrounded by a lipid envelope; viruses gain entry by attaching to specific membrane proteins on the host cell.

vital dye *n.* a dye that is excluded by living cells but not by dead cells. Trypan blue is widely used in light microscopy, whereas propidium iodide (which is fluorescent) is widely used in flow cytometry.

viviparous *a.* the condition of giving birth to live young; all mammals except monotremes (lay eggs) and a number of exceptions in other classes of vertebrates.

Western blot the use of antibody to detect specific antigen among complex mixtures by first resolving the proteins according to molecular weight by polyacrylamide gel electrophoresis, followed by transfer ('blotting') to a nitrocellulose membrane; primary antibody is used to detect the antigen band and visualized using a secondary antibody conjugated to a chromogenic enzyme.

white pulp *n.* lymphoid tissue surrounding the arterioles in the spleen.

xenoantiserum *n.* antiserum raised in one species of animal against the cells from another species (e.g. in mice against human cells).

xenogeneic *a.* two genetically dissimilar individuals of different species.

Xenopus *n.* a genus of anuran amphibians, including *X. laevis* (the African clawed toad).

ZAP 70 an acronym for zeta associated protein (70 kDa); a tyrosine kinase part of the T lymphocyte receptor complex related to Syk in B lymphocytes.

Appendix A: CD molecules

Phenotyping of human leucocytes is based on the presence or absence of particular cell-surface molecules that are identified using specific antibodies. These cell-surface molecules are also known as **differentiation antigens** or **markers** because their expression can be used to identify cells of different lineages or cells in the same lineage at different stages of maturation or differentiation. For example, T and B lymphocytes can be distinguished by T lymphocyte-specific markers CD3 or Thy-1 (CD90), or by the B lymphocyte-specific markers CD19 or CD20. Mature T lymphocytes are conventionally subtyped according to the mutually exclusive expression of CD4 and CD8, which overlap well with helper and cytotoxic functions, respectively. T lymphocytes recently activated by antigen express new antigens, such as the high molecular weight isoform of CD45. The 'proper' functions of the majority of these molecules are known, although some have as yet no known function.

The CD system, administered by the *International Workshop on Human Leucocyte Differentiation Antigens,* is a standardized naming system designed to prevent the same surface antigen that is identified by different antibodies from being given different names. The CD system also allows homologous antigens from species other than human to be named. To analyse and compare the antibodies, panels of identical cells and tissues are distributed around the world to researchers in a blind trial. Antibodies are tested on each of the samples and the results collated by the Workshop. Different antibodies that recognize the same antigen have similar patterns of reactivity, and cluster together in the data analysis. The antibodies are then assigned a CD (**cluster of differentiation**) number, although the CD number is used synonymously with the antigen recognized. Antigen receptors and classical MHC molecules have not been assigned CD numbers. See Table A1.

Table A1 Mammalian CD antigens

CD number	Description	Cellular expression	Function
CD1a–e	Class I MHC-like molecules; Ig superfamily member	Various antigen-presenting cells	? Presentation of antigen to γδ T lymphocytes
CD2	LFA-2; Ig superfamily member.	T lymphocytes, thymocytes and NK cells.	Ligand of LFA-3 (CD58) on antigen-presenting cells and cytotoxic T lymphocyte targets; ?T lymphocyte activation.
CD3	TCR complex gamma (γ), delta (δ), epsilon (ε), zeta (ζ) and eta (η) chains. The γ, δ, and ε chains are Ig superfamily members.	T lymphocytes	Signal transduction through the αβ TCR
CD4	TCR coreceptor; Ig superfamily member.	Class II MHC-restricted T lymphocytes	Ligand of class II MHC molecule; adhesion and signal transduction; also used as receptor by HIV to infect T lymphocytes.
CD5		T and some B lymphocytes.	? Role in activation
CD6		T lymphocytes	?
CD7	Early T lymphocyte marker; Ig superfamily member.	T lymphocyte stem cells	?
CD8	TCR coreceptor (heterodimer of α and β chains)	Class I MHC-restricted T lymphocytes	Ligand of class I MHC molecule; adhesion and signal transduction
CD9		Platelets, early B lymphocytes	? Platelet activation
CD10	Cell-surface proteinase	Various cell types	?
CD11a	Integrin αL chain (α chain of LFA-1); forms heterodimer with CD18	Leucocytes; elevated expression on memory T lymphocytes.	Adhesion molecule; ligand of ICAM-1 (CD54), ICAM-2 (CD102) and ICAM-3 (CD50)
CD11b	Integrin αM chain (α chain of CR3); forms heterodimer with CD18	Myeloid cells, NK cells	Adhesion to extracellular matrix; receptor for microorganisms opsonized with iC3b
CD11c	Integrin αX chain (α chain of CR4); forms heterodimer with CD18	Granulocytes, monocytes, NK cells	Adhesion to fibrinogen
CD12		Monocytes, granulocytes, platelets	?
CD13		granulocytes, monocytes and epithelial cells	?
CD14		Monocytes, macrophages, some granulocytes	? Receptor for bacteria opsonized with LPS-binding protein.

CD	Molecule	Cellular distribution	Function
CD15	Lewis X; a branched polysaccharide	Neutrophils, eosinophils and monocytes	When sialated (sial-Lewis X) is a ligand of selectins (CD62)
CD16	FcγRIII; Ig superfamily member	Neutrophils, macrophages, NK cells	Low affinity IgG Fc receptor
CD17		neutrophils, monocytes, platelets	?
CD18	Integrin β chain of LFA-1 (with CD11a), CR3 (with CD11b) and CR4 (CD11c)	Leucocytes	β chain of 'β$_2$-family' of integrins; see CD11
CD19	B lymphocyte marker, Ig superfamily member	B lymphocytes and B lymphocyte progenitors	Signal transduction, regulation of B lymphocyte proliferation
CD20		B lymphocytes	Signal transduction, ?activation.
CD21	CR2	B lymphocytes, pharyngeal epithelial cells	Receptor for complement C3d; used by Epstein–Barr virus to infect B lymphocytes and nasopharyngeal epithelium.
CD22a	BCR complex α chain (Igα); Ig superfamily member	B lymphocytes, follicular dendritic cells, some epithelial cells	Signal transduction through antigen receptor
CD22b	BCR complex α chain (Igβ); Ig superfamily member	B lymphocytes	Signal transduction through antigen receptor
CD23	FcεRIII; C-type lectin	B lymphocytes, other leucocytes	Low affinity receptor for IgE Fc
CD24		B lymphocytes	Signal transduction, ? differentiation
CD25	IL-2 receptor α chain (Tac antigen)	Activated T and B lymphocytes, monocytes	Signal transduction; IL-2 has many roles in cell-mediated immune responses
CD26	Cell-surface proteinase	Epithelial cells, endothelial cells, T lymphocytes	?
CD27		T lymphocytes	?
CD28	Receptor for costimulatory signals; Ig superfamily member	T lymphocytes	Ligand for CD80 on antigen-presenting cells; signal transduction
CD29	Integrin β chain; forms heterodimer with CD49 to form very late antigen (VLA) proteins	Many cell types	β chain of 'β$_1$ family' of integrins; adhesion molecules; see CD49
CD30		Activated T and B lymphocytes	?
CD31	Platelet/endothelial cell adhesion molecule (PECAM-1); Ig superfamily member	Platelets, monocytes, granulocytes and endothelial cells	Adhesion of platelets to vascular endothelium
CD32	FcγRII; Ig superfamily member	B lymphocytes, neutrophils, monocytes	Low affinity IgG Fc receptor II
CD33	Myeloid cell marker; Ig superfamily member	Myeloid stem cells	?
CD34	Vascular addressin	Haemopoietic stem cells, endothelial cells	Ligand for L-selectin (CD62L)

Table A1 Continued

CD number	Description	Cellular expression	Function
CD35	CR1	B lymphocytes, eosinophils, neutrophils, erythrocytes, monocytes, follicular dendritic cells	Receptor for microorganisms opsonized by complement C3b, C4b
CD36	Platelet glycoprotein (gpIIb)	Platelets, monocytes	?; receptor used by *Plasmodium* to penetrate erythrocytes
CD37		Naive B lymphocytes	?
CD38		Immature and activated lymphocytes	A cell-surface enzyme; ?involved in B lymphocyte activation.
CD39		Many leucocytes	?
CD40	Receptor for costimulation	B lymphocytes	Receptor of costimulatory molecule, CD40L, on activated helper T lymphocytes
CD41	Integrin glycoprotein IIb (gpIIb) chain that forms heterodimer with CD61	Platelets	Platelet adhesion to fibrinogen and fibronectin
CD42	Platelet glycoprotein	Platelets and megakaryocytes	Platelet adhesion
CD43		Thymocytes and T lymphocytes	Ligand of ICAM-1 (CD54)
CD44	Leucocyte homing receptor	Leucocytes, erythrocytes	T lymphocyte homing to high endothelial venule cells
CD45	Leucocyte common antigen	Leucocytes	T lymphocyte activation (receptor phosphatase activity)
CD45R	Isoforms of CD45 with restricted cellular distribution	CD45RA: naïve T lymphocytes CD45RO: recently activated/memory T lymphocytes	?
CD46	Membrane cofactor protein	Many leucocytes, endothelial and epithelial cells	Protection of self surfaces against complement; binds C3b or C4b to permit their degradation by the serum proteinase, factor I
CD47		Haemopoietic cells	?
CD48		Mostly T and B lymphocytes.	? Signal transduction
CD49a	Integrin α_1 chain of VLA-1 (heterodimer with CD29)	T lymphocytes, monocytes	Adhesion to laminin and collagen
CD49b	Integrin α_2 chain of VLA-2 (heterodimer with CD29)	B lymphocytes, monocytes, platelets	Adhesion to laminin, collagen
CD49c	Integrin α_3 chain of VLA-3 (heterodimer with CD29)	B lymphocytes	Adhesion to laminin, fibronectin

CD	Molecule	Cellular distribution	Function
CD49d	Integrin α_4 chain of VLA-4 (heterodimer with CD29)	Thymocytes and naive T lymphocytes, B lymphocytes	Adhesion to fibronectin, Peyer's patch HEV; also ligand of VCAM-1 on inflamed endothelial cells
CD49e	Integrin α_5 chain of VLA-5 (heterodimer with CD29)	Activated T lymphocytes, monocytes, platelets	Adhesion to fibronectin
CD49f	Integrin α_6 chain of VLA-6 (heterodimer with CD29)	Thymocytes, activated T lymphocytes, monocytes, platelets, epithelial cells	Adhesion to laminin
CD50	ICAM 3	T and B lymphocytes, monocytes	?
CD51	Integrin αV chain of vitronectin receptor (heterodimer with CD61)	Platelets and megakaryocytes	Adhesion of platelets to vitronectin, components of blood clotting cascade
CD52	CAMPATH-1 antigen	Various leucocytes	?
CD53		Various leucocytes	?
CD54	ICAM-1; Ig superfamily member	Inducible on many haemopoietic and non-haemopoietic cell types	Cell-to-cell adhesion; ligand of LFA-1 (CD11a/CD18) and Mac-1 (CD11b/CD18)
CD55	Decay accelerating factor (DAF) of complement	On many cell types	Protects self surfaces from complement
CD56	NK cell isoform of neural cell adhesion molecule (N-CAM); Ig superfamily member	NK cells	Ligand of N-CAM (homotypic adhesion) involved in platelet aggregation
CD57	NK cell molecules (HNK-1)	NK cells	?
CD58	LFA-3; Ig superfamily member	On many cell types	Ligand of CD2; mediates adhesion of T lymphocytes to many cell types
CD59	MAC inhibition factor	On many cell types	Protects self surfaces from membrane attack complex of complement
CD60		Monocytes, platelets and some T lymphocytes	?
CD61	Integrin glycoprotein IIIa (gpIIIa) chain that forms heterodimer with CD41	Platelets, megakaryocytes and macrophages	Platelet aggregation and adhesion to fibrinogen and fibronectin
CD62L	L-selectin	T lymphocytes	Lymph node homing receptor
CD62E	E-selectin	Vascular endothelial cells	Adhesion of leucocytes to endothelium
CD62P	P-selectin	Vascular endothelial cells, platelets	Adhesion of leucocytes to endothelium, platelets
CD63	Lysosome-associated membrane glycoprotein 3 (LAMP-3)	Activated platelets, monocytes, macrophages	Becomes transported to cell surface
CD64	FcγRI; Ig superfamily member	Monocytes, macrophages, activated neutrophils	High affinity IgG Fc receptor
CD65		Myeloid cells	?
CD66		Mostly neutrophils	?

Table A1 Continued

CD number	Description	Cellular expression	Function
CD67		Granulocytes	?
CD68		Various leucocytes	?
CD69		Activated T and B lymphocytes	?
CD70		Naïve and activated B lymphocytes, naïve T lymphocytes	?
CD71	Transferrin receptor	Activated leucocytes	Transports serum iron via receptor-mediated endocytosis.
CD72	C-type lectin	B lymphocytes	Ligand for CD5; ?activation and differentiation of B lymphocytes
CD73	A cell-surface nucleotidease	Some T and B lymphocytes	Dephosphorylates nucleoside monophosphates prior to endocytosis
CD74	Invariant (or γ) chain of MHC class II	Class II-positive cells, mostly intracellularly	Molecular chaperone for nascent class II MHC molecules
CD75		Mostly B lymphocytes	?
CD76		B lymphocytes, some T lymphocytes	?
CD77	Burkitt's lymphoma-associated antigen	Some activated B lymphocytes	?
CD78		B lymphocytes	?
CD79a,b	BCR complex α chain (Igα) and β chain (Igβ)	B lymphocytes	Signal transduction through membrane Ig
CD80	B7.1; costimulatory molecule for T lymphocytes	Professional antigen-presenting cells (dendritic cells, macrophages activated B lymphocytes)	Ligand for CD28 and CTLA-4 on T lymphocytes; provides costimulatory signals required for activation through the TCR
CD81	Target of anti-proliferative antibody (TAPA-1)	Various cell types	?
CD86	B7.2; costimulatory molecule for T lymphocytes	Professional antigen-presenting cells (dendritic cells, macrophages, B lymphocytes)	
CD87	Urokinase plasminogen activator surface receptor		
CD88	Receptor for complement C5a	Neutrophils, macrophages, mast cells, smooth muscle	C5a is chemotactic peptide for neutrophils and macrophages, and stimulates degranulation of mast cells and macrophages
CD89	FcαR	Most leucocytes	IgA Fc receptor

CD	Name	Cellular distribution	Function
CD90	Thy-1; Ig superfamily member	Thymocytes, T lymphocytes, brain tissue	?
CD91	α_2-macroglobulin receptor	Macrophages, monocytes	? Clearance of serum proteinases inactivated by α_2-macroglobulin
CD95	Apoptosis-mediating surface antigen (APO-1; Fas antigen in the mouse)	Tumour cells	Ligand for molecule on cytotoxic T lymphocytes that triggers target cell apoptosis
CD102	ICAM-2; Ig superfamily member	Naïve lymphocytes, monocytes, vascular endothelial cells	Ligand for integrin LFA-1 (CD11a/CD18) but not Mac-1 [see ICAM-1 (CD54)]; adhesion of leucocytes to endothelium
CD103	Integrin αE chain		
CD104	Integrin β_4 chain	Endothelial cells, activated macrophages	The β chain of 'β_4 family' of integrins
CD105	Endoglin		?
CD106	Vascular cell adhesion molecule 1 (VCAM-1)	Endothelial cells, macrophages, follicular dendritic cells	Receptor for VLA-4 (CD29/49d); cellular adhesion molecule.
CD107	Lysosome-associated membrane glycoprotein 1, 2 (LAMP-1, -2)	Various cell types	?
CD115	M-CSF receptor; Ig superfamily member	Monocytes, macrophages and their progenitors	Signal transduction; M-CSF stimulates survival, proliferation and differentiation of monocytes, macrophages and their progenitors
CDw116	GM-CSF receptor α chain	Monocytes, neutrophils, eosinophils, fibroblasts, endothelial cells	Signal transduction; GM-CSF stimulates the growth of myeloid progenitor cells
CDw119	IFN-γ receptor	Macrophages, monocytes, B lymphocytes, fibroblasts, epithelial cells, endothelium	Signal transduction; IFN-γ is a multifunctional immunomodulatory cytokine
CD120a	Type I TNF receptor	Mostly epithelial cell types	Signal transduction; TNF-α induces production of inflammatory mediators, TNF-β enhances activity of phagocytes
CD120b	Type II TNF receptor	Mostly myeloid cell types	see above
CDw121a	Type I IL-1 receptor	T lymphocytes and many other non-haemopoietic cell types	Signal transduction; IL-1 has several roles in inflammation
CDw121b	Type II IL-1 receptor	B lymphocytes, macrophages and monocytes	see above
CD122	IL-2 receptor β chain	see CD25	
CD123	IL-3 receptor α chain	Myeloid progenitor cells	Signal transduction; Il-3 stimulates growth of haemopoietic cells

Table A1 Continued

CD number	Description	Cellular expression	Function
CD124	IL-4 receptor α chain	T and B lymphocytes and several non-haemopoietic cells	Signal transduction; IL-4 is a growth factor for B lymphocytes and for Th2 cells
CD125	IL-5 receptor α chain	Eosinophils and basophils, B lymphocytes	Signal transduction; IL-5 needed for B lymphocyte class switching
CD126	IL-6 receptor α chain	Activated B lymphocytes	Signal transduction; IL-6 needed for B lymphocyte differentiation into plasma cells; also has inflammatory effects
CD127	IL-7 receptor α chain	Lymphoid stem cells	Signal transduction; differentiation of T and B lymphocytes from stem cells
CDw128	High affinity IL-8 receptor	Various leucocytes	Signal transduction; induces chemotaxis of neutrophils, basophils and T lymphocytes
CDw130	IL-6 receptor β chain	see CD126	

Abbreviations used: BCR, B lymphocyte antigen receptor; CR, complement receptor; ICAM, intercellular adhesion molecule; Ig, immunoglobulin; HEV, high endothelial venule cell; NK, natural killer cell; Th, helper T lymphocyte; LFA, leucocyte function associated antigen; MHC, major histocompatibility complex; TCR, T lymphocyte receptor.

Information in this table has been obtained mostly from Barclay, N.A. et al. (1993) *The Leucocyte Antigen Facts Book*. Academic Press, London and Schlossman, S.F., *et al.* (1994) CD antigens 1993. *Immunology Today*, **15(3)**: 88–89.

Appendix B: Commonly used immunoassays

Western blotting (Figure 1)

Complex mixtures of protein, which contain the antigen to be detected, are first separated according to size by electrophoresis through a polyacrylamide gel. An additional sample containing proteins of known molecular weight is also separated on the gel. This will enable the weight of antigens detected by the technique to be calculated. Protein samples are first dissolved by boiling in a buffer containing sodium dodecyl sulphate (SDS) and mercaptoethanol. Boiling ensures the proteins are fully unfolded and SDS, a strong cationic detergent, coats both hydrophobic and hydrophilic regions of proteins, and imparts a strong negative charge. Any natural charge the protein has is overwhelmed by SDS. Mercaptoethanol is a reducing agent that can be included to break disulphide bonds between and within polypeptide chains. During electrophoresis proteins migrate towards the anode; the speed of migration of a given protein through the gel is dependent on its size. Progress of larger proteins is hindered more by the matrix of acrylamide and they consequently migrate more slowly than smaller proteins.

In the next step, the proteins are transferred horizontally from the gel by electrophoresis onto a sheet of nitrocellulose paper ('blotted') upon which they are immobilized. The vacant areas of the nitrocellulose are blocked with an irrelevant protein (casein or dried milk powder is often used). The blot is then incubated with primary antibody chosen to bind to its specific antigen on the blot. The blot is then washed and incubated with a secondary antibody labelled for detection of the bound primary antigen. For example, if the primary antigen is of mouse origin, a secondary antibody might be rabbit anti-mouse immunoglobulin coupled to peroxidase. Bound second-

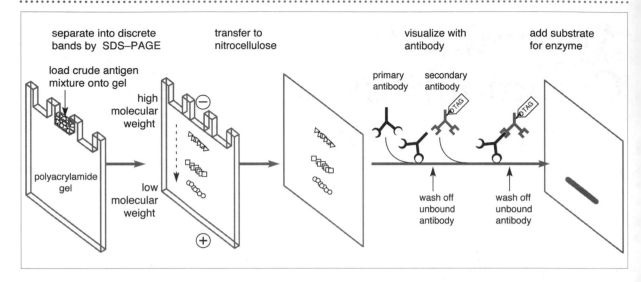

separate into discrete
bands by SDS–PAGE

load crude antigen
mixture onto gel

high
molecular
weight

polyacrylamide
gel

low
molecular
weight

transfer to
nitrocellulose

⊖

⊕

visualize with
antibody

primary   secondary
antibody   antibody

TAG   TAG

wash off   wash off
unbound   unbound
antibody   antibody

add substrate
for enzyme

Figure 1 Western blotting.

ary antibody is visualized by incubating the blot in a substrate that is converted into a coloured dye by the peroxidase.

This technique has the advantage that the molecular weight of the antigen can be determined, which often helps to verify the specificity of the antibody. A disadvantage is that the antigen is denatured prior to binding by antibody, which reduces the chance that the antigen is recognized by antibody elicited to native antigen.

Immunoprecipitation (Figure 2)

This technique also involves the separation of complex mixtures of antigens, such as cell lysates, by SDS polyacrylamide gel electrophoresis. This enables the molecular weights of antigens detected by antibodies to be determined, which helps verify the specificity of the interaction. In contrast to the Western blot, the antigen–antibody interaction occurs *prior* to electrophoresis and consequently with *native* antigen. This is achieved by adding the antibodies to the complex antigen mixture. To allow recovery of the immune complexes, the primary or secondary antibodies are first 'immobilized' on the surface of inert macroscopic beads. The beads provide a convenient means to recover the primary antibody and any bound antigens by centrifugation. Importantly, this approach allows any other proteins that are naturally associated with the antigen to be co-precipitated. The precipitate is then dissolved in SDS and mercaptoethanol and subjected to polyacrylamide gel electrophoresis as described. To visualize the antigen, the proteins in the original

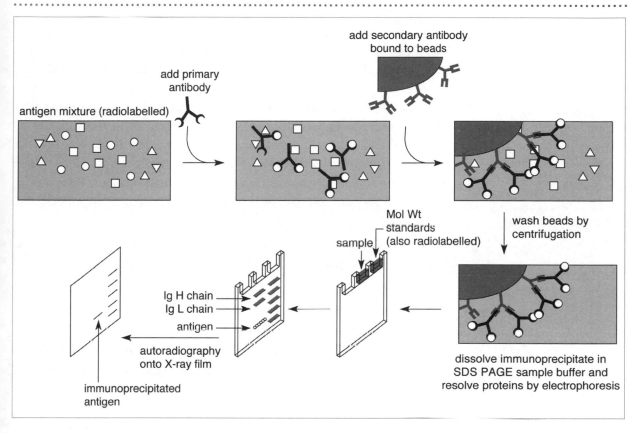

mixture are usually pre-labelled with a radioisotope prior to immunoprecipitation and the radiolabelled band on the gel visualized by
autoradiography. This technique is used when preservation of the
antigen's native conformation is important and when information con-
cerning other proteins that associate with the antigen is sought.

Figure 2 Immunoprecipitation.

Immunofluorescence and flow cytometry

Antibodies are widely used for determining the phenotype of cells. Each of
the CD molecules listed in Appendix A, for example, is defined by different
antibodies that each recognize a specific cell-surface molecule. Antibodies
can also be used to label intracellular antigens, such as cytoskeletal filaments,
if the cells are first allowed to adhere to a glass microscope slide and
permeabilized to allow the antibodies to pass through the plasma membrane. Bound primary antibodies are visualized with secondary antibodies
coupled to a fluorochrome, such as fluorescein or rhodamine. When excited
by the appropriate wavelength of light (in the ultraviolet range) these emit

at a lower wavelength in the visible spectrum – fluorescein emits yellow/green light and rhodamine emits red light.

Labelled cells can be inspected in a fluorescence light microscope. Alternatively, quantitative data, such as the intensity of fluorescence for each cell, can be obtained using a **flow cytometer**. This machine passes a stream of cells through a beam of light from a laser to excite the fluorescent label and the light emitted is measured by a detector. The intensity of several thousand cells can be measured in a few seconds. By using different primary antibodies conjugated directly to different fluorochromes, two or more different surface antigens can be detected and measured simultaneously.

> **flow cytometer** *n.* a machine for quantitating the level of fluorescence on individual cells in suspension stained by immunofluorescence (see **fluorescence activated cell sorter**).

Immunocytochemistry and immunohistochemistry (Figure 3)

Like classical cytochemistry and histochemistry, these techniques are used to study the organization and composition of cells and tissues by light microscopy. Whereas the former use chemical stains that react according to the chemical properties of the specimen, the latter use defined antibodies that bind to a specific antigen. Cells (containing the antigen) are immobilized on a transparent surface, such as a glass microscope slide. Cell-surface antigens can be visualized without further treatment whereas intracellular antigens require the cells to be permeabilized. Whole tissues are too thick to be studied by light microscopy and must first be sliced into thin sections on a microtome. This requires the specimen to be embedded in a supporting material that prevents the specimen being distorted as the sections are cut. For conventional histology, paraffin wax is used although for immunohistochemistry, a water-soluble embedding material must be used. Often, specimens are embedded in a water-soluble compound and then frozen, and sections cut in a refrigerated microtome (a cryostat). After the cells or sections have been immobilized, antigens are then detected by primary antibody and visualized by a fluorochrome or chromogen-tagged secondary antibody. The ultrastructural distribution of an antigen can be probed using immunoelectron microscopy. Ultrathin sections of cells are stained with antibodies tagged with a label that is visible in the electron microscope such as colloidal gold particles.

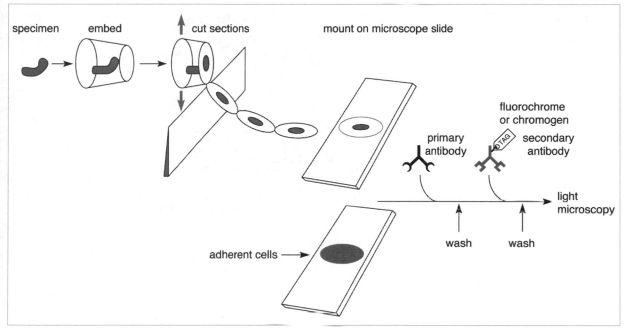

specimen   embed   cut sections   mount on microscope slide

fluorochrome
or chromogen

primary
antibody   secondary
antibody

light
microscopy

adherent cells

wash   wash

Figure 3 Immunohistochemistry
(above) and
immunocytochemistry.

Enzyme-linked immunosorbent assay (ELISA; Figure 4).

This is a widely used quantitative assay that exists in several different configurations. At its simplest, antigen is allowed to adsorb onto a plastic surface. ELISAs are conventionally performed in a plastic ELISA 'microtitre plate' which contains 96 small wells (arranged 12 × 8) each able to hold approximately 200 μl of buffer. After the antigen has adsorbed, unbound antigen is washed out and unoccupied plastic surface blocked with an irrelevant protein. The primary antibody is then placed into the wells at different dilutions, and after washing, antibody bound to the antigen is visualized with a secondary antibody labelled with a chromogenic enzyme. Upon adding the substrate for the enzyme the buffer changes colour, the density of the colour being directly proportional to the amount of bound primary antibody. This can be quantitated by measuring the optical density of each well in a special spectrophotometer (or 'ELISA plate reader'). This technique uses antigen in its native, rather than denatured, conformation and so has this advantage over the Western blot. However, it is impossible to establish the molecular weight of the antigen to verify the specificity of the reaction. Thus ELISAs are usually performed with defined antigens (purified proteins, synthetic peptides) rather than complex mixtures and are usually used to estimate the titre of specific antibody in biological fluids such as serum or hybridoma culture supernatants.

Figure 4 Principle of the ELISA.

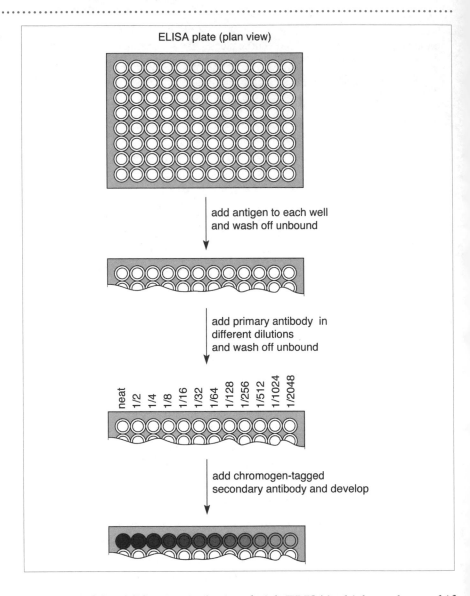

ELISA plate (plan view)

add antigen to each well
and wash off unbound

add primary antibody in
different dilutions
and wash off unbound

neat 1/2 1/4 1/8 1/16 1/32 1/64 1/128 1/256 1/512 1/1024 1/2048

add chromogen-tagged
secondary antibody and develop

One useful modification is the 'sandwich ELISA' which can be used if the antigen is impure. Also required is a monospecific (affinity purified) or monoclonal antibody to the antigen in question. This antibody is absorbed to the plate first and then used to capture the antigen from the complex mixture. The plate is then washed. The plate is then ready to be used to assay for antibodies to the antigen as in a conventional ELISA. Sandwich ELISAs (so called because the antigen is sandwiched between antibodies) are also used if the antigen is bulky, such as a virus, and may be distorted into a non-native conformation if it is adsorbed onto the ELISA plate directly.

Further reading

Harlow, E. and Lane, D. (1988) *Antibodies: A Laboratory Manual.* Cold Spring Harbor Laboratory Press, Cold Spring Harbor, NY.

Hudson, L. and Hay, F.C. (1989) *Practical Immunology,* 3rd edn. Blackwell Scientific Publications, Oxford.

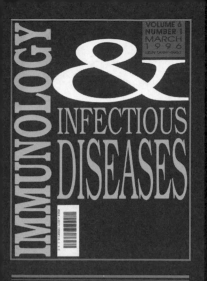

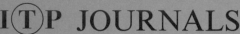

Index

Page numbers in **bold** refer to figures and page numbers in *italic* refer to tables.